Douglas R. Stinson

Combinatorial Designs

Constructions and Analysis

Springer

Douglas R. Stinson
School of Computer Science
University of Waterloo
Waterloo ON N2L 3G1
Canada
dstinson@waterloo.ca

Library of Congress Cataloging-in-Publication Data
Stinson, Douglas R. (Douglas Robert), 1956–
Combinatorial designs : constructions and analysis / Douglas R. Stinson.
p. cm.
Includes bibliographical references and index.
ISBN 0-387-95487-2
1. Combinatorial designs and configurations. I. Title
QA166.25.S75 2003
511'.6—dc21 2003052964
ISBN 0-387-95487-2

Springer-Verlag New York Berlin Heidelberg

Stinson: Combinatorial Designs: Constructions & Analysis

First Indian Reprint 2010

ISBN 978-81-8489-705-0

This edition is published by Springer (India) Private Limited, (a part of Springer Science+Business Media), Registered Office: 3rd Floor, Gandharva Mahavidyalaya, 212, Deen Dayal Upadhyaya Marg, New Delhi – 110 002, India

Printed in India by Rashtriya Printers, Delhi.

Combinatorial Designs

To Ron Mullin, who taught me design theory

Foreword

The evolution of combinatorial design theory has been one of remarkable successes, unanticipated applications, deep connections with fundamental mathematics, and the desire to produce order from apparent chaos. While some of its celebrated successes date from the eighteenth and nineteenth centuries in the research of Euler, Kirkman, Cayley, Hamilton, Sylvester, Moore, and others, not until the twentieth century did the study of combinatorial designs emerge as an academic subject in its own right. When Fisher and his colleagues developed the mathematics of experimental design in the 1920s, combinatorial design theory was born as a field intimately linked to its applications. Beginning in the 1930s, Bose and his school laid the foundations, embedding the nascent field firmly as a mathematical discipline by developing deep connections with finite geometry, number theory, finite fields, and group theory; however, Bose accomplished much more. His foundation entwined deep mathematics with its applications in experimental design and in recreational problems and anticipated its fundamental importance in the theory of error-correcting codes.

The rapid advances in design theory can be attributed in large degree to its impetus from applications in coding theory and communications and its continued deep interactions with geometry, algebra, and number theory. The last fifty years have witnessed not only the emergence of certain combinatorial designs (balanced incomplete block designs, Hadamard matrices, pairwise balanced designs, and orthogonal arrays, for example) as central, but also powerful combinatorial and computational techniques for their construction. Indeed the field grew so far and so fast that its historical connection to applications was strained.

Yet, in the last twenty years, combinatorial design theory has emerged again as a field rich in current and practical applications. The fundamental connections with algebra, number theory, and finite geometry remain and flourish. The applications in experimental design and coding theory have developed a breadth and depth that defy brief explanation. Yet combinatorial design theory has matured into more than this through applications in cryptography, optical communications, storage system design, communication protocols, algorithm design and analysis, and wireless communications, to mention just a few areas.

Combinatorial design theory is mature and widely applied today because it has respected and advanced its mathematical heritage while finding genuine new applications. I am honored to write this foreword for two reasons. Doug Stinson has for twenty-five years been the epitome of a researcher and expositor who has advanced combinatorial design theory as a marriage of mathematics and applications. But more than that, the book you hold in your hands presents design theory as a seamless interaction of deep mathematics and challenging applications. By providing an accessible introduction, it serves as an invitation to those in applications areas to appreciate and employ beautiful mathematics and concurrently invites mathematicians to learn from the applications themselves.

In which directions will combinatorial design theory evolve in the next century? We cannot yet know. We can know, however, that new mathematical truths will be found and that unanticipated applications will arise. Our challenge is to seek both and to know that each profits from the other.

Phoenix, Arizona
April, 2003

Charles J. Colbourn

Preface

Overview and Goals

Combinatorial design theory is one of the most beautiful areas of mathematics. Design theory has its roots in recreational mathematics, but it evolved in the twentieth century into a full-fledged mathematical discipline with diverse applications in statistics and computer science. The fundamental problems in design theory are simple enough that they can be explained to non-mathematicians, yet the solutions of those problems have involved the development of innovative new combinatorial techniques as well as ingenious applications of methods from other areas of mathematics such as algebra and number theory. Many classical problems remain unsolved to this day as well.

This book is intended primarily to be a textbook for study at the senior undergraduate or beginning graduate level. Courses in mathematics or computer science can be based on this book. Regardless of the audience, however, it requires a certain amount of "mathematical maturity" to study design theory. The main technical prerequisites are some familiarity with basic abstract algebra (group theory, in particular), linear algebra (matrices and vector spaces), and some number-theoretic fundamentals (e.g., modular arithmetic and congruences).

Topic Coverage and Organization

The first seven chapters of this book provide a thorough treatment of the classical core of the subject of combinatorial designs. These chapters concern symmetric BIBDs, difference sets, Hadamard matrices, resolvable BIBDs, Latin squares, and pairwise balanced designs. A one-semester course can cover most of this material. For example, when I have taught courses on designs, I have based my lectures on material selected from the following chapters and sections:

- Chapter 1: Sections 1.1–1.3, Section 1.4 (optional), Sections 1.5–1.6
- Chapter 2: Sections 2.1–2.4
- Chapter 3: Sections 3.1–3.4
- Chapter 4: Sections 4.1–4.4, Section 4.5 (optional), Section 4.6
- Chapter 5: Sections 5.1–5.2, Section 5.3 (optional)
- Chapter 6: Sections 6.1, Section 6.2 (optional), Sections 6.3–6.8
- Chapter 7: Sections 7.1–7.3

There are many variations possible, of course. Typically, I would provide a complete proof of the Bruck-Ryser-Chowla Theorem or the Multiplier Theorem, but not both. It is possible to omit Wilson's Construction for MOLS in order to spend more time on pairwise balanced designs. Another option is to include the optional Section 6.2 and omit some of the material in Chapter 7. Yet another possibility is to present an introduction to t-designs (incorporating some material from Chapter 9, Sections 9.1 and 9.2) and delete some of the optional sections listed above.

More advanced or specialized material is covered in the last four chapters as well as in some later sections of the first seven chapters. The main topics in the last four chapters are minimal pairwise balanced designs, t-designs, orthogonal arrays and codes, and four selected applications of designs (in the last chapter).

Key Features

There are several features of this book that will make it useful as a textbook. Complete, carefully written proofs of most major results are given. There are many examples provided throughout in order to illustrate the definitions, concepts, and theorems. Numerous and varied exercises are provided at the end of each chapter. As well, certain mathematical threads flow through this book:

- The linear algebraic method of proving Fisher's Inequality reappears several times.
- The theme of Boolean functions is introduced in the study of bent functions and revisited in the discussion of Reed-Muller codes and a brief treatment of resilient functions.
- The use of permutation groups as a construction technique is pervasive.
- Elegant combinatorial arguments are used in many places in preference to alternative proofs that employ heavier mathematical machinery.
- Finite fields are used throughout the book. For this reason, some background material on finite fields is summarized in an Appendix. However, another option for an instructor is to specialize constructions utilizing finite fields $\mathbb{F}_q$ to the more familiar fields $\mathbb{Z}_p$, where p is a prime.

As mentioned earlier, there are a variety of advanced or specialized topics that are discussed in the book. Highlights include the following:

- regular Hadamard matrices and excess of Hadamard matrices;
- bent functions;
- bounds and constructions for minimal pairwise balanced designs;
- the Ryser-Woodall Theorem;
- constructions and bounds for t-wise balanced designs, including a proof of the Kramer Conjecture;
- a survey of the combinatorial connections between orthogonal arrays, codes, and designs;
- constructions and bounds for various classes of optimal codes and orthogonal arrays;
- Reed-Muller codes;
- resilient functions;
- four selected applications of designs: authentication codes, threshold schemes, group testing, and two-point sampling.

It must be recognized that design theory is an enormous subject, and any choice of optional material in a 300 page book is dependent on the whim of the author! Thus there are many interesting or important areas of design theory that are not discussed in the book. I hope, however, that readers of the book will find a fascinating mix of topics that serve to illustrate the breadth and beauty of design theory.

Audience

As mentioned above, this book is primarily intended to be a textbook. In addition, all of the material in this book is suitable for self-study by graduate students, who will find it provides helpful background information concerning research topics in design theory. Researchers may also find that some of the sections on advanced topics provide a useful reference for material that is not easily accessible in textbook form.

Acknowledgments

I have benefitted from the suggestions, comments and encouragement of many people while this book was being written. In particular, I would like to thank Charlie Colbourn, Don Kreher, and Brett Stevens. Special thanks goes to Dameng Deng for his help with proofreading. Also, I appreciate the assistance and advice of Wayne Yuhasz and Wayne Wheeler from Springer during this project.

Waterloo, Ontario, Canada
July, 2003

Douglas R. Stinson

Contents

1

Introduction to Balanced Incomplete Block Designs

1.1 What Is Design Theory?

Combinatorial design theory concerns questions about whether it is possible to arrange elements of a finite set into subsets so that certain "balance" properties are satisfied. Types of designs that we will discuss include balanced incomplete block designs, t-designs, pairwise balanced designs, orthogonal Latin squares, and many more. Many of the fundamental questions are existence questions: Does a design of a specified type exist? Modern design theory includes many existence results as well as nonexistence results. However, there remain many open problems concerning the existence of certain types of designs.

Design theory has its roots in recreational mathematics. Many types of designs that are studied today were first considered in the context of mathematical puzzles or brain-teasers in the eighteenth and nineteenth centuries. The study of design theory as a mathematical discipline really began in the twentieth century due to applications in the design and analysis of statistical experiments. Designs have many other applications as well, such as tournament scheduling, lotteries, mathematical biology, algorithm design and analysis, networking, group testing, and cryptography.

This work will provide a mathematical treatment of the most important "classical" results in design theory. This roughly covers the period from 1940 to 1980. In addition, we cover some selected recent topics in design theory that have applications in other areas, such as bent functions and resilient functions.

Design theory makes use of tools from linear algebra, groups, rings and fields, and number theory, as well as combinatorics. The basic concepts of design theory are quite simple, but the mathematics used to study designs is varied, rich, and ingenious.

1.2 Basic Definitions and Properties

Definition 1.1. *A* design *is a pair* $(X, \mathcal{A})$ *such that the following properties are satisfied:*

1. *X is a set of elements called* points, *and*
2. $\mathcal{A}$ *is a collection (i.e., multiset) of nonempty subsets of X called* blocks.

If two blocks in a design are identical, they are said to be *repeated blocks*. This is why we refer to $\mathcal{A}$ as a *multiset* of blocks rather than a set. A design is said to be a *simple design* if it does not contain repeated blocks.

If we want to list the elements in a multiset (with their multiplicities), we will use the notation [ ]. If all elements of a multiset have multiplicity one, then the multiset is a set. For example, we have that $[1,2,5] = \{1,2,5\}$, but $[1,2,5,2] \neq \{1,2,5,2\} = \{1,2,5\}$. The order of the elements in a multiset is irrelevant, as with a set.

Balanced incomplete block designs are probably the most-studied type of design. The study of balanced incomplete block designs was begun in the 1930s by Fisher and Yates. Here is a definition:

Definition 1.2. *Let* $v, k,$ *and* λ *be positive integers such that* $v > k \geq 2$. *A* (v, k, λ)-balanced incomplete block design *(which we abbreviate to* (v, k, λ)-BIBD*) is a design* $(X, \mathcal{A})$ *such that the following properties are satisfied:*

1. $|X| = v$,
2. *each block contains exactly k points, and*
3. *every pair of distinct points is contained in exactly* λ *blocks.*

Property 3 in the definition above is the "balance" property. A BIBD is called an *incomplete block design* because $k < v$, and hence all its blocks are *incomplete blocks*.

A BIBD may possibly contain repeated blocks if $\lambda > 1$. The use of the letter "v" to denote the number of points is an artifact of the original motivation for studying BIBDs, namely to facilitate the design of agricultural experiments. "v" was an abbreviation for "varieties", as in "varieties of wheat".

We give a few examples of BIBDs now. To save space, we write blocks in the form *abc* rather than $\{a, b, c\}$.

Example 1.3. A (7,3,1)-BIBD.

$$X = \{1,2,3,4,5,6,7\}, \quad \text{and}$$
$$\mathcal{A} = \{123, 145, 167, 246, 257, 347, 356\}.$$

This BIBD has a nice diagrammatic representation; see Figure 1.1. The blocks of the BIBD are the six lines and the circle in this diagram. ∎

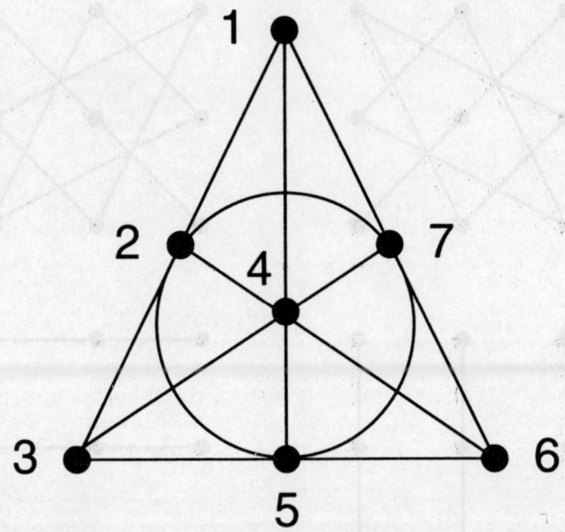

Fig. 1.1. The Fano Plane: A $(7, 3, 1)$-BIBD

Example 1.4. A $(9, 3, 1)$-BIBD.

$$X = \{1, 2, 3, 4, 5, 6, 7, 8, 9\}, \quad \text{and}$$
$$\mathcal{A} = \{123, 456, 789, 147, 258, 369, 159, 267, 348, 168, 249, 357\}.$$

This BIBD can also be presented diagrammatically; see Figure 1.2. The 12 blocks of the BIBD are depicted as eight lines and four triangles. Observe that the blocks can be separated into four sets of three, where each of these four sets covers every point in the BIBD. ▮

Example 1.5. A $(10, 4, 2)$-BIBD.

$$X = \{0, 1, 2, 3, 4, 5, 6, 7, 8, 9\}, \quad \text{and}$$
$$\begin{aligned}\mathcal{A} = \{&0123, 0145, 0246, 0378, 0579, 0689, 1278, 1369, 1479, 1568,\\ &2359, 2489, 2567, 3458, 3467\}.\end{aligned}$$

▮

Example 1.6. Let $\mathcal{A}$ consist of all k-subsets of X. Then $(X, \mathcal{A})$ is a $\left(v, k, \binom{v-2}{k-2}\right)$-BIBD. ▮

Example 1.7. A $(7, 3, 2)$-BIBD containing a repeated block.

$$X = \{0, 1, 2, 3, 4, 5, 6\}, \quad \text{and}$$
$$\begin{aligned}\mathcal{A} = [&123, 145, 167, 246, 257, 347, 356,\\ &123, 147, 156, 245, 267, 346, 357].\end{aligned}$$

▮

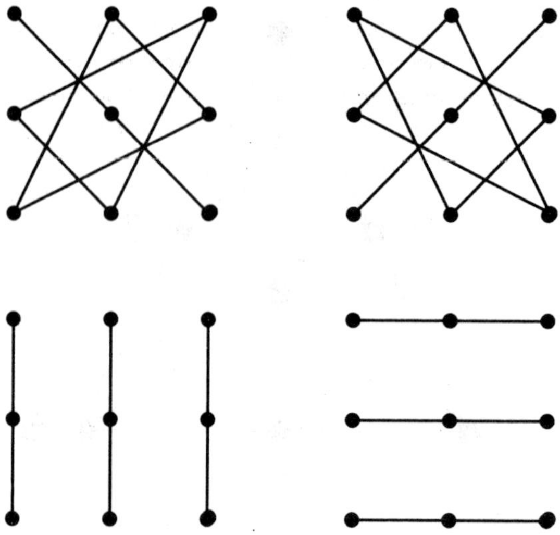

Fig. 1.2. A $(9,3,1)$-BIBD

We now state and prove two basic properties of BIBDs.

Theorem 1.8. *In a (v,k,λ)-BIBD, every point occurs in exactly*

$$r = \frac{\lambda(v-1)}{k-1}$$

blocks.

Proof. Let $(X, \mathcal{A})$ be a (v,k,λ)-BIBD. Suppose $x \in X$, and let r_x denote the number of blocks containing x. Define a set

$$I = \{(y, A) : y \in X, y \neq x, A \in \mathcal{A}, \{x,y\} \subseteq A\}.$$

We will compute $|I|$ in two different ways.

First, there are $v-1$ ways to choose $y \in X$ such that $y \neq x$. For each such y, there are λ blocks A such that $\{x,y\} \subseteq A$. Hence,

$$|I| = \lambda(v-1).$$

On the other hand, there are r_x ways to choose a block A such that $x \in A$. For each choice of A, there are $k-1$ ways to choose $y \in A$, $y \neq x$. Hence,

$$|I| = r_x(k-1).$$

Combining these two equations, we see that

$$\lambda(v-1) = r_x(k-1).$$

Hence $r_x = \lambda(v-1)/(k-1)$ is independent of x, and the result follows. □

The value r is often called the *replication number* of the BIBD.

Theorem 1.9. *A (v, k, λ)-BIBD has exactly*

$$b = \frac{vr}{k} = \frac{\lambda(v^2 - v)}{k^2 - k}$$

blocks.

Proof. Let $(X, \mathcal{A})$ be a (v, k, λ)-BIBD, and let $b = |\mathcal{A}|$. Define a set

$$I = \{(x, A) : x \in X, A \in \mathcal{A}, x \in A\}.$$

We will compute $|I|$ in two different ways.

First, there are v ways to choose $x \in X$. For each such x, there are r blocks A such that $x \in A$. Hence,

$$|I| = vr.$$

On the other hand, there are b ways to choose a block $A \in \mathcal{A}$. For each choice of A, there are k ways to choose $x \in A$. Hence,

$$|I| = bk.$$

Combining these two equations, we see that

$$bk = vr,$$

as desired. □

Sometimes we will use the notation (v, b, r, k, λ)-BIBD if we want to record the values of all five parameters.

Since b and r must be integers, these two theorems allow us to conclude that BIBDs with certain parameter sets do not exist. We state the following obvious corollary of Theorems 1.8 and 1.9.

Corollary 1.10. *If a (v, k, λ)-BIBD exists, then $\lambda(v - 1) \equiv 0 \pmod{k - 1}$ and $\lambda v(v - 1) \equiv 0 \pmod{k(k - 1)}$.*

For example, an $(8, 3, 1)$-BIBD does not exist because $\lambda(v - 1) = 7 \not\equiv 0 \pmod 2$. As another example, let us consider the parameter set $(19, 4, 1)$. Here, we see that $\lambda v(v - 1) = 342 \not\equiv 0 \pmod{12}$. Hence a $(19, 4, 1)$-BIBD cannot exist.

A more general use of Corollary 1.10 is to determine necessary conditions for families of BIBDs with fixed values of k and λ. For example, it is not hard to show that a $(v, 3, 1)$-BIBD exists only if $v \equiv 1, 3 \pmod 6$.

One of the main goals of combinatorial design theory is to determine necessary and sufficient conditions for the existence of a (v, k, λ)-BIBD. This is a very difficult problem in general, and there are many parameter sets where the answer is not yet known. For example, it is currently unknown if there exists a $(22, 8, 4)$-BIBD (such a BIBD would have $r = 12$ and $b = 33$). On the other hand, there are many known constructions for infinite classes of BIBDs as well as some other necessary conditions that we will discuss a bit later.

1.3 Incidence Matrices

It is often convenient to represent a BIBD by means of an incidence matrix. This is especially useful for computer programs. We give the definition of an incidence matrix now.

Definition 1.11. *Let* $(X, \mathcal{A})$ *be a design where* $X = \{x_1, \ldots, x_v\}$ *and* $\mathcal{A} = \{A_1, \ldots, A_b\}$. *The* incidence matrix *of* $(X, \mathcal{A})$ *is the* $v \times b$ $0-1$ *matrix* $M = (m_{i,j})$ *defined by the rule*

$$m_{i,j} = \begin{cases} 1 \text{ if } x_i \in A_j \\ 0 \text{ if } x_i \notin A_j. \end{cases}$$

The incidence matrix, M, of a (v, b, r, k, λ)-BIBD satisfies the following properties:

1. every column of M contains exactly k "1"s;
2. every row of M contains exactly r "1"s;
3. two distinct rows of M both contain "1"s in exactly λ columns.

Example 1.12. Consider the $(9, 3, 1)$-BIBD presented in Example 1.4. The incidence matrix of this design is the following 9×12 matrix:

$$M = \begin{pmatrix} 1&0&0&1&0&0&1&0&0&1&0&0 \\ 1&0&0&0&1&0&0&1&0&0&1&0 \\ 1&0&0&0&0&1&0&0&1&0&0&1 \\ 0&1&0&1&0&0&0&0&1&0&1&0 \\ 0&1&0&0&1&0&1&0&0&0&0&1 \\ 0&1&0&0&0&1&0&1&0&1&0&0 \\ 0&0&1&1&0&0&0&1&0&0&0&1 \\ 0&0&1&0&1&0&0&0&1&1&0&0 \\ 0&0&1&0&0&1&1&0&0&0&1&0 \end{pmatrix}.$$

∎

We need a few more definitions before stating the next theorem. Suppose I_n denotes an $n \times n$ identity matrix, J_n denotes the $n \times n$ matrix in which every entry is a "1", and $\mathbf{u}_n$ denotes the vector of length n in which every coordinate is a "1". Finally, for a matrix $M = (m_{i,j})$, define the *transpose* of M, denoted M^T, to be the matrix whose (j, i) entry is $m_{i,j}$.

Theorem 1.13. *Let* M *be a* $v \times b$ $0-1$ *matrix and let* $2 \leq k < v$. *Then* M *is the incidence matrix of a* (v, b, r, k, λ)-BIBD *if and only if* $MM^T = \lambda J_v + (r - \lambda) I_v$ *and* $\mathbf{u}_v M = k\mathbf{u}_b$.

Proof. First, suppose $(X, \mathcal{A})$ is a (v, k, λ)-BIBD, where $X = \{x_1, \ldots, x_v\}$ and $\mathcal{A} = \{A_1, \ldots, A_b\}$. Let M be its incidence matrix. The (i, j)-entry of MM^T is

$$\sum_{h=1}^{b} m_{i,h}m_{j,h} = \begin{cases} r \text{ if } i = j \\ \lambda \text{ if } i \neq j. \end{cases}$$

Hence, from properties 2 and 3 enumerated above, every entry on the main diagonal of the matrix MM^T is equal to r, and every off-diagonal entry is equal to λ, so $MM^T = \lambda J_v + (r - \lambda)I_v$.

Furthermore, the ith entry of $\mathbf{u}_v M$ is equal to the number of "1"s in column i of M. By property 1, this equals k. Hence, $\mathbf{u}_v M = k\mathbf{u}_b$.

Conversely, suppose that M is a $v \times b$ $0-1$ matrix such that $MM^T = \lambda J_v + (r-\lambda)I_v$ and $\mathbf{u}_v M = k\mathbf{u}_b$. Let $(X, \mathcal{A})$ be the design whose incidence matrix is M. Clearly we have $|X| = v$ and $|\mathcal{A}| = b$. From the equation $\mathbf{u}_v M = k\mathbf{u}_b$, it follows that every block in $\mathcal{A}$ contains k points. From the equation $MM^T = \lambda J_v + (r-\lambda)I_v$, it follows that every pair of points occurs in exactly λ blocks, and every point occurs in r blocks. Hence, $(X, \mathcal{A})$ is a (v, b, r, k, λ)-BIBD. □

We will show that the converse part of the theorem above does not hold if the second condition is omitted. Incidence matrices satisfying the first condition are equivalent to a certain type of design, which we define now.

Definition 1.14. *A* pairwise balanced design *(or* PBD*) is a design $(X, \mathcal{A})$ such that every pair of distinct points is contained in exactly λ blocks, where λ is a positive integer. Furthermore, $(X, \mathcal{A})$ is a* regular pairwise balanced design *if every point $x \in X$ occurs in exactly r blocks $A \in \mathcal{A}$, where r is a positive integer.*

A PBD $(X, \mathcal{A})$ is allowed to contain blocks of size $|X|$ (i.e., complete blocks*). If $(X, \mathcal{A})$ consists only of complete blocks, it is said to be a* trivial pairwise balanced design*. If $(X, \mathcal{A})$ contains no complete blocks, it is said to be a* proper pairwise balanced design .

We state the following variation of Theorem 1.13 without proof.

Theorem 1.15. *Let M be a $v \times b$ $0-1$ matrix. Then M is the incidence matrix of a regular pairwise balanced design having v points and b blocks if and only if there exist positive integers r and λ such that $MM^T = \lambda J_v + (r-\lambda)I_v$.*

Here is an example to illustrate Theorem 1.15.

Example 1.16. Consider the following 6×11 matrix:

$$M = \begin{pmatrix} 1&0&1&1&1&0&0&0&0&0&0 \\ 1&0&0&0&0&1&1&1&0&0&0 \\ 1&0&0&0&0&0&0&0&1&1&1 \\ 0&1&1&0&0&1&0&0&1&0&0 \\ 0&1&0&1&0&0&1&0&0&1&0 \\ 0&1&0&0&1&0&0&1&0&0&1 \end{pmatrix}.$$

This matrix M is the incidence matrix of the following regular pairwise balanced design:

$$X = \{1,2,3,4,5,6\}, \quad \text{and}$$
$$\mathcal{A} = \{123,456,14,15,16,24,25,26,34,35,36\}.$$

Here $v = 6$, $b = 11$, $r = 4$, and $\lambda = 1$. The design is not a BIBD because the blocks do not all have the same size—there are two blocks of size three and nine blocks of size two.

It is easily verified that $MM^T = J_v + 3I_v = \lambda J_v + (r - \lambda)I_v$. However,

$$\mathbf{u}_6 M = (3,3,2,2,2,2,2,2,2,2,2),$$

so $\mathbf{u}_6 M \neq k\mathbf{u}_b$ for any integer k. ∎

Suppose that $(X, \mathcal{A})$ is a design with $|X| = v$ and $|\mathcal{A}| = b$. Let M be the $v \times b$ incidence matrix of $(X, \mathcal{A})$. The design having incidence matrix M^T is called the *dual design* of $(X, \mathcal{A})$. Suppose that $(Y, \mathcal{B})$ is the dual design of $(X, \mathcal{A})$; then $|Y| = |\mathcal{A}| = b$ and $|\mathcal{B}| = |X| = v$. Properties of dual designs of BIBDs are summarized in the following theorem.

Theorem 1.17. *Suppose that $(X, \mathcal{A})$ is a (v, b, r, k, λ)-BIBD, and let $(Y, \mathcal{B})$ be the dual design of $(X, \mathcal{A})$. Then the following properties hold:*

1. *every block in $\mathcal{B}$ has size r,*
2. *every point in Y occurs in exactly k blocks in $\mathcal{B}$, and*
3. *any two distinct blocks $B_i, B_j \in \mathcal{B}$ intersect in exactly λ points.*

Example 1.18. Suppose that $(X, \mathcal{A})$ is the $(9, 3, 1)$-BIBD presented in Example 1.4. Then $(Y, \mathcal{B})$ is the dual design of $(X, \mathcal{A})$, where

$$Y = \{1,2,3,4,5,6,7,8,9,T,E,V\}, \quad \text{and}$$
$$\mathcal{B} = \{147T,158E,169V,248E,257V,268T,348V,359T,367E\}.$$

It is easy to verify that every block in $\mathcal{B}$ has size four, every point in Y occurs in exactly three blocks in $\mathcal{B}$, and every pair of distinct blocks in $\mathcal{B}$ intersect in exactly one point. ∎

1.4 Isomorphisms and Automorphisms

We begin with a definition.

Definition 1.19. *Suppose $(X, \mathcal{A})$ and $(Y, \mathcal{B})$ are two designs with $|X| = |Y|$. $(X, \mathcal{A})$ and $(Y, \mathcal{B})$ are* isomorphic *if there exists a bijection $\alpha : X \to Y$ such that*

$$[\{\alpha(x) : x \in A\} : A \in \mathcal{A}] \models \mathcal{B}.$$

In other words, if we rename every point $x \in X$ by $\alpha(x)$, then the collection of blocks $\mathcal{A}$ is transformed into $\mathcal{B}$. The bijection α is called an isomorphism.

Example 1.20. Here are two (7,3,1)-BIBDs, $(X, \mathcal{A})$ and $(Y, \mathcal{B})$:

$$X = \{1,2,3,4,5,6,7\}, \quad \text{and}$$
$$\mathcal{A} = \{123, 145, 167, 246, 257, 347, 356\};$$

$$Y = \{a,b,c,d,e,f,g\}, \quad \text{and}$$
$$\mathcal{B} = \{abd, bce, cdf, deg, aef, bfg, acg\}.$$

Suppose we define the bijection α as $\alpha(1) = a$, $\alpha(2) = b$, $\alpha(3) = d$, $\alpha(4) = c$, $\alpha(5) = g$, $\alpha(6) = e$ and $\alpha(7) = f$. Then, when we relabel the points in X using α, the blocks of $\mathcal{A}$ become the following:

$$123 \to abd$$
$$145 \to acg$$
$$167 \to aef$$
$$246 \to bce$$
$$257 \to bfg$$
$$347 \to cdf$$
$$356 \to deg.$$

Thus α is an isomorphism of the two BIBDs. ∎

We need to clarify how isomorphisms affect BIBDs having repeated blocks. Suppose that $(X, \mathcal{A})$ and $(Y, \mathcal{B})$ are two (v,k,λ)-BIBDs, and suppose that $\alpha : X \to Y$ is an isomorphism of these two designs. Suppose further that $(X, \mathcal{A})$ contains c copies of the block A. Then it must also be the case that $(Y, \mathcal{B})$ contains c copies of the block $\{\alpha(x) : x \in A\}$.

We can describe isomorphism of designs in terms of incidence matrices as follows.

Theorem 1.21. *Suppose $M = (m_{i,j})$ and $N = (n_{i,j})$ are both $v \times b$ incidence matrices of designs. Then the two designs are isomorphic if and only if there exists a permutation γ of $\{1,\ldots,v\}$ and a permutation β of $\{1,\ldots,b\}$ such that*

$$m_{i,j} = n_{\gamma(i),\beta(j)}$$

for all $1 \le i \le v$, $1 \le j \le b$.

Proof. Suppose that $(X, \mathcal{A})$ and $(Y, \mathcal{B})$ are designs having $v \times b$ incidence matrices M and N, respectively. Suppose that $X = \{x_1,\ldots,x_v\}$, $Y = \{y_1,\ldots,y_v\}$, $\mathcal{A} = \{A_1,\ldots,A_b\}$, and $\mathcal{B} = \{B_1,\ldots,B_b\}$.

Suppose first that $(X, \mathcal{A})$ and $(Y, \mathcal{B})$ are isomorphic. Then, there exists a bijection $\alpha : X \to Y$ such that $[\{\alpha(x) : x \in A\} : A \in \mathcal{A}] = \mathcal{B}$. For $1 \le i \le v$, define

$$\gamma(i) = j \text{ if and only if } \alpha(x_i) = y_j.$$

Since α is a bijection of X and Y, it follows that γ is a permutation of $\{1,\ldots,v\}$.

Next, there exists a permutation β of $\{1,\ldots,b\}$ that has the property that

$$\{\alpha(x) : x \in A_j\} = B_{\beta(j)}$$

for $1 \leq j \leq b$. Such a permutation exists because α is an isomorphism of $(X, \mathcal{A})$ and $(Y, \mathcal{B})$.

Now, we have

$$\begin{aligned} m_{i,j} = 1 &\Leftrightarrow x_i \in A_j \\ &\Rightarrow y_{\gamma(i)} \in B_{\beta(j)} \\ &\Leftrightarrow n_{\gamma(i),\beta(j)} = 1. \end{aligned}$$

Conversely, suppose we have permutations γ and β such that $m_{i,j} = n_{\gamma(i),\beta(j)}$ for all i, j. Define $\alpha : X \to Y$ by the rule

$$\alpha(x_i) = y_j \text{ if and only if } \gamma(i) = j.$$

Then it is easily seen that

$$\{\alpha(x) : x \in A_j\} = B_{\beta(j)}$$

for $1 \leq j \leq b$. Hence, α defines an isomorphism of $(X, \mathcal{A})$ and $(Y, \mathcal{B})$. □

A *permutation matrix* is a $0-1$ matrix in which every row and every column contain exactly one entry equal to "1". The following corollary of Theorem 1.21 provides an alternate characterization of isomorphic designs. The proof is left to the reader.

Corollary 1.22. *Suppose M and N are incidence matrices of two (v, b, r, k, λ)-*BIBDs. *Then the two BIBDs are isomorphic if and only if there exists a $v \times v$ permutation matrix, say P, and a $b \times b$ permutation matrix, say Q, such that $M = PNQ$.*

In general, determining whether or not two designs are isomorphic is a difficult computational problem. There are $v!$ possible bijections between two sets of cardinality v. To show that two designs are not isomorphic, it must be shown that none of the $v!$ possible bijections constitutes an isomorphism. Since $v!$ grows exponentially quickly as a function of v, it soon becomes impractical to actually test every possible bijection. Fortunately, there are more sophisticated algorithms than testing every possibility exhaustively, and isomorphism testing is practical for relatively large designs.

Suppose $(X, \mathcal{A})$ is a design. An *automorphism* of $(X, \mathcal{A})$ is an isomorphism of this design with itself. In this case, the bijection α is a *permutation* of X such that

$$[\{\alpha(x) : x \in A\} : A \in \mathcal{A}] = \mathcal{A}.$$

Of course, the identity mapping on X is always a (trivial) automorphism, but a design may have other, nontrivial automorphisms.

Example 1.23. Let $(X, \mathcal{A})$ be the following (7,3,1)-BIBD:

$$X = \{1,2,3,4,5,6,7\}, \quad \text{and}$$
$$\mathcal{A} = \{123, 145, 167, 246, 257, 347, 356\}.$$

Suppose we define the permutation α as follows: $\alpha(1) = 1$, $\alpha(2) = 2$, $\alpha(3) = 3$, $\alpha(4) = 5$, $\alpha(5) = 4$, $\alpha(6) = 7$, and $\alpha(7) = 6$. Then, when we relabel the points in X using α, the blocks of $\mathcal{A}$ become the following:

$$123 \rightarrow 123$$
$$145 \rightarrow 145$$
$$167 \rightarrow 167$$
$$246 \rightarrow 257$$
$$257 \rightarrow 246$$
$$347 \rightarrow 356$$
$$356 \rightarrow 347.$$

Thus α is an automorphism of the BIBD. ∎

It is often convenient to present a permutation α on a set X using the *disjoint cycle representation*. Each cycle in this representation has the form

$$(x \; \alpha(x) \; \alpha(\alpha(x)) \; \cdots)$$

for some $x \in X$. Eventually, we get back to x, creating a cycle. The cycles thus obtained are disjoint, and they have lengths that sum to $|X|$. The *order* of the permutation α is the least common multiple of the lengths of the cycles in the disjoint cycle representation. A *fixed point* of α is a point x such that $\alpha(x) = x$; note that fixed points of α correspond to cycles of length one in the disjoint cycle representation of α.

The permutation α in the example above has the disjoint cycle representation (1)(2)(3)(4 5)(6 7). It is a permutation of order 2 that contains three fixed points.

It is easy to show that the set of all automorphisms of a BIBD $(X, \mathcal{A})$ forms a group under the operation of composition of permutations. This group is called the *automorphism group* of the BIBD and is denoted $\mathsf{Aut}(X, \mathcal{A})$. $\mathsf{Aut}(X, \mathcal{A})$ is a subgroup of the *symmetric group* $S_{|X|}$ (where S_v is the group consisting of all $v!$ permutations on a set of v elements). Note that a subgroup of S_v is called a *permutation group*, so automorphism groups of designs are examples of permutation groups.

Example 1.24. The (7,3,1)-BIBD $(X, \mathcal{A})$ in the previous example has another automorphism, $\beta = (1\,2\,4\,3\,6\,7\,5)$. The composition $\gamma = \alpha \circ \beta$ is defined as $\gamma(x) = \beta(\alpha(x))$ for all $x \in X$. It can be checked that $\gamma = (1\,2\,4)(3\,6\,5)(7)$. Thus γ is an automorphism of the BIBD because it is the composition of two automorphisms.

$(X, \mathcal{A})$ has many other automorphisms. In fact, it turns out that $\mathsf{Aut}(X, \mathcal{A})$ is a group of order 168. ∎

1.4.1 Constructing BIBDs with Specified Automorphisms

In this section, we describe a method that can often be used to determine the existence or nonexistence of a (v, k, λ)-BIBD having specified automorphisms.

Let S_v denote the symmetric group on a v-set, say X. For a positive integer $j \leq v$, let $\binom{X}{j}$ denote the set of all $\binom{v}{j}$ j-subsets of X. For a subset $Y \subseteq X$ and for a permutation $\beta \in S_v$, define

$$\beta(Y) = \{\beta(x) : x \in Y\}.$$

Suppose that G is a subgroup of S_v. Let $j \leq v$ be a positive integer, and for $A, B \in \binom{X}{j}$, define $A \sim_j B$ if $\beta(A) = B$ for some $\beta \in G$. It is not hard to prove that $\sim_j$ is an equivalence relation on $\binom{X}{j}$. The equivalence classes of this relation are called the *j-orbits* of X with respect to the group G. The j-orbits comprise a partition of the set $\binom{X}{j}$, and $\beta(A) = B$ for some $\beta \in G$ if and only if A and B are in the same *orbit* of G.

The well-known Cauchy-Frobenius-Burnside Lemma provides a method of computing the number of j-orbits of X. For each $\beta \in G$, define

$$\mathsf{fix}(\beta) = \left| \left\{ A \in \binom{X}{j} : \beta(A) = A \right\} \right|.$$

We state the following lemma without proof.

Lemma 1.25 (Cauchy-Frobenius-Burnside Lemma). *The number of j-orbits of X with respect to the group G is exactly*

$$\frac{1}{|G|} \sum_{\beta \in G} \mathsf{fix}(\beta).$$

Suppose that $\mathcal{O}_1, \ldots, \mathcal{O}_n$ are the k-orbits, and $\mathcal{P}_1, \ldots, \mathcal{P}_m$ are the 2-orbits of X with respect to the group G. We define an $n \times m$ matrix, denoted $A_{k,2}$, as follows. For $1 \leq j \leq m$, choose any 2-subset $Y_j \in \mathcal{P}_j$. Then, for $1 \leq i \leq n$, the i, j entry of $A_{k,2}$, denoted $a_{i,j}$, is defined as follows:

$$a_{i,j} = |\{A \in \mathcal{O}_i : Y_j \subseteq A\}|.$$

It can be shown that the definition of $a_{i,j}$ does not depend on the particular *orbit representatives* Y_j that are chosen; this follows immediately from the next lemma.

Lemma 1.26. *Suppose that $\mathcal{O}_1, \ldots, \mathcal{O}_n$ are the k-orbits, and $\mathcal{P}_1, \ldots, \mathcal{P}_m$ are the 2-orbits of X with respect to the group G. Suppose that $Y, Y' \in \mathcal{P}_j$ for some j, and suppose $1 \leq i \leq n$. Then*

$$|\{A \in \mathcal{O}_i : Y \subseteq A\}| = |\{A \in \mathcal{O}_i : Y' \subseteq A\}|.$$

Proof. There exists $\beta \in G$ such that $\beta(Y) = Y'$. For each $A \in \mathcal{O}_i$ such that $Y \subsetneq A$, it holds that $Y' \subseteq \beta(A)$. β is a permutation, so $\beta(A) \neq \beta(B)$ if $A \neq B$. Therefore, for each $A \in \mathcal{O}_i$ such that $Y \subseteq A$, we obtain a block $A' = \beta(A) \in \mathcal{O}_i$ such that $Y' \subseteq A'$, and the blocks $\beta(A)$, where $A \in \mathcal{O}_i$ and $Y \subseteq A$, are all distinct. Therefore

$$|\{A \in \mathcal{O}_i : Y \subseteq A\}| \leq |\{A \in \mathcal{O}_i : Y' \subseteq A\}|.$$

The inequality in the opposite direction follows by interchanging the roles of Y and Y', and replacing β by β^{-1}. Combining the two inequalities, the desired result is proven. □

Here now is the main result of this section.

Theorem 1.27 (Kramer-Mesner Theorem). *There exists a* (v,k,λ)-BIBD *having G as a subgroup of its automorphism group if and only if there exists a solution* $\mathbf{z} \in \mathbb{Z}^n$ *to the matrix equation*

$$\mathbf{z}A_{k,2} = \lambda \mathbf{u}_m, \tag{1.1}$$

where $\mathbf{z}$ *has nonnegative entries.*

Proof. We give a sketch of the proof. First, suppose that $\mathbf{z} = (z_1, \ldots, z_n)$ is a nonnegative integral solution to equation (1.1). Define

$$\mathcal{A} = \bigcup_{i=1}^{n} z_i \mathcal{O}_i.$$

The notation above is a multiset union; it means that $\mathcal{A}$ is formed by taking z_i copies of every block in $\mathcal{O}_i$ for $1 \leq i \leq n$. It is easy to see that $(X, \mathcal{A})$ is a (v,k,λ)-BIBD having G as a subgroup of its automorphism group.

Conversely, suppose that $(X, \mathcal{A})$ is the desired BIBD. Then $\mathcal{A}$ necessarily must consist of a multiset union of the orbits $\mathcal{O}_i$, $1 \leq i \leq n$. Let z_i denote the number of times each of the blocks of the orbit $\mathcal{O}_i$ occurs in $\mathcal{A}$; then $\mathbf{z} = (z_1, \ldots, z_n)$ is a nonnegative integral solution to equation (1.1). □

As an additional remark, we observe that the BIBD in Theorem 1.27 is simple if and only if the vector $\mathbf{z} \in \{0,1\}^n$.

Example 1.28. We use the technique described above to construct a $(6,3,2)$-BIBD having an automorphism of order 5. Suppose that $\alpha = (0\ 1\ 2\ 3\ 4)(5)$ and $G = \{\alpha^i : 0 \leq i \leq 4\}$. It is easy to see that there are three 2-orbits of $X = \{0,1,2,3,4,5\}$, namely

$$\mathcal{P}_1 = \{01, 12, 23, 34, 40\},$$
$$\mathcal{P}_2 = \{02, 13, 24, 30, 41\}, \quad \text{and}$$
$$\mathcal{P}_3 = \{05, 15, 25, 35, 45\}.$$

Also, there are four 3-orbits:

$$\begin{aligned}\mathcal{O}_1 &= \{012, 123, 234, 340, 401\},\\ \mathcal{O}_2 &= \{013, 124, 230, 341, 402\},\\ \mathcal{O}_3 &= \{015, 125, 235, 345, 405\}, \quad \text{and}\\ \mathcal{O}_4 &= \{025, 135, 245, 305, 415\}.\end{aligned}$$

The matrix $A_{3,2}$ is as follows:

$$A_{3,2} = \begin{pmatrix} 2\,1\,0 \\ 1\,2\,0 \\ 1\,0\,2 \\ 0\,1\,2 \end{pmatrix}.$$

The equation $\mathbf{z}A_{3,2} = 2\mathbf{u}_3$ has exactly two nonnegative integral solutions: $\mathbf{z} = (1,0,0,1)$ and $\mathbf{z} = (0,1,1,0)$. Each of these solutions yields a $(6,3,2)$-BIBD having α as an automorphism. ∎

Here is a more interesting example, in which the orbits do not all have the same size.

Example 1.29. We construct a $(9,3,1)$-BIBD having a certain automorphism of order six. Suppose that $\alpha = (0\,1\,2\,3\,4\,5)(6\,7\,8)$ and $G = \{\alpha^i : 0 \le i \le 5\}$. The permutations in G are as follows:

$$\begin{aligned}\alpha &= (0\,1\,2\,3\,4\,5)(6\,7\,8),\\ \alpha^2 &= (0\,2\,4)(1\,3\,5)(6\,8\,7),\\ \alpha^3 &= (0\,3)(1\,4)(2\,5)(6)(7)(8),\\ \alpha^4 &= (0\,4\,2)(1\,5\,3)(6\,7\,8),\\ \alpha^5 &= (0\,5\,4\,3\,2\,1)(6\,8\,7), \quad \text{and}\\ \alpha^0 &= (0)(1)(2)(3)(4)(5)(6)(7)(8).\end{aligned}$$

Lemma 1.25 can be used to compute the number of 2- and 3-orbits. First we consider 2-orbits. It is not hard to see that $\mathsf{fix}(\alpha) = \mathsf{fix}(\alpha^2) = \mathsf{fix}(\alpha^4) = \mathsf{fix}(\alpha^5) = 0$, $\mathsf{fix}(\alpha^3) = 6$, and $\mathsf{fix}(\alpha^0) = \binom{9}{2} = 36$. Therefore, the number of 2-orbits is $(36+6)/6 = 7$.

Now we turn to 3-orbits. It is not hard to check that $\mathsf{fix}(\alpha) = \mathsf{fix}(\alpha^5) = 1$, $\mathsf{fix}(\alpha^2) = \mathsf{fix}(\alpha^4) = 3$, $\mathsf{fix}(\alpha^3) = 10$, and $\mathsf{fix}(\alpha^0) = \binom{9}{3} = 84$. Therefore, the number of 3-orbits is $(84+10+2(3)+2(1))/6 = 17$.

We leave it as an exercise for the reader to construct the $A_{3,2}$ matrix and solve the matrix equation. It turns out that there is a solution; the following $(9,3,1)$-BIBD, consisting of four of the 3-orbits, has α as an automorphism:

orbit	orbit size
018 126 237 348 456 507	6
036 147 258	3
024 135	2
678	1

The total number of blocks is 12, as it must be. ∎

It is, in general, a nontrivial task to construct an $A_{k,2}$ matrix if the set X is even of moderate size. It is a considerably more difficult problem to find the desired integral solution to the matrix equation (and of course there is no guarantee that the sought-after solution even exists). The known algorithms to find nonnegative integral solutions of matrix equations have exponential complexity and may require enormous amounts of computing time to run to completion. Nevertheless, this approach to finding designs having specified automorphisms has been very useful in practice in discovering previously unknown designs.

1.5 New BIBDs from Old

In this section, we give two simple methods of constructing new BIBDs from old. The first construction can be called the "sum construction". Given two BIBDs on the same point set, it involves forming the collection of all the blocks in both designs.

Theorem 1.30 (Sum Construction). *Suppose there exists a* (v,k,λ_1)-BIBD *and a* (v,k,λ_2)-BIBD. *Then there exists a* $(v,k,\lambda_1+\lambda_2)$-BIBD.

Corollary 1.31. *Suppose there exists a* (v,k,λ)-BIBD. *Then there exists a* $(v,k,s\lambda)$-BIBD *for all integers* $s \geq 1$.

Note that the BIBDs produced by Corollary 1.31 with $s \geq 2$ are not simple designs, even if the initial (v,k,λ)-BIBD is simple. For $\lambda > 1$, construction of simple BIBDs is, in general, more difficult than construction of BIBDs with repeated blocks.

To illustrate an application of the sum construction, let us consider $(16,6,\lambda)$-BIBDs. We will see in the next section that there does not exist a $(16,6,1)$-BIBD. However, both a $(16,6,2)$-BIBD and a $(16,6,3)$-BIBD are known to exist. By application of the sum construction, it then follows that there exists a $(16,6,\lambda)$-BIBD if and only if $\lambda > 1$.

The second construction is called "block complementation". Suppose $(X,\mathcal{A})$ is a BIBD, and we replace every block $A \in \mathcal{A}$ by $X \backslash A$. The result is again a BIBD, as stated in the following theorem.

Theorem 1.32 (Block Complementation). *Suppose there exists a* (v,b,r,k,λ)-BIBD, *where* $k \leq v-2$. *Then there also exists a* $(v,b,b-r,v-k,b-2r+\lambda)$-BIBD.

Proof. Suppose $(X,\mathcal{A})$ is a (v,b,r,k,λ)-BIBD. We will show that

$$(X,\{X\backslash A : A \in \mathcal{A}\})$$

is a BIBD. Clearly, this design has v points and b blocks, every block contains $v-k \geq 2$ points, and every point occurs in $b-r$ blocks. Hence, we just need to show that every pair of points occurs in exactly $b-2r+\lambda$ blocks.

Let $x, y \in X$, $x \neq y$. Define

$$\begin{aligned} a_1 &= |\{A \in \mathcal{A} : x, y \in A\}|, \\ a_2 &= |\{A \in \mathcal{A} : x \in A, y \notin A\}|, \\ a_3 &= |\{A \in \mathcal{A} : x \notin A, y \in A\}|, \quad \text{and} \\ a_4 &= |\{A \in \mathcal{A} : x, y \notin A\}|. \end{aligned}$$

Then it is easy to see that

$$\begin{aligned} a_1 &= \lambda, \\ a_1 + a_2 &= r, \\ a_1 + a_3 &= r, \quad \text{and} \\ a_1 + a_2 + a_3 + a_4 &= b. \end{aligned}$$

These four equations may be solved easily to obtain

$$a_4 = b - 2r + \lambda,$$

as desired. □

For example, the complement of a $(7,3,1)$-BIBD is a $(7,4,2)$-BIBD, and the complement of a $(9,3,1)$-BIBD is a $(9,6,5)$-BIBD. In view of Theorem 1.32, it suffices to study BIBDs with $k \leq v/2$.

1.6 Fisher's Inequality

We have already discussed two necessary conditions for the existence of a (v,k,λ)-BIBD, namely Theorems 1.8 and 1.9. Another important necessary condition is known as "Fisher's Inequality".

Theorem 1.33 (Fisher's Inequality). *In any (v,b,r,k,λ)-BIBD, $b \geq v$.*

Proof. Let $(X, \mathcal{A})$ be a (v,b,r,k,λ)-BIBD, where $X = \{x_1, \ldots, x_v\}$ and $\mathcal{A} = \{A_1, \ldots, A_b\}$. Let M be the incidence matrix of this BIBD, and define $\mathbf{s}_j$ to be the jth row of M^T (equivalently, $\mathbf{s}_j{}^T$ is the jth column of M). Note that $\mathbf{s}_1, \ldots, \mathbf{s}_b$ are all v-dimensional vectors in the real vector space $\mathbb{R}^v$.

Define $S = \{\mathbf{s}_j : 1 \leq j \leq b\}$ and define $\mathbf{S} = \mathsf{span}(\mathbf{s}_j : 1 \leq j \leq b)$. $\mathbf{S}$ is the subspace of $\mathbb{R}^v$ spanned by the $\mathbf{s}_j$'s; it consists of the following vectors:

$$\mathbf{S} = \left\{ \sum_{j=1}^{b} \alpha_j \mathbf{s}_j : \alpha_1, \ldots, \alpha_b \in \mathbb{R} \right\}.$$

In other words, $\mathbf{S}$ consists of all linear combinations of the vectors $\mathbf{s}_1, \ldots, \mathbf{s}_b$.

We will prove that $\mathbf{S} = \mathbb{R}^v$; i.e., the b vectors in S span the vector space $\mathbb{R}^v$. Since $\mathbb{R}^v$ has dimension v and is spanned by a set of b vectors, it must be the case that $b \geq v$.

Our task is thus to show that $\mathbf{S} = \mathbb{R}^v$. For $1 \leq i \leq v$, define $\mathbf{e}_i \in \mathbb{R}^v$ to be the vector with a "1" in the ith coordinate and "0"s in all other coordinates. The vectors $\mathbf{e}_1, \ldots, \mathbf{e}_v$ form a basis for $\mathbb{R}^v$, so every vector in $\mathbb{R}^v$ can be expressed as a linear combination of these v vectors. Therefore, to show that $\mathbf{S} = \mathbb{R}^v$, it suffices to show that $\mathbf{e}_i \in \mathbf{S}$ for $1 \leq i \leq v$ (i.e., that each basis vector $\mathbf{e}_i$ can be expressed as a linear combination of vectors in S).

First, we observe that

$$\sum_{j=1}^{b} \mathbf{s}_j = (r, \ldots, r), \tag{1.2}$$

from which it follows that

$$\sum_{j=1}^{b} \frac{1}{r} \mathbf{s}_j = (1, \ldots, 1). \tag{1.3}$$

Next, fix a value i, $1 \leq i \leq v$. Then we have

$$\sum_{\{j : x_i \in A_j\}} \mathbf{s}_j = (r - \lambda)\mathbf{e}_i + (\lambda, \ldots, \lambda). \tag{1.4}$$

Since $\lambda(v-1) = r(k-1)$ and $v > k$, it follows that $r > \lambda$, and hence $r - \lambda \neq 0$. Then we can combine equations (1.3) and (1.4) to obtain

$$\mathbf{e}_i = \sum_{\{j : x_i \in A_j\}} \frac{1}{r - \lambda} \mathbf{s}_j - \sum_{j=1}^{b} \frac{\lambda}{r(r - \lambda)} \mathbf{s}_j. \tag{1.5}$$

Equation (1.5) gives a formula expressing $\mathbf{e}_i$ as a linear combination of $\mathbf{s}_1, \ldots, \mathbf{s}_b$, as desired. □

Note that the conclusion of Theorem 1.33, $b \geq v$, can be stated in other, equivalent ways, such as $r \geq k$ and $\lambda(v-1) \geq k^2 - k$.

As an example, consider the parameter set $(16, 6, 1)$. In a $(16, 6, 1)$-BIBD, we would have $r = 3$, but it would then be the case that $r < k$, which is impossible. Hence, a $(16, 6, 1)$-BIBD does not exist.

Theorem 1.33 can easily be generalized to regular pairwise balanced designs. We have the following.

Theorem 1.34. *In any nontrivial regular pairwise balanced design, $b \geq v$.*

Proof. By examining the proof of Theorem 1.33, it can be seen that the fact that all blocks have the same size is not used in the proof. Therefore, Fisher's Inequality holds for regular pairwise balanced designs in which $r > \lambda$. It is easy to see that a regular PBD has $r > \lambda$ if and only if it is not a trivial PBD. Therefore we conclude that Fisher's Inequality is valid for all nontrivial regular PBDs. □

In fact, Fisher's Inequality holds for all nontrivial pairwise balanced designs (not just the regular ones), but a slightly different proof is required. We will return to this topic in Chapter 8.

1.7 Notes and References

Fisher's Inequality was first proven in 1940 by the famous statistician Ronald Fisher [45]. There are many proofs of this result; we have chosen to employ a linear-algebraic proof technique that will be used to prove several other results later in this book.

The Kramer-Mesner Theorem was proven in 1975 in [71]. It has since been used to find many previously unknown designs. For a nice survey of computational techniques in design theory, see Gibbons [47].

There are several reference books and textbooks on combinatorial design theory. The book "Combinatorial Designs" by Wallis [115] is a fairly easy-to-read general introduction. Two other good introductory textbooks are "Combinatorial Designs and Tournaments" by Anderson [2] and "Design Theory" by Lindner and Rodger [77]. A more advanced book that contains a great deal of useful information is the two-volume work also entitled "Design Theory" by Beth, Jungnickel, and Lenz [9, 10]. The reader can also profitably consult "Design Theory" by Hughes and Piper [61] and "Combinatorics of Experimental Design" [107] by Street and Street (however, these two books are currently out of print).

The "CRC Handbook of Combinatorial Designs", edited by Colbourn and Dinitz [27], is an enormous, encyclopedic reference work that is a valuable resource for researchers. This book also has an on-line Web page located at the following URL: `http://www.emba.uvm.edu/~dinitz/hcd.html`. "Contemporary Design Theory, A Collection of Surveys", edited by Dinitz and Stinson [41], is a collection of twelve surveys on various topics in design theory.

Two books that explore the links between combinatorial design theory and other branches of combinatorial mathematics are "Designs, Codes, Graphs and Their Links" by Cameron and van Lint [20] and "Combinatorial Configurations: Designs, Codes, Graphs" by Tonchev [110].

Several "general" combinatorics textbooks contain one or more sections on designs. Three books that are worth consulting are "Combinatorics: Topics, Techniques, Algorithms", by Cameron [19]; "Combinatorial Theory (Second Edition)", by Hall [53]; and "A Course in Combinatorics (Second Edition)", by Van Lint and Wilson [79].

Much recent research on combinatorial designs can be found in the *Journal of Combinatorial Designs*, which has been published by John Wiley & Sons since 1993.

1.8 Exercises

1.1 What is the value of b in a $(46, 6, 1)$-BIBD (if it exists)?

1.2 What is the value of r in a $(65, 5, 1)$-BIBD?

1.3 For all integers k and v such that $3 \leq k \leq v/2$ and $v \leq 10$, determine the smallest integer λ such that the parameter set (v, k, λ) satisfies the necessary conditions stated in Corollary 1.10.

1.4 For an integer $k \geq 2$, let $\lambda^*(k)$ denote the minimum integer such that the conditions stated in Corollary 1.10 are satisfied for all integers $v > k$.

(a) Compute $\lambda^*(k)$ for $k = 3, 4, 5$ and 6.

(b) Prove that

$$\lambda^*(k) = \begin{cases} \binom{k}{2} & \text{if } k \text{ is even} \\ k(k-1) & \text{if } k \text{ is odd.} \end{cases}$$

1.5 Let M be the incidence matrix of a $(v, b, r, k, 1)$-BIBD and define $N = M^T M$. Denote $N = (n_{i,j})$. Prove that

$$n_{i,j} = \begin{cases} k & \text{if } i = j \\ 0 \text{ or } 1 & \text{if } i \neq j. \end{cases}$$

1.6 Construct a regular pairwise balanced design on six points that contains exactly four blocks of size three.

1.7 Give a complete proof of Theorem 1.15.

1.8 Give a complete proof of Theorem 1.17.

1.9 (a) Prove that no $(6, 3, 2)$-BIBD can contain repeated blocks.

(b) Prove that all $(6, 3, 2)$-BIBDs are isomorphic.

1.10 Give a complete proof of Corollary 1.22.

1.11 Show that all $(7, 3, 1)$-BIBDs are isomorphic by the following method. (Fill in the details of the proof.)

(a) Without loss of generality, we can take the points to be $\{1, \ldots, 7\}$, and let the blocks containing the point 1 be $\{1, 2, 3\}, \{1, 4, 5\}$, and $\{1, 6, 7\}$.

(b) Find all ways to complete this structure to a $(7, 3, 1)$-BIBD.

(c) Then show that all the designs obtained are isomorphic.

1.12 Find an isomorphism π of the two $(9, 3, 1)$-BIBDs $(X, \mathcal{A})$ and $(Y, \mathcal{B})$, and give a complete verification that the two BIBDs are isomorphic.

$$X = \{1, 2, 3, 4, 5, 6, 7, 8, 9\}$$
$$\mathcal{A} = \{123, 147, 159, 168, 258, 267, 249, 369, 348, 357, 456, 789\}$$
$$Y = \{a, b, c, d, e, f, g, h, i\}$$
$$\mathcal{B} = \{abe, acd, afi, agh, bcf, bdg, bhi, ceh, cgi, dfh, dei, efg\}.$$

Hint: Observe that if $\pi(x) = \alpha$, $\pi(y) = \beta$, $\{x, y, z\} \in \mathcal{A}$, and $\{\alpha, \beta, \gamma\} \in \mathcal{B}$, then it must be the case that $\pi(z) = \gamma$.

1.13 Suppose we arrange the elements of a set $X = \{0, \ldots, 15\}$ in a 4×4 array A as follows:

$$A = \begin{pmatrix} 0 & 1 & 2 & 3 \\ 4 & 5 & 6 & 7 \\ 8 & 9 & 10 & 11 \\ 12 & 13 & 14 & 15 \end{pmatrix}.$$

For each x, $0 \leq x \leq 15$, suppose we define a block B_x consisting of the elements in the same row or column of A as x, excluding x. Then define a set of blocks $\mathcal{B} = \{B_x : 0 \leq x \leq 15\}$. We are going to study the design $(X, \mathcal{B})$.

(a) Prove that this design is a $(16, 6, 2)$-BIBD.
(b) Construct the incidence matrix of this BIBD.
(c) Prove that the mapping $\alpha(x) = (x + 4) \bmod 16$ is an automorphism of this BIBD.
(d) Prove that this BIBD has automorphisms of orders 2, 3, and 4.

1.14 Suppose that α is an automorphism of order p of a $(v, k, 1)$-BIBD, where p is prime. Let α_f denote the number of fixed points in α.

(a) Prove that $\alpha_f \equiv v \pmod{p}$.
(b) Suppose that $2 \leq \alpha_f \leq k - 1$. Prove that $k \geq p + 2$.
(c) As a corollary, prove that a $(7, 3, 1)$-BIBD cannot have an automorphism of order 5.

1.15 Let G be the permutation group of order 3 on the set $X = \{1, \ldots, 7\}$ that is generated by the permutation $\alpha = (1\,2\,3)(4\,5\,6)(7)$.

(a) Use Lemma 1.25 to compute the number of 2- and 3-orbits of X with respect to G.
(b) Use Theorem 1.27 to find all $(7, 3, 1)$-BIBDs having α as an autormorphism.

1.16 Referring to Example 1.29, carry out the following computations.

(a) Construct all the 2-orbits and 3-orbits.
(b) Construct the $A_{3,2}$ matrix.
(c) Find all solutions to the matrix equation $\mathbf{z}A_{3,2} = \mathbf{u}_7$.

1.17 Construct $(9, 3, 1)$-BIBDs having the following permutations as automorphisms.

(a) $(1)(2\,3\,4\,5\,6\,7\,8\,9)$.
(b) $(1)(2)(3)(4\,5\,6)(7\,8\,9)$.
(c) $(1)(2)(3)(4\,5)(6\,7)(8\,9)$.

1.18 (a) Construct a $(7, 4, 2)$-BIBD.
(b) Determine the incidence matrix of this BIBD.
(c) For the incidence matrix that you have computed, express the vector $\mathbf{e}_3$ as a linear combination of the vectors $\mathbf{s}_1, \ldots, \mathbf{s}_7$ using (1.5). Then verify that the resulting linear combination indeed yields the vector $\mathbf{e}_3$.

1.19 Let B_0 be a block in a $(v, k, 1)$-BIBD, say $(X, \mathcal{B})$.

(a) Find a formula for the number of blocks $B \in \mathcal{B}$ such that $|B \cap B_0| = 1$.

(b) Use your formula to show that $b \geq k(r-1)+1$ if a $(v, k, 1)$-BIBD exists.

(c) Using the facts that $vr = bk$ and $v = r(k-1)+1$, deduce that $(r-k)(r-1)(k-1) \geq 1$, and hence $r \geq k$, which implies Fisher's Inequality.

1.20 Let B_0 be a block in a $(v, k, 1)$-BIBD, say $(X, \mathcal{B})$. Let $x \in X \backslash B_0$, and show that there are at least k blocks that contain x and intersect B_0. From this, deduce that $r \geq k$, which implies Fisher's Inequality.

2

Symmetric BIBDs

2.1 An Intersection Property

Definition 2.1. *A BIBD in which $b = v$ (or, equivalently, $r = k$ or $\lambda(v-1) = k^2 - k$) is called a* symmetric BIBD. *(Note that this terminology does not mean that the incidence matrix is a symmetric matrix.)*

A simple but rather trivial family of symmetric BIBDs can be obtained from Example 1.6 when $v = k + 1$. These are symmetric $(v, v-1, v-2)$-BIBDs. We will see many examples of more interesting symmetric BIBDs later in this and other chapters.

In the next three chapters, we study various properties and constructions of symmetric BIBDs. We begin by stating and proving an important theorem about the intersections of blocks in a symmetric BIBD.

Theorem 2.2. *Suppose that $(X, \mathcal{A})$ is a symmetric (v, k, λ)-BIBD and denote $\mathcal{A} = \{A_1, \ldots, A_v\}$. Suppose that $1 \leq i, j \leq v$, $i \neq j$. Then $|A_i \cap A_j| = \lambda$.*

Proof. We use the same notation as in the proof of Theorem 1.33 (Fisher's Inequality). Fix a value h, $1 \leq h \leq b$. Applying equations (1.2) and (1.4), we have the following:

$$\begin{aligned}
\sum_{\{i:x_i \in A_h\}} \sum_{\{j:x_i \in A_j\}} \mathbf{s}_j &= \sum_{\{i:x_i \in A_h\}} ((r-\lambda)\mathbf{e}_i + (\lambda, \ldots, \lambda)) \\
&= (r-\lambda)\mathbf{s}_h + k(\lambda, \ldots, \lambda) \\
&= (r-\lambda)\mathbf{s}_h + \sum_{j=1}^{b} \frac{\lambda k}{r} \mathbf{s}_j.
\end{aligned}$$

On the other hand, we can compute this double sum in a different way by interchanging the order of summation:

$$\sum_{\{i:x_i\in A_h\}} \sum_{\{j:x_i\in A_j\}} \mathbf{s}_j = \sum_{j=1}^{b} \sum_{\{i:x_i\in A_h\cap A_j\}} \mathbf{s}_j$$
$$= \sum_{j=1}^{b} |A_h \cap A_j| \mathbf{s}_j.$$

Hence, we have that

$$(r-\lambda)\mathbf{s}_h + \sum_{j=1}^{b} \frac{\lambda k}{r} \mathbf{s}_j = \sum_{j=1}^{b} |A_h \cap A_j| \mathbf{s}_j. \tag{2.1}$$

Since $b = v$ and $r = k$, we can rewrite equation (2.1) as

$$(r-\lambda)\mathbf{s}_h + \sum_{j=1}^{v} \lambda \mathbf{s}_j = \sum_{j=1}^{v} |A_h \cap A_j| \mathbf{s}_j. \tag{2.2}$$

In the proof of Theorem 1.33, we showed that $\mathbf{S} = \mathbb{R}^v$, where

$$\mathbf{S} = \left\{ \sum_{j=1}^{b} \alpha_j \mathbf{s}_j : \alpha_1, \ldots, \alpha_b \in \mathbb{R} \right\}.$$

Since we are now assuming that $b = v$, it must be the case that S is a basis for $\mathbb{R}^v$. Since S is a basis for $\mathbb{R}^v$, the coefficients of any $\mathbf{s}_j$ on the left and right sides of equation (2.2) must be equal. Therefore,

$$|A_h \cap A_j| = \lambda$$

for all $j \neq h$. Since h was chosen arbitrarily, it follows that $|A \cap A'| = \lambda$ for any two blocks $A \neq A'$. □

We observed in Theorem 1.34 that Fisher's Inequality also holds for nontrivial regular pairwise balanced designs. The next theorem shows that nontrivial regular pairwise balanced designs with $b = v$ are, in fact, symmetric BIBDs.

Theorem 2.3. *Suppose that $(X, \mathcal{A})$ is a nontrivial regular pairwise balanced design with $b = v$. Then $(X, \mathcal{A})$ is a (symmetric) (v, k, λ)-BIBD.*

Proof. We compute the sum

$$\sum_{i=1}^{v} \sum_{\{j:x_i\in A_j\}} \mathbf{s}_j$$

in two ways. First, we have that

$$\begin{aligned}\sum_{i=1}^{v} \sum_{\{j: x_i \in A_j\}} \mathbf{s}_j &= \sum_{i=1}^{v} ((r-\lambda)\mathbf{e}_i + (\lambda, \ldots, \lambda)) \\ &= (r - \lambda + \lambda v)(1, \ldots, 1) \\ &= \frac{\lambda(v-1)+r}{r} \sum_{j=1}^{b} \mathbf{s}_j.\end{aligned}$$

On the other hand, we can compute

$$\begin{aligned}\sum_{i=1}^{v} \sum_{\{j: x_i \in A_j\}} \mathbf{s}_j &= \sum_{j=1}^{b} \sum_{\{i: x_i \in A_j\}} \mathbf{s}_j \\ &= \sum_{j=1}^{b} |A_j|\, \mathbf{s}_j.\end{aligned}$$

Now, using the facts that $b = v$ and S is a basis for $\mathbb{R}^v$, it follows that

$$|A_j| = \frac{\lambda(v-1)+r}{r}$$

for $1 \leq j \leq b$. Hence, $(X, \mathcal{A})$ is a (v, k, λ)-BIBD, where $k = (\lambda(v-1)+r)/r$. □

The next result is an immediate consequence of Theorems 1.17 and 2.2.

Corollary 2.4. *Suppose M is the incidence matrix of a symmetric (v, k, λ)-BIBD. Then M^T is also the incidence matrix of a (symmetric) (v, k, λ)-BIBD.*

Corollary 2.4 says that the dual of a symmetric BIBD is again a symmetric BIBD. We note that these two BIBDs need not be identical or even isomorphic.

Here is another corollary of the results of this section. This result is a converse to Theorem 2.2.

Corollary 2.5. *Suppose that μ is a positive integer and $(X, \mathcal{A})$ is a (v, b, r, k, λ)-BIBD such that $|A \cap A'| = \mu$ for all $A, A' \in \mathcal{A}$. Then $(X, \mathcal{A})$ is a symmetric BIBD and $\mu = \lambda$.*

Proof. Theorem 1.17 ensures that the dual of $(X, \mathcal{A})$ is a (b, v, k, r, μ)-BIBD. Fisher's Inequality (for $(X, \mathcal{A})$) implies that $b \geq v$, and Fisher's Inequality (for the dual design) implies that $v \geq b$. Hence $b = v$, and then $\mu = \lambda$ follows from Theorem 2.2. □

2.2 Residual and Derived BIBDs

Recall that Theorem 2.2 states that any two blocks of a symmetric BIBD contain λ common points. This result provides another method of constructing new BIBDs from old.

Definition 2.6. *Suppose that* $(X, \mathcal{A})$ *is a symmetric* (v, k, λ)-BIBD, *and let* $A_0 \in \mathcal{A}$. *Define*

$$\mathsf{Der}(X, \mathcal{A}, A_0) = (A_0, \{A \cap A_0 : A \in \mathcal{A}, A \neq A_0\})$$

and define

$$\mathsf{Res}(X, \mathcal{A}, A_0) = (X \backslash A_0, \{A \backslash A_0 : A \in \mathcal{A}, A \neq A_0\}).$$

$\mathsf{Der}(X, \mathcal{A}, A_0)$ *is called a* derived BIBD, *and* $\mathsf{Res}(X, \mathcal{A}, A_0)$ *is called a* residual BIBD.

We form a derived design by deleting all the points not in a given block A_0 and then deleting A_0. The residual design is constructed by deleting all points in A_0.

It is clear that the derived and residual designs are BIBDs, provided that the block sizes are at least two, and at most the number of points minus one.

Theorem 2.7. *Suppose that* $(X, \mathcal{A})$ *is a symmetric* (v, k, λ)-BIBD, *and let* $A_0 \in \mathcal{A}$. *Then* $\mathsf{Der}(X, \mathcal{A}, A_0)$ *is a* $(k, v-1, k-1, \lambda, \lambda-1)$-BIBD *provided that* $\lambda \geq 2$. *Furthermore,* $\mathsf{Res}(X, \mathcal{A}, A_0)$ *is a* $(v-k, v-1, k, k-\lambda, \lambda)$-BIBD *provided that* $k \geq \lambda + 2$.

Proof. $\mathsf{Der}(X, \mathcal{A}, A_0)$ is a BIBD with the stated parameters provided that $k > \lambda \geq 2$ (k is the number of points in the derived design, and the blocks have size λ). However, $k > \lambda$ in any symmetric BIBD because $\lambda(v-1) = k(k-1)$ and $v > k$, so this condition is superfluous.

$\mathsf{Res}(X, \mathcal{A}, A_0)$ is a BIBD with the stated parameters provided that $v - k > k - \lambda \geq 2$ ($v - k$ is the number of points in the residual design, and the blocks have size $k - \lambda$). We now prove that $v - k > k - \lambda$ in a symmetric BIBD. Suppose that $v \leq 2k - \lambda$; then we have $k(k-1) = \lambda(v-1) \leq \lambda(2k - \lambda - 1)$. This is equivalent to $(k-\lambda)(k-\lambda-1) \leq 0$. But k and λ are integers, so this last inequality holds if and only if $k = \lambda$ or $k = \lambda + 1$. We are assuming that $k \geq \lambda + 2$, so we have a contradiction. Therefore the condition $v - k > k - \lambda$ is superfluous. □

Let's consider an example:

Example 2.8. An $(11, 5, 2)$-BIBD is symmetric because $2(11-1) = 5(5-1)$. A residual BIBD is a $(6, 3, 2)$-BIBD, and a derived BIBD is a $(5, 2, 1)$-BIBD. In Figure 2.1, we have written out the 11 blocks in an $(11, 5, 2)$-BIBD. The block $A_0 = \{1, 3, 4, 5, 9\}$. The remaining 10 blocks are each partitioned into two parts, which form a $(5, 2, 1)$-BIBD on point set $\{1, 3, 4, 5, 9\}$ and a $(6, 3, 2)$-BIBD on point set $\{0, 2, 6, 7, 8, 10\}$. ∎

Suppose we write the parameters $(v-k, v-1, k, k-\lambda, \lambda)$ of a residual BIBD as $(v', b', r', k', \lambda')$. These parameters satisfy the numerical condition $r' = k' + \lambda'$. A (v, b, r, k, λ)-BIBD with $r = k + \lambda$ is called a *quasiresidual BIBD*. A quasiresidual (v, b, r, k, λ)-BIBD can be constructed as the residual BIBD

1 3	4 5 9
4 5	2 6 10
3 5	6 7 0
1 4	6 7 8
5 9	2 7 8
3 9	6 8 10
4 9	0 7 10
1 5	0 8 10
1 9	2 6 0
1 3	2 7 10
3 4	0 2 8

Fig. 2.1. Derived and Residual BIBDS of a Symmetric (11, 5, 2)-BIBD

of a symmetric $(v+r, r, \lambda)$-BIBD, provided that this symmetric BIBD exists. (The numerical condition $\lambda(v+r-1) = r(r-1)$ necessarily holds when $r = k + \lambda$, but this does not guarantee existence of the symmetric BIBD.)

Similarly, we can write the parameters $(k, v-1, k-1, \lambda, \lambda-1)$ of a derived BIBD as $(v', b', r', k', \lambda')$. These parameters satisfy the numerical condition $k' = \lambda' + 1$. Any (v, b, r, k, λ)-BIBD with $k = \lambda + 1$ is called a *quasiderived BIBD*. A quasiderived (v, b, r, k, λ)-BIBD can be constructed as the derived BIBD of a symmetric $(b+1, r+1, \lambda+1)$-BIBD, provided that this symmetric BIBD exists. (Again, the numerical condition $(\lambda+1)b = r(r+1)$ necessarily holds when $k = \lambda + 1$, but this does not guarantee existence of the symmetric BIBD.)

Here are a couple of examples. The parameter set $(10, 15, 6, 4, 2)$ is quasiresidual because $6 = 4 + 2$. Therefore a $(10, 15, 6, 4, 2)$-BIBD exists if a (symmetric) $(16, 6, 2)$-BIBD exists. The parameter set $(9, 19, 8, 4, 3)$ is quasiderived because $4 = 3 + 1$. Therefore a $(9, 18, 8, 4, 3)$-BIBD exists if a (symmetric) $(19, 9, 4)$-BIBD exists. Both of these symmetric BIBDs exist, so it follows from Theorem 2.7 that a $(10, 15, 6, 4, 2)$-BIBD and a $(9, 18, 8, 4, 3)$-BIBD both exist.

It is clear from the definitions that a residual BIBD is quasiresidual and a derived BIBD is quasiderived. The converse is, in general, not true. However, we will show in Theorem 5.10 that every quasiresidual BIBD with $\lambda = 1$ is residual. (It is also true that any quasiresidual BIBD with $\lambda = 2$ is residual, but this is much harder to prove.)

2.3 Projective Planes and Geometries

Definition 2.9. *An* $(n^2+n+1, n+1, 1)$-BIBD *with* $n \geq 2$ *is called a* projective plane *of* order n.

Observe that a $(3, 2, 1)$-BIBD certainly exists. For technical reasons, however, this BIBD is not regarded as being a projective plane of order 1. Noting that $1(n^2+n)=(n+1)n$, we see that projective planes are symmetric BIBDs. Therefore, from Theorem 2.2, every point occurs in $n+1$ blocks and every pair of blocks intersects in a unique point.

We now prove that a projective plane of order q exists whenever q is a prime power. Suppose q is a prime power. Let $\mathbb{F}_q$ be the finite field of order q, and let V denote the three-dimensional vector space over $\mathbb{F}_q$. (To save space, we will write vectors $(x_1,x_2,x_3) \in V$ in the form $x_1x_2x_3$.)

Let $\mathcal{V}_1$ consist of all the one-dimensional subspaces of V, and let $\mathcal{V}_2$ consist of all the two-dimensional subspaces of V. For each $B \in \mathcal{V}_2$, define a block

$$A_B = \{C \in \mathcal{V}_1 : C \subseteq B\}.$$

Finally, define

$$\mathcal{A} = \{A_B : B \in \mathcal{V}_2\}.$$

We claim that $(\mathcal{V}_1, \mathcal{A})$ is a projective plane of order q.

First, observe that $|C| = q$ and $000 \in C$ for all $C \in \mathcal{V}_1$. The sets $C\backslash\{000\}$, $C \in \mathcal{V}_1$, form a partition of $V\backslash\{000\}$. Hence,

$$|\mathcal{V}_1| = \frac{q^3-1}{q-1} = q^2+q+1.$$

Next, let $B \in \mathcal{V}_2$. Clearly $|B| = q^2$. The sets $C\backslash\{000\}$ such that $C \in \mathcal{V}_1$ and $C \subseteq B$ partition the set $B\backslash\{000\}$. Hence, it follows that

$$|A_B| = \frac{q^2-1}{q-1} = q+1.$$

Finally, let $C, D \in \mathcal{V}_1$, $C \neq D$. Clearly there is a unique two-dimensional subspace B containing the one-dimensional subspaces C and D. This subspace determines the unique block A_B containing the points C and D.

The discussion above establishes the following theorem.

Theorem 2.10. *For every prime power $q \geq 2$, there exists a (symmetric) $(q^2+q+1, q+1, 1)$-BIBD (i.e., a projective plane of order q).*

The $(7,3,1)$-BIBD presented in Example 1.3 is a projective plane of order 2. We give another example of a projective plane now.

Example 2.11. We construct a $(13,4,1)$-BIBD, which is a projective plane of order 3. The construction takes place in the finite field $\mathbb{Z}_3$. The one-dimensional and two-dimensional subspaces of $(\mathbb{Z}_3)^3$ are listed in Figure 2.2 and the 13 blocks of the projective plane are presented in Figure 2.3. ∎

The question of the existence of a projective plane of nonprime power order is one of the most celebrated open questions in design theory. We will see later in this section that projective planes of certain (nonprime power) orders can be proven not to exist. There is no known example at present of any projective plane of nonprime power order, and there are infinitely many orders where the existence question has not yet been answered.

$$\begin{array}{ll}
C_1 = \{000, 001, 002\} & B_1 = \{000, 001, 002, 010, 020, 011, 012, 021, 022\} \\
C_2 = \{000, 010, 020\} & B_2 = \{000, 001, 002, 100, 200, 101, 102, 201, 202\} \\
C_3 = \{000, 011, 022\} & B_3 = \{000, 001, 002, 110, 220, 111, 112, 221, 222\} \\
C_4 = \{000, 012, 021\} & B_4 = \{000, 001, 002, 120, 210, 121, 122, 211, 212\} \\
C_5 = \{000, 100, 200\} & B_5 = \{000, 010, 020, 100, 200, 110, 120, 210, 220\} \\
C_6 = \{000, 101, 202\} & B_6 = \{000, 010, 020, 101, 202, 111, 121, 212, 222\} \\
C_7 = \{000, 102, 201\} & B_7 = \{000, 010, 020, 102, 201, 112, 122, 211, 221\} \\
C_8 = \{000, 110, 220\} & B_8 = \{000, 011, 022, 100, 200, 111, 122, 211, 222\} \\
C_9 = \{000, 111, 222\} & B_9 = \{000, 011, 022, 101, 202, 112, 120, 210, 221\} \\
C_{10} = \{000, 112, 221\} & B_{10} = \{000, 011, 022, 102, 201, 110, 121, 212, 220\} \\
C_{11} = \{000, 120, 210\} & B_{11} = \{000, 012, 021, 100, 200, 112, 121, 212, 221\} \\
C_{12} = \{000, 122, 211\} & B_{12} = \{000, 012, 021, 101, 202, 110, 122, 211, 220\} \\
C_{13} = \{000, 121, 212\} & B_{13} = \{000, 012, 021, 102, 201, 111, 120, 210, 222\}.
\end{array}$$

Fig. 2.2. The One-dimensional and Two-dimensional Subspaces of $(\mathbb{Z}_3)^3$

$$\begin{aligned}
A_{B_1} &= \{C_1, C_2, C_3, C_4\} \\
A_{B_2} &= \{C_1, C_5, C_6, C_7\} \\
A_{B_3} &= \{C_1, C_8, C_9, C_{10}\} \\
A_{B_4} &= \{C_1, C_{11}, C_{12}, C_{13}\} \\
A_{B_5} &= \{C_2, C_5, C_8, C_{11}\} \\
A_{B_6} &= \{C_2, C_6, C_9, C_{13}\} \\
A_{B_7} &= \{C_2, C_7, C_{10}, C_{12}\} \\
A_{B_8} &= \{C_3, C_5, C_9, C_{12}\} \\
A_{B_9} &= \{C_3, C_6, C_{10}, C_{11}\} \\
A_{B_{10}} &= \{C_3, C_7, C_8, C_{13}\} \\
A_{B_{11}} &= \{C_4, C_5, C_{10}, C_{13}\} \\
A_{B_{12}} &= \{C_4, C_6, C_8, C_{12}\} \\
A_{B_{13}} &= \{C_4, C_7, C_9, C_{11}\}.
\end{aligned}$$

Fig. 2.3. The Blocks of the Projective Plane of Order 3

Definition 2.12. *Let* $n \geq 2$. *An* $(n^2, n^2 + n, n + 1, n, 1)$-BIBD *is called an* affine plane *of* order n.

It is easy to verify that the residual design of a projective plane of order n is an affine plane of order n. Therefore the following is an immediate consequence of Theorems 2.7 and 2.10.

Theorem 2.13. *For every prime power* $q \geq 2$, *there exists a* $(q^2, q, 1)$-BIBD *(i.e., an affine plane of order* q*).*

Note that the derived design of a projective plane has block size equal to one, and so it is not a BIBD.

The projective planes we have constructed are usually denoted $\mathsf{PG}_2(q)$. They are regarded as two-dimensional projective geometries. A straightforward generalization to higher dimensions is given in the next theorem.

Theorem 2.14. *Suppose $q \geq 2$ is a prime power and $d \geq 2$ is an integer. Then there exists a symmetric*

$$\left(\frac{q^{d+1}-1}{q-1}, \frac{q^d-1}{q-1}, \frac{q^{d-1}-1}{q-1}\right)\text{-BIBD}.$$

Proof. Let $V = (\mathbb{F}_q)^{d+1}$, let $\mathcal{V}_1$ consist of all one-dimensional subspaces of V, and let $\mathcal{V}_d$ consist of all d-dimensional subspaces of V. Each d-dimensional subspace gives rise to a block, as before. □

Note that Theorem 2.10 is the special case $d = 2$ of Theorem 2.14. The points and blocks of the BIBD constructed in Theorem 2.14 correspond to the points and hyperplanes of the d-dimensional *projective geometry*, $\mathsf{PG}_d(q)$.

We can obtain residual BIBDs from the symmetric BIBDs constructed in Theorem 2.14. We get derived BIBDs as well when $d > 2$. These BIBDs have parameters as stated in the following Corollary.

Corollary 2.15. *Suppose $q \geq 2$ is a prime power and $d \geq 2$ is an integer. Then there exists a*

$$\left(q^d, q^{d-1}, \frac{q^{d-1}-1}{q-1}\right)\text{-BIBD}.$$

Furthermore, if $d > 2$, there is a

$$\left(\frac{q^d-1}{q-1}, \frac{q^{d-1}-1}{q-1}, \frac{q(q^{d-2}-1)}{q-1}\right)\text{-BIBD}.$$

Observe that the second BIBD in Corollary 2.15 has the same parameters as q copies of $\mathsf{PG}_{d-1}(q)$.

2.4 The Bruck-Ryser-Chowla Theorem

We now look at two necessary existence conditions for symmetric BIBDs, which are known (together) as the "Bruck-Ryser-Chowla Theorem".

Theorem 2.16 (Bruck-Ryser-Chowla Theorem, v even). *Suppose there exists a symmetric (v, k, λ)-BIBD with v even. Then $k - \lambda$ is a perfect square.*

Proof. Let M be the incidence matrix of a symmetric (v, k, λ)-BIBD with v even. Then, from Theorem 1.13, and using the fact that $r = k$, we have that $MM^T = \lambda J_v + (k - \lambda) I_v$. Since $b = v$, the matrices M and M^T are v by v matrices. Let det() denote the determinant of a square matrix. Since

$$\det(MM^T) = (\det M)(\det M^T) = (\det M)^2$$

for any square matrix M, it follows that

$$(\det M)^2 = \det(\lambda J_v + (k - \lambda) I_v).$$

We proceed to compute $\det(\lambda J_v + (k-\lambda) I_v)$ by performing elementary row and column operations. (Recall that elementary row and column operations do not affect the value of the determinant.) The matrix $\lambda J_v + (k-\lambda) I_v$ looks like

$$\begin{pmatrix} k & \lambda & \lambda & \cdots & \lambda \\ \lambda & k & \lambda & \cdots & \lambda \\ \lambda & \lambda & k & \cdots & \lambda \\ \vdots & \vdots & \vdots & \ddots & \vdots \\ \lambda & \lambda & \lambda & \cdots & k \end{pmatrix}.$$

If we subtract the first row from every other row, then we obtain the matrix

$$\begin{pmatrix} k & \lambda & \lambda & \cdots & \lambda \\ \lambda-k & k-\lambda & 0 & \cdots & 0 \\ \lambda-k & 0 & k-\lambda & \cdots & 0 \\ \vdots & \vdots & \vdots & \ddots & \vdots \\ \lambda-k & 0 & 0 & \cdots & k-\lambda \end{pmatrix}.$$

Now add columns 2 through v to the first column, obtaining the following:

$$\begin{pmatrix} k+(v-1)\lambda & \lambda & \lambda & \cdots & \lambda \\ 0 & k-\lambda & 0 & \cdots & 0 \\ 0 & 0 & k-\lambda & \cdots & 0 \\ \vdots & \vdots & \vdots & \ddots & \vdots \\ 0 & 0 & 0 & \cdots & k-\lambda \end{pmatrix}.$$

This matrix is an upper triangular matrix, so its determinant is the product of the entries on the main diagonal. Hence, we see that

$$(\det M)^2 = (k+(v-1)\lambda)(k-\lambda)^{v-1} = k^2(k-\lambda)^{v-1},$$

where we use the fact that $(v-1)\lambda = k(k-1)$ in a symmetric BIBD. The matrix M has integer entries, so $\det M$ is an integer. Therefore, if v is even, then it must be the case that $k-\lambda$ is a perfect square. □

As an example, we use Theorem 2.16 to show that a $(22,7,2)$-BIBD cannot exist. First, if this BIBD were to exist, it would be symmetric, because $2(22-1) = 7(7-1)$. However, 22 is even and $7-2=5$ is not a perfect square, so we can conclude that the BIBD does not exist.

Before stating and proving the second part of the Bruck-Ryser-Chowla Theorem, we record a couple of other results that are needed in the proof. The first is a well-known theorem from number theory, which we do not prove here.

Lemma 2.17. *For any integer $n \geq 0$, there exist integers $a_0, a_1, a_2, a_3 \geq 0$ such that $n = {a_0}^2 + {a_1}^2 + {a_2}^2 + {a_3}^2$.*

The next lemma is easily verified.

Lemma 2.18. *Suppose that*

$$C = \begin{pmatrix} a_0 & a_1 & a_2 & a_3 \\ -a_1 & a_0 & -a_3 & a_2 \\ -a_2 & a_3 & a_0 & -a_1 \\ -a_3 & -a_2 & a_1 & a_0 \end{pmatrix}$$

and let $n = {a_0}^2 + {a_1}^2 + {a_2}^2 + {a_3}^2$. *Then* $C^{-1} = \frac{1}{n}C^T$.

Now we proceed to the Bruck-Ryser-Chowla Theorem in the case when v is odd.

Theorem 2.19 (Bruck-Ryser-Chowla Theorem, v odd). *Suppose there exists a symmetric* (v, k, λ)-BIBD *with* v *odd. Then there exist integers* x, y, *and* z *(not all* 0*) such that*

$$x^2 = (k - \lambda)y^2 + (-1)^{(v-1)/2}\lambda z^2. \tag{2.3}$$

Proof. First, we suppose that $v \equiv 1 \pmod 4$, and we denote $v = 4w + 1$.

Let M be the incidence matrix of a symmetric (v, k, λ)-BIBD. Let $x_1, \ldots, x_v$ be indeterminates. For $1 \le i \le v$, define

$$L_i = \sum_{j=1}^{v} m_{j,i}x_j.$$

Each L_i is a linear function of the x_j's having integral coefficients.

With a bit of simple algebra, it can be shown that

$$\sum_{i=1}^{v} {L_i}^2 = \lambda \left(\sum_{j=1}^{v} x_j \right)^2 + (k - \lambda) \sum_{j=1}^{v} {x_j}^2. \tag{2.4}$$

We prove that the equation above holds as follows. First, we have that

$${L_i}^2 = \sum_{j=1}^{v} \sum_{h=1}^{v} m_{j,i}m_{h,i}x_jx_h.$$

Then, we have

$$\begin{aligned} \sum_{i=1}^{v} {L_i}^2 &= \sum_{i=1}^{v} \sum_{j=1}^{v} \sum_{h=1}^{v} m_{j,i}m_{h,i}x_jx_h \\ &= \sum_{j=1}^{v} \sum_{h=1}^{v} \left(\sum_{i=1}^{v} m_{j,i}m_{h,i} \right) x_jx_h. \end{aligned}$$

Now, from Theorem 1.13, noting that $r = k$, it follows that

$$\sum_{i=1}^{v} m_{j,i} m_{h,i} = \begin{cases} \lambda & \text{if } j \neq h \\ k & \text{if } j = h. \end{cases}$$

Substituting into the equation above, we have that

$$\begin{aligned}\sum_{i=1}^{v} {L_i}^2 &= \sum_{\{j,h:j\neq h\}} \lambda x_j x_h + \sum_{j=1}^{v} k{x_j}^2 \\ &= \sum_{j=1}^{v}\sum_{h=1}^{v} \lambda x_j x_h + \sum_{j=1}^{v} (k-\lambda){x_j}^2 \\ &= \lambda\left(\sum_{j=1}^{v} x_j\right)^2 + (k-\lambda)\sum_{j=1}^{v} {x_j}^2,\end{aligned}$$

as desired.

Equation (2.4) is an identity in the variables $x_1, \ldots, x_v$ in which all the coefficients are integers. Next, we transform the variables $x_1, \ldots, x_v$ into new variables $y_1, \ldots, y_v$, where each y_i is a certain integral linear combination of the x_j's. Let a_0, a_1, a_2, a_3 be integers such that ${a_0}^2 + {a_1}^2 + {a_2}^2 + {a_3}^2 = k - \lambda$; these exist by Lemma 2.17. Let the matrix C be defined as in Lemma 2.18. Then, for $1 \leq h \leq w$, let

$$(y_{4h-3}, y_{4h-2}, y_{4h-1}, y_{4h}) = (x_{4h-3}, x_{4h-2}, x_{4h-1}, x_{4h})C.$$

Finally, let $y_v = x_v$ and let

$$y_0 = \sum_{i=1}^{v} x_i.$$

It is easy to see, using Lemma 2.18, that

$$\sum_{j=1}^{v-1} {y_j}^2 = (k-\lambda)\sum_{j=1}^{v-1} {x_j}^2.$$

This follows from the following equations, which hold for $1 \leq h \leq w$:

$$\begin{aligned}&{y_{4h-3}}^2 + {y_{4h-2}}^2 + {y_{4h-1}}^2 + {y_{4h}}^2 \\ &= (y_{4h-3}, y_{4h-2}, y_{4h-1}, y_{4h})(y_{4h-3}, y_{4h-2}, y_{4h-1}, y_{4h})^T \\ &= (x_{4h-3}, x_{4h-2}, x_{4h-1}, x_{4h})C((x_{4h-3}, x_{4h-2}, x_{4h-1}, x_{4h})C)^T \\ &= (x_{4h-3}, x_{4h-2}, x_{4h-1}, x_{4h})CC^T(x_{4h-3}, x_{4h-2}, x_{4h-1}, x_{4h})^T \\ &= (x_{4h-3}, x_{4h-2}, x_{4h-1}, x_{4h})(k-\lambda)I_4(x_{4h-3}, x_{4h-2}, x_{4h-1}, x_{4h})^T \\ &= (k-\lambda)({x_{4h-3}}^2 + {x_{4h-2}}^2 + {x_{4h-1}}^2 + {x_{4h}}^2).\end{aligned}$$

Hence, it follows that

$$\sum_{i=1}^{v} L_i{}^2 = \lambda y_0{}^2 + \sum_{j=1}^{v-1} y_j{}^2 + (k-\lambda) y_v{}^2. \tag{2.5}$$

The L_i's were defined as integral linear combinations of the x_j's. However, by virtue of Lemma 2.18, we can express each x_j as a rational linear combination of $y_1, \ldots, y_v$. Similarly, y_0 is a linear combination of $y_1, \ldots, y_v$ having rational coefficients.

In view of the observations above, equation (2.5) can be regarded as an identity in the indeterminates $y_1, \ldots, y_v$ in which all the coefficients are rational numbers. It is possible to specialize this identity by expressing any of the indeterminates as a rational combination of the remaining indeterminates, and the result will be an identity in the remaining indeterminates in which the coefficients are (still) all rational.

First, suppose that

$$L_1 = \sum_{i=1}^{v} e_i y_i.$$

If $e_1 \neq 1$, then let $y_1 = L_1$, and if $e_1 = 1$, then let $y_1 = -L_1$. We have expressed y_1 as a rational linear combination of $y_2, \ldots, y_v$ in such a way that $L_1{}^2 = y_1{}^2$. Then equation (2.5) is transformed into the following identity in $y_2, \ldots, y_v$:

$$\sum_{i=2}^{v} L_i{}^2 = \lambda y_0{}^2 + \sum_{j=2}^{v-1} y_j{}^2 + (k-\lambda) y_v{}^2. \tag{2.6}$$

We continue in this fashion, eliminating the variables $y_2, \ldots, y_{v-1}$ one at a time, making sure that each y_j is a rational linear combination of $y_{j+1}, \ldots, y_v$ such that $y_j{}^2 = L_j{}^2$ for all such j. We end up with the following equation:

$$L_v{}^2 = \lambda y_0{}^2 + (k-\lambda) y_v{}^2. \tag{2.7}$$

In this equation, L_v and y_0 are rational multiples of y_v. Suppose that $L_v = s y_v$ and $y_0 = t y_v$, where $s, t \in \mathbb{Q}$. Let $y_v = 1$; then

$$s^2 = \lambda t^2 + k - \lambda.$$

Now, we can write $s = s_1/s_2$ and $t = t_1/t_2$, where $s_1, s_2, t_1, t_2 \in \mathbb{Z}$ and $s_2, t_2 \neq 0$. Our equation becomes

$$(s_1 t_2)^2 = \lambda (s_2 t_1)^2 + (k-\lambda)(s_2 t_2)^2.$$

If we let $x = s_1 t_2$, $y = s_2 t_2$, and $z = s_2 t_1$, then we have an integral solution to the equation $x^2 = (k-\lambda) y^2 + (-1)^{(v-1)/2} \lambda z^2$ in which at least one of $x, y,$ and z is nonzero (note also that $(-1)^{(v-1)/2} = 1$ because $v \equiv 1 \pmod 4$).

There remains the case $v \equiv 3 \pmod 4$ to consider. It is similar to the previous case, with a few modifications. Denote $v = 4w - 1$. Introduce a new indeterminate, x_{v+1}, and add $(k-\lambda)x_{v+1}{}^2$ to both sides of equation (2.4), producing the following:

$$\sum_{i=1}^{v} L_i{}^2 + (k-\lambda)x_{v+1}{}^2 = \lambda\left(\sum_{j=1}^{v} x_j\right)^2 + (k-\lambda)\sum_{j=1}^{v+1} x_j{}^2. \tag{2.8}$$

Then, for $1 \le h \le w$, let

$$(y_{4h-3}, y_{4h-2}, y_{4h-1}, y_{4h}) = (x_{4h-3}, x_{4h-2}, x_{4h-1}, x_{4h})C.$$

Finally, let

$$y_0 = \sum_{i=1}^{v} x_i.$$

Then we have that

$$\sum_{i=1}^{v} L_i{}^2 + (k-\lambda)x_{v+1}{}^2 = \lambda y_0{}^2 + \sum_{j=1}^{v+1} y_j{}^2. \tag{2.9}$$

Proceed as in the case $v \equiv 1 \pmod 4$, eliminating all the L_i's. The following equation results:

$$(k-\lambda)x_{v+1}{}^2 = \lambda y_0{}^2 + y_{v+1}{}^2.$$

We end up with a solution to the equation $x^2 = (k-\lambda)y^2 + (-1)^{(v-1)/2}\lambda z^2$ in which at least one of x, y, and z is nonzero (note that $(-1)^{(v-1)/2} = -1$ when $v \equiv 3 \pmod 4$). □

Theorem 2.19 is more difficult to apply than Theorem 2.16 because it involves determining if a certain diophantine equation has a nontrivial solution. Here is an example to illustrate this:

Example 2.20. We will show that a (symmetric) $(43, 7, 1)$-BIBD does not exist. Theorem 2.16 tells us that if this BIBD exists, then the equation

$$x^2 + z^2 = 6y^2 \tag{2.10}$$

has a solution in integers, not all of which are zero. Let us assume that (x, y, z) is an integral solution to equation (2.10). Reducing this equation modulo 3, it follows that $x^2 + z^2 \equiv 0 \pmod 3$. Since $x^2 \equiv 0, 1 \pmod 3$ for any integer x, the only way that we can have $x^2 + z^2 \equiv 0 \pmod 3$ is if $x \equiv 0 \pmod 3$ and $z \equiv 0 \pmod 3$. Let us write $x = 3x_1$ and $z = 3z_1$, where x_1 and z_1 are integers. Then equation (2.10) becomes

$$(3x_1)^2 + (3z_1)^2 = 6y^2,$$

or

$$3x_1{}^2 + 3z_1{}^2 = 2y^2.$$

The left side of this equation is divisible by 3, so it must be the case that $y \equiv 0 \pmod 3$. Writing $y = 3y_1$, we have

$$3{x_1}^2 + 3{z_1}^2 = 2(3y_1)^2,$$

or

$${x_1}^2 + {z_1}^2 = 6{y_1}^2.$$

We have shown that if (x, y, z) is any integral solution to equation (2.10), then $\left(\frac{x}{3}, \frac{y}{3}, \frac{z}{3}\right)$ is also an integral solution to equation (2.10). This process can be repeated infinitely often, which is a contradiction unless $(x, y, z) = (0, 0, 0)$. We conclude that the only solution to equation (2.10) is $(0, 0, 0)$, and therefore a $(43, 7, 1)$-BIBD does not exist. ∎

The example above was a bit tedious. It is worthwhile to use some results from number theory to establish a more general result. Let us first consider the situation of a projective plane of arbitrary order n. We will give a complete analysis of the Bruck-Ryser-Chowla conditions in this situation.

First, suppose that $n \equiv 0, 3 \pmod 4$. In this case, equation (2.3) reduces to $x^2 = ny^2 + z^2$. This always has the nontrivial solution $x = z = 1$, $y = 0$. Therefore the Bruck-Ryser-Chowla Theorem does not yield any non-existence results for $(n^2 + n + 1, n + 1, 1)$-BIBDs when $n \equiv 0, 3 \pmod 4$.

Now we turn to the case where $n \equiv 1, 2 \pmod 4$. For such integers n, we have that $(n^2 + n)/2$ is odd, so the equation to be solved is $x^2 = ny^2 - z^2$, or

$$x^2 + z^2 = ny^2. \tag{2.11}$$

We are interested in determining the conditions under which equation (2.11) has an integral solution (x, y, z) not all of which are zero. Although we do not give the proof here, it is possible to show that equation (2.11) has a solution of the desired type if and only if

$$x^2 + z^2 = n \tag{2.12}$$

has an integral solution (x, z). Furthermore, it is known precisely when equation (2.12) has an integral solution. The following is a famous result from number theory.

Theorem 2.21. *A positive integer n can be expressed as the sum of two integral squares if and only if there does not exist a prime $p \equiv 3 \pmod 4$ such that the largest power of p that divides n is odd.*

Summarizing the previous discussion, we obtain the following result.

Theorem 2.22. *Suppose that $n \equiv 1, 2 \pmod 4$, and there exists a prime $p \equiv 3 \pmod 4$ such that the largest power of p that divides n is odd. Then a projective plane of order n does not exist.*

The first few values of n for which Theorem 2.22 can be applied are $n = 6, 14, 21, 22$, and 30. Hence, projective planes of these orders do not exist.

We now turn to the situation of arbitrary λ, where we derive an easy-to-use corollary of the Bruck-Ryser-Chowla Theorem. Before proceeding to our main result, we define the concept of a quadratic residue. Suppose that $m \geq 2$ is an integer and a is any integer. Then we say that a is a *quadratic residue* modulo m if the congruence $x^2 \equiv a \pmod{m}$ has a solution $x \in \mathbb{Z}_m \backslash \{0\}$. For future reference, we record the following well-known result, which is known as *Euler's Criterion*.

Theorem 2.23 (Euler's Criterion). *An integer a is a quadratic residue modulo the odd prime p if and only if $a^{(p-1)/2} \equiv 1 \pmod{p}$.*

A positive integer is said to be *square-free* provided that it is not divisible by j^2 for any integer $j > 1$. Any positive integer n can be written uniquely in the form $n = A^2 n_1$ where A is a positive integer and n_1 is square-free (note that we allow $A = 1$ and/or $n_1 = 1$). The integer n_1 is called the *square-free part* of n.

Theorem 2.24. *Suppose that v, k and λ are positive integers such that $\lambda(v-1) = k(k-1)$ and $v > k \geq 2$. Let λ_1 be the square-free part of λ and let n_1 be the square-free part of $k - \lambda$. Suppose that p is an odd prime such that $n_1 \equiv 0 \pmod{p}$, $\lambda_1 \not\equiv 0 \pmod{p}$, and $(-1)^{(v-1)/2}\lambda_1$ is not a quadratic residue modulo p. Then there does not exist a (v, k, λ)-*BIBD.

Proof. We will prove that equation (2.3) does not have an integral solution $(x, y, z) \neq (0, 0, 0)$. Assuming that it does, we will derive a contradiction.

First, we have that $\lambda = B^2\lambda_1$ and $k - \lambda = A^2 n_1$, where A and B are positive integers. Then

$$x^2 = n_1(Ay)^2 + (-1)^{(v-1)/2}\lambda_1(Bz)^2.$$

Letting $y_1 = Ay$ and $z_1 = Bz$, the equation

$$x^2 = n_1{y_1}^2 + (-1)^{(v-1)/2}\lambda_1{z_1}^2 \tag{2.13}$$

has a solution $(x, y_1, z_1) \neq (0, 0, 0)$. We can assume that $\gcd(x, y_1, z_1) = 1$ (for if $\gcd(x, y_1, z_1) = d > 1$, then we can divide each of x, y_1, and z_1 by d, obtaining a solution in which the gcd is equal to 1).

Suppose that $z_1 \equiv 0 \pmod{p}$. Then $x \equiv 0 \pmod{p}$ because $n_1 \equiv 0 \pmod{p}$. But if z_1 and x are both divisible by p, then ${z_1}^2$ and x^2 are both divisible by p^2, and hence $n_1{y_1}^2$ is divisible by p^2. n_1 is square-free, so it is not divisible by p^2. Therefore y_1 is divisible by p. But then $\gcd(x, y_1, z_1) \geq p$, which is a contradiction. We conclude that $z_1 \not\equiv 0 \pmod{p}$.

Now we reduce equation (2.13) modulo p. We obtain the following congruence:

$$x^2 \equiv (-1)^{(v-1)/2}\lambda_1 {z_1}^2 \pmod{p}.$$

We proved above that $z_1 \not\equiv 0 \pmod{p}$. Since p is prime, there exists a multiplicative inverse ${z_1}^{-1}$ mod p. Then

$$(x{z_1}^{-1})^2 \equiv (-1)^{(v-1)/2}\lambda_1 \pmod{p}.$$

This means that $(-1)^{(v-1)/2}\lambda_1$ is a quadratic residue modulo p, which contradicts the hypotheses of the theorem.

We conclude that equation (2.3) does not have a solution $(x, y, z) \neq (0, 0, 0)$. Hence, from Theorem 2.19, there does not exist a (v, k, λ)-BIBD. □

We illustrate the application of the theorem above in the following example.

Example 2.25. Consider the parameter set $(v, k, \lambda) = (67, 12, 2)$. We compute $2 \times 66 = 12 \times 11$, so it is conceivable that a (symmetric) $(67, 12, 2)$-BIBD exists. We show that this is not the case using Theorem 2.24.

We have $\lambda_1 = 2$ and $n_1 = 10$, so we will take $p = 5$. We compute $(-1)^{(v-1)/2}\lambda_1 \equiv 3 \pmod{5}$, and it is easily verified that 3 is not a quadratic residue modulo 5. Therefore we conclude from Theorem 2.24 that a $(67, 12, 2)$-BIBD does not exist. ▌

As another example, we show that Theorem 2.22 can be derived as a corollary of Theorem 2.24.

Example 2.26. Suppose that $n \equiv 1, 2 \pmod{4}$ and there exists a prime $p \equiv 3 \pmod{4}$ such that the largest power of p that divides n is odd. We want to show, using Theorem 2.24, that an $(n^2 + n + 1, n + 1, 1)$-BIBD does not exist. Clearly we have $\lambda_1 = \lambda = 1$, $k - \lambda = n$, and $\lambda_1 \not\equiv 0 \pmod{p}$. Using the fact that the largest power of p that divides n is odd, it follows that $n_1 \equiv 0 \pmod{p}$.

We need to verify that $(-1)^{(v-1)/2}\lambda_1$ is not a quadratic residue modulo p. As observed previously, $(-1)^{(v-1)/2} = (-1)^{(n^2+n)/2} = -1$ when $n \equiv 1, 2 \pmod{4}$. Therefore $(-1)^{(v-1)/2}\lambda_1 = -1$. However, using Euler's Criterion, it is immediate that -1 is not a quadratic residue modulo p if $p \equiv 3 \pmod{4}$.

It therefore follows from Theorem 2.24 that a projective plane of order n does not exist if the given hypotheses hold. ▌

The Bruck-Ryser-Chowla Theorem was proven over fifty years ago. It is remarkable that no general necessary conditions for existence of symmetric BIBDs have been proven since then. In fact, the only nonexistence result for any symmetric BIBD, other than those ruled out by the Bruck-Ryser-Chowla theorem, is that a projective plane of order 10 does not exist. This was proven in 1989 using a computer.

2.5 Notes and References

Lander [75] is a 1983 monograph devoted to symmetric designs. Tran [113] is a more recent survey.

Dembowski [39] is a standard reference on projective geometries. Hughes and Piper [60] is a specialized study of projective planes.

Most of the results in Section 2.1 (including Theorem 2.2) were proven in Ryser [89] and Chowla and Ryser [23].

The result that a quasiresidual BIBD with $\lambda = 2$ is residual is known as the "Hall-Connor Theorem" and was proven in [54]. There are quite a number of constructions for quasiresidual BIBDs that are not residual. Tran [112] gave an extensive treatment of this subject in 1990; see also Ionin and Mackenzie-Fleming [62] (and the references found therein) for more recent results.

The theorem known as the Bruck-Ryser-Chowla Theorem was proven (for odd v) by Bruck and Ryser [18] and by Chowla and Ryser [23]. The part of the theorem pertaining to even v was first obtained by Schützenberger [91].

The proof of the nonexistence of a projective plane of order 10 is due to Lam, Thiel, and Swiercz [74].

2.6 Exercises

2.1 Give a proof of Theorem 2.2 in the special case $\lambda = 1$ using the technique of Exercise 1.19.

2.2 Suppose that there is a symmetric (v, k, λ)-BIBD, say $(X, \mathcal{A})$, and denote $n = k - \lambda$. n is called the *order* of the symmetric BIBD $(X, \mathcal{A})$.
   (a) Prove that the block complement of $(X, \mathcal{A})$ has order n.
   (b) Prove that $\lambda^2 + (2n - v)\lambda + n^2 - n = 0$.
   (c) Solve this quadratic equation for λ.
   (d) Using the fact that $\lambda \geq 1$, deduce that $v \leq n^2 + n + 1$.
   (e) Prove that $v \geq 4n - 1$.

2.3 Let $(X, \mathcal{A})$ be a symmetric (v, k, λ)-BIBD having order $n = k - \lambda$.
   (a) If $v = n^2 + n + 1$, prove that $(X, \mathcal{A})$ is a projective plane of order n (or its block-complement).
   (b) If $v = 4n - 1$, prove that $(v, k, \lambda) = (4n - 1, 2n - 1, n - 1)$ or $(4n - 1, 2n, n)$.
   (c) If $v = 4n$, prove that $(v, k, \lambda) = (4u^2, 2u^2 \pm u, u^2 \pm u)$ for some positive integer u.
   Hint: Use the Bruck-Ryser Theorem and Exercise 2.2.

2.4 Suppose that (v, b, r, k, λ) are parameters of a BIBD.
   (a) Prove that $\lambda(v + r - 1) = r(r - 1)$ whenever $r = k + \lambda$.
   (b) Prove that $(\lambda + 1)b = r(r + 1)$ whenever $k = \lambda + 1$.

2.5 (a) State the parameters (v_1, k_1, λ_1) of the residual BIBD of a symmetric (v, k, λ)-BIBD.
(b) State the parameters (v_2, k_2, λ_2) of the derived BIBD of the block-complement of a symmetric (v, k, λ)-BIBD.
(c) Prove that the parameter triples (v_1, k_1, λ_1) and (v_2, k_2, λ_2) are identical if and only if $k = 2\lambda + 1$ and $v = 4\lambda + 3$.

2.6 Suppose that a (v, k, λ)-BIBD is both a derived and a residual BIBD. Prove that $v = 2\lambda + 2$.

2.7 Construct a projective plane of order 4 using the technique of Example 2.11.
Note: The finite field $\mathbb{F}_4 = \mathbb{Z}_2[x]/(x^2 + x + 1)$.

2.8 Construct a $(15, 7, 3)$-BIBD using the method described in Theorem 2.14.

2.9 The following triples (v, k, λ) all satisfy the condition $\lambda(v-1) = k(k-1)$, so they could be parameters of a symmetric BIBD. For each triple, investigate the Bruck-Ryser-Chowla conditions. You should either prove that the Diophantine equation

$$x^2 = (k - \lambda)y^2 + (-1)^{(v-1)/2}\lambda z^2$$

has no integral solution $(x, y, z) \neq (0, 0, 0)$ (which implies that the BIBD does not exist) or find a solution $(x, y, z) \neq (0, 0, 0)$ by trial and error (you are not required to try to construct the BIBD in this situation). The parameter triples are as follows.
(a) $(29, 8, 2)$.
(b) $(53, 13, 3)$.
(c) $(43, 15, 5)$.
(d) $(81, 16, 3)$.
(e) $(77, 20, 5)$.
(f) $(85, 28, 9)$.

2.10 A $W(n, w)$ is an $n \times n$ matrix whose entries are elements of the set $\{0, 1, -1\}$ such that $WW^T = wI_n$. Prove the following Bruck-Ryser-Chowla type theorems for the existence of these matrices.
(a) Suppose that a $W(n, w)$ exists, where n is odd. Then prove that w is a perfect square.
(b) Suppose that a $W(n, w)$ exists, where $n \equiv 2 \pmod 4$. Then prove that w is the sum of two integral squares.
Hint: Eventually, you should obtain an equation of the form ${L_1}^2 + {L_2}^2 = w({y_{v-1}}^2 + {y_v}^2)$. Set $y_{v-1} = 1$ and $y_v = 0$, and make use of the fact (which you are not required to prove) that an integer is the sum of two integral squares if and only if it is the sum of two rational squares.

3

Difference Sets and Automorphisms of Designs

3.1 Difference Sets and Automorphisms

We now study an important construction method for symmetric BIBDs.

Definition 3.1. *Suppose $(G, +)$ is a finite group of order v in which the identity element is denoted "0". Unless explicitly stated, we will not require that G be an Abelian group. (In many examples, however, we will take $G = (\mathbb{Z}_v, +)$, the integers modulo v.) Let k and λ be positive integers such that $2 \le k < v$. A (v, k, λ)-*difference set *in $(G, +)$ is a subset $D \subseteq G$ that satisfies the following properties:*

1. *$|D| = k$,*
2. *the multiset $[x - y : x, y \in D, x \neq y]$ contains every element in $G \setminus \{0\}$ exactly λ times.*

Note that $\lambda(v-1) = k(k-1)$ if a (v, k, λ)-difference set exists.

Example 3.2. A $(21, 5, 1)$-difference set in $(\mathbb{Z}_{21}, +)$:

$$D = \{0, 1, 6, 8, 18\}.$$

If we compute the differences (modulo 21) we get from pairs of distinct elements in D, we obtain the following:

$$\begin{array}{ll}
1 - 0 = 1 & 0 - 1 = 20 \\
6 - 0 = 6 & 0 - 6 = 15 \\
8 - 0 = 8 & 0 - 8 = 13 \\
18 - 0 = 18 & 0 - 18 = 3 \\
6 - 1 = 5 & 1 - 6 = 16 \\
8 - 1 = 7 & 1 - 8 = 14 \\
18 - 1 = 17 & 1 - 18 = 4 \\
8 - 6 = 2 & 6 - 8 = 19 \\
18 - 6 = 12 & 6 - 18 = 9 \\
18 - 8 = 10 & 8 - 18 = 11.
\end{array}$$

So we get every element of $\mathbb{Z}_{21}\backslash\{0\}$ exactly once as a difference of two elements in D. ∎

Example 3.3. A $(15,7,3)$-difference set in $(\mathbb{Z}_{15},+)$:

$$D = \{0,1,2,4,5,8,10\}.$$

∎

Example 3.4. A $(16,6,2)$-difference set in $(\mathbb{Z}_4 \times \mathbb{Z}_4,+)$:

$$D = \{(0,1),(0,2),(0,3),(1,0),(2,0),(3,0)\}.$$

(**Note**: This example is particularly interesting in view of the fact that there does not exist a $(16,6,2)$-difference set in $(\mathbb{Z}_{16},+)$.) ∎

Example 3.5. A $(45,12,3)$-difference set in $(\mathbb{Z}_5 \times \mathbb{Z}_3 \times \mathbb{Z}_3,+)$:

$$D = \left\{ \begin{array}{l} (0,0,0),(0,0,1),(0,0,2),(1,0,0),(1,1,0),(1,2,0), \\ (2,0,0),(2,1,1),(2,2,2),(3,0,0),(3,1,2),(3,2,1) \end{array} \right\}.$$

∎

Example 3.6. A $(36,15,6)$-difference set in $(\mathbb{Z}_6 \times \mathbb{Z}_6,+)$:

$$D = \{(0,i) : 1 \le i \le 5\} \cup \{(i,0) : 1 \le i \le 5\} \cup \{(i,i) : 1 \le i \le 5\}.$$

∎

Example 3.7. We give an example of a difference set in a non-Abelian group. Consider the following group (written multiplicatively):

$$G = \{a^i b^j : a^3 = b^7 = 1, ba = ab^4\}.$$

It can be shown that G is a non-Abelian group of order 21. The set $D = \{a,a^2,b,b^2,b^4\}$ is a $(21,5,1)$-difference set in $(G,\cdot)$. Because the group is written multiplicatively, what we mean by this is that

$$\{xy^{-1} : x,y \in D, x \neq y\} = G\backslash\{1\}.$$

∎

Difference sets can be used to construct symmetric BIBDs as follows. Let D be a (v,k,λ)-difference set in a group $(G,+)$. For any $g \in G$, define

$$D+g = \{x+g : x \in D\}.$$

Any set $D+g$ is called a *translate* of D. Then, define $\mathsf{Dev}(D)$ to be the collection of all v translates of D. $\mathsf{Dev}(D)$ is called the *development* of D.

Theorem 3.8. *Let D be a (v,k,λ)-difference set in an Abelian group $(G,+)$. Then $(G,\mathsf{Dev}(D))$ is a symmetric (v,k,λ)-BIBD.*

Proof. Suppose $x,y \in G$, $x \neq y$. We first prove that there are exactly λ elements $g \in G$ such that $\{x,y\} \subseteq D+g$.

Denote $x-y=d$. There are exactly λ ordered pairs (x',y') such that $x',y' \in D$ and $x'-y'=d$. Let these ordered pairs be denoted (x_i,y_i), $1 \leq i \leq \lambda$. For $1 \leq i \leq \lambda$, define $g_i = -x_i + x$. Then $g_i = -y_i + y$ and $\{x,y\} = \{x_i+g_i, y_i+g_i\} \subseteq D+g_i$. The g_i's are distinct because the x_i's are distinct, so this shows that there are at least λ values of g such that $\{x,y\} \subseteq D+g$.

Conversely, suppose that there are exactly ℓ values of g such that $\{x,y\} \subseteq D+g$, namely $g = h_1,\ldots,h_\ell$. (We have shown above that $\ell \geq \lambda$.) Then $(x-h_i)+(h_i-y) = x-y = d$ for $1 \leq i \leq \ell$. Also, $\{x-h_i, y-h_i\} \subseteq D$ for $1 \leq i \leq \ell$. The h_i's are distinct, so we have found ℓ ordered pairs $(x',y') \in D$ such that $x'-y'=d$. There are exactly λ such ordered pairs, however, so $\ell \leq \lambda$.

We have therefore proven that $\ell = \lambda$. Every block $D+g$ contains k points, so the collection of v blocks $D+g$ $(g \in G)$ is a symmetric (v,k,λ)-BIBD. □

Corollary 3.9. *Suppose D is a (v,k,λ)-difference set in an Abelian group $(G,+)$. Then $\mathsf{Dev}(D)$ consists of v distinct blocks.*

Proof. Suppose that $D+g_1 = D+g_2$, where $g_1 \neq g_2$. Then the symmetric BIBD $(G,\mathsf{Dev}(D))$ contains two blocks that intersect in k points. However, Theorem 2.2 states that any two blocks in a symmetric (v,k,λ)-BIBD intersect in λ points. The result follows. □

Thus, for example, the $(21,5,1)$-BIBD developed from the difference set of Example 3.2 has 21 distinct blocks:

$$\{0,1,6,8,18\},\{1,2,7,9,19\},\ldots,\{0,5,7,17,20\}.$$

The next result establishes the existence of nontrivial automorphisms of the symmetric BIBDs constructed from difference sets.

Theorem 3.10. *Suppose $(G,\mathsf{Dev}(D))$ is the symmetric BIBD constructed from a (v,k,λ)-difference set D in a group $(G,+)$. Then $\mathsf{Aut}(G,\mathsf{Dev}(D))$ contains a subgroup $\widehat{G}$ that is isomorphic to G.*

Proof. For every $g \in G$, define the mapping $\widehat{g} : G \rightarrow G$ as follows:

$$\widehat{g}(x) = x+g$$

for all $x \in G$. It is clear that each $\widehat{g}$ is one-to-one and onto and therefore a permutation of G. Define $\widehat{G} = \{\widehat{g} : g \in G\}$. $\widehat{G}$ is a permutation group, and it is known as the *permutation representation* of G.

We will prove the following:

1. $(G,+)$ is isomorphic to $(\widehat{G},\circ)$, where the group operation "$\circ$" denotes composition of permutations; and
2. $(\widehat{G},\circ)$ is a subgroup of $\mathsf{Aut}(G,\mathsf{Dev}(D))$.

To prove the first assertion, we exhibit an isomorphism between $(G,+)$ and $(\widehat{G},\circ)$. Define $\alpha : G \to \widehat{G}$ in the obvious way: $\alpha(g) = \widehat{g}$ for all $g \in G$. First, α is a group homomorphism because

$$\begin{aligned}(\alpha(g)\circ\alpha(h))(x) &= (\widehat{g}\circ\widehat{h})(x)\\ &= \widehat{h}(\widehat{g}(x))\\ &= \widehat{h}(x+g)\\ &= x+g+h\\ &= \widehat{g+h}(x)\\ &= \alpha(g+h)(x)\end{aligned}$$

holds for all $g,h,x \in G$, and hence $\alpha(g)\circ\alpha(h) = \alpha(g+h)$ holds for all $g,h \in G$. Next, it is clear that α is surjective. We also have that α is injective since $\widehat{g} = \widehat{h}$ if and only if $g = h$. Hence α is a group isomorphism.

To prove the second statement, we observe that

$$\begin{aligned}\widehat{g}(D+h) &= \{\widehat{g}(x) : x \in D+h\}\\ &= \{x+g : x \in D+h\}\\ &= \{x+g+h : x \in D\}\\ &= D+h+g.\end{aligned}$$

Hence, for any permutation $\widehat{g} \in \widehat{G}$ and for any block $D+h \in \mathsf{Dev}(D)$, it holds that $\widehat{g}(D+h) \in \mathsf{Dev}(D)$. That is, every $\widehat{g} \in \widehat{G}$ is an automorphism of $(G,\mathsf{Dev}(D))$. Since $\widehat{G}$ is a group, it is a subgroup of $\mathsf{Aut}(G,\mathsf{Dev}(D))$. □

Example 3.11. Consider the symmetric $(7,3,1)$-BIBD developed from the difference set $\{1,2,4\}$ in $(\mathbb{Z}_7,+)$. The blocks of the BIBD are 124, 235, 346, 450, 561, 602, and 013.

The elements g of the group $G = (\mathbb{Z}_7,+)$, and the elements $\widehat{g}$ in its permutation representation, $\widehat{G}$, are as follows:

g	$\widehat{g}$
0	(0)(1)(2)(3)(4)(5)(6)
1	(0 1 2 3 4 5 6)
2	(0 2 4 6 1 3 5)
3	(0 3 6 2 5 1 4)
4	(0 4 1 5 2 6 3)
5	(0 5 3 1 6 4 2)
6	(0 6 5 4 3 2 1)

It is easy to verify that every permutation in $\widehat{G}$ is an automorphism of the BIBD. ∎

Under suitable conditions, a converse to Theorem 3.10 holds. We will prove a result of this type for the special case of a difference set in a cyclic group. However, before doing so, we state and prove some preliminary results that will be required.

Suppose that $(X, \mathcal{A})$ is a symmetric (v, k, λ)-BIBD, and suppose that $\alpha \in \mathsf{Aut}(X, \mathcal{A})$. α is a permutation of X, and therefore it follows from Section 1.4 that α consists of a union of disjoint cycles whose lengths sum to v. The *cycle type* of α is the collection (i.e., multiset) of the sizes of the cycles in the disjoint cycle representation of α. Recall that a fixed point of α is a point x such that $\alpha(x) = x$.

As an example, consider the permutation α of $\{0, \ldots, 8\}$ defined as follows: $\alpha(0) = 3$, $\alpha(1) = 4$, $\alpha(2) = 2$, $\alpha(3) = 0$, $\alpha(4) = 5$, $\alpha(5) = 1$, $\alpha(6) = 8$, $\alpha(7) = 7$, and $\alpha(8) = 6$. If we write α as a union of disjoint cycles, then we have

$$\alpha = (0\ 3)(1\ 4\ 5)(2)(6\ 8)(7).$$

The cycle type of α, written as a list of nondecreasing integers, is $[1, 1, 2, 2, 3]$. Note that α has two fixed points, namely 2 and 7.

Any automorphism α of a symmetric BIBD, say $(X, \mathcal{A})$, will permute the blocks in the set $\mathcal{A}$. Hence, we can consider the permutation of $\mathcal{A}$ induced by α and define the cycle type of this permutation in the obvious way. A fixed "point" of this permutation is a block $A \in \mathcal{A}$ that is fixed setwise by α; i.e., $\{\alpha(x) : x \in A\} = A$. We refer to such a block as a *fixed block* to avoid confusion.

We now state and prove a useful combinatorial lemma.

Lemma 3.12. *Suppose that* $(X, \mathcal{A})$ *is a symmetric* (v, k, λ)-BIBD, *and suppose that* $\alpha \in \mathsf{Aut}(X, \mathcal{A})$ *has exactly* f *fixed points. Then* α *fixes exactly* f *blocks in* $\mathcal{A}$.

Proof. Suppose that α fixes exactly F blocks. Define

$$I = \{(x, A) : x \in X, A \in \mathcal{A}, \{x, \alpha(x)\} \subseteq A\}.$$

We will compute $|I|$ in two different ways. First, we have

$$\begin{aligned} |I| &= \sum_{x \in X} |\{A \in \mathcal{A} : \{x, \alpha(x)\} \subseteq A\}| \\ &= \sum_{\{x \in X : \alpha(x) = x\}} |\{A \in \mathcal{A} : \{x, \alpha(x)\} \subseteq A\}| \\ &\qquad + \sum_{\{x \in X : \alpha(x) \neq x\}} |\{A \in \mathcal{A} : \{x, \alpha(x)\} \subseteq A\}| \\ &= fk + (v - f)\lambda. \end{aligned}$$

On the other hand, we have

$$\begin{aligned}
|I| &= \sum_{A \in \mathcal{A}} |\{x \in X : \{x, \alpha(x)\} \subseteq A\}| \\
&= \sum_{\{A \in \mathcal{A}: \alpha(A) = A\}} |\{x \in X : \{x, \alpha(x)\} \subseteq A\}| \\
&\quad + \sum_{\{A \in \mathcal{A}: \alpha(A) \neq A\}} |\{x \in X : \{x, \alpha(x)\} \subseteq A\}|.
\end{aligned}$$

Now, if $\alpha(A) = A$, then $\alpha(x) \in A$ for all $x \in A$, and it is easily seen that

$$\{x \in X : \{x, \alpha(x)\} \subseteq A\} = A.$$

Therefore,

$$|\{x \in X : \{x, \alpha(x)\} \subseteq A\}| = k.$$

Now assume that $\alpha(A) \neq A$. Clearly, $\{x, \alpha(x)\} \subseteq A$ if and only if $x \in A \cap \alpha^{-1}(A)$. $A \neq \alpha^{-1}(A)$, and hence, applying Theorem 2.2, we have that $|A \cap \alpha^{-1}(A)| = \lambda$. Therefore

$$|\{x \in X : \{x, \alpha(x)\} \subseteq A\}| = \lambda,$$

and hence

$$|I| = Fk + (v - F)\lambda.$$

Equating the two expressions we have derived for $|I|$, we have that

$$fk + (v - f)\lambda = Fk + (v - F)\lambda.$$

This implies that

$$(f - F)(k - \lambda) = 0.$$

In a symmetric BIBD, it holds that $k \neq \lambda$, and hence we conclude that $f = F$. □

The proof of our next theorem will make use of a combinatorial technique known as the "Möbius Inversion Formula". This interesting formula involves the *Möbius function*, denoted μ, which is defined on the positive integers as follows:

$$\mu(n) = \begin{cases} 1 & \text{if } n = 1 \\ (-1)^k & \text{if } n = p_1 \times \cdots \times p_k, \text{ where the } p_i\text{'s are distinct primes} \\ 0 & \text{if } n \text{ is divisible by } p^2 \text{ for some prime } p. \end{cases}$$

We now state the Möbius Inversion Formula.

Theorem 3.13 (Möbius Inversion Formula). *Suppose that $f, g : \mathbb{Z}^+ \to \mathbb{R}$ are functions, and suppose that the following equation holds for all positive integers j:*

$$f(j) = \sum_{i | j} g(i).$$

Then the following equation holds for all positive integers i:

$$g(i) = \sum_{j|i} \mu\left(\frac{i}{j}\right) f(j).$$

Theorem 3.14. *Suppose that $(X, \mathcal{A})$ is a symmetric (v, k, λ)-BIBD, and suppose that $\alpha \in \mathsf{Aut}(X, \mathcal{A})$. Then the cycle type of the permutation of X induced by α is the same as the cycle type of the permutation of $\mathcal{A}$ induced by α.*

Proof. Suppose a permutation α of a finite set S has exactly c_i cycles of length i, for $1 \leq i \leq |S|$. Let f_j denote the number of fixed points of the permutation α^j. It is not hard to see that a point $x \in S$ is fixed by the permutation α^j if and only if x occurs in a cycle of length $i|j$ in the permutation α. Hence the following equation holds:

$$f_j = \sum_{i|j} ic_i. \tag{3.1}$$

The Möbius Inversion Formula can be used to solve for the c_i's in terms of the f_j's. This is easily done by defining $g(i) = ic_i$ and applying Theorem 3.13. The following formula is the result:

$$c_i = \frac{1}{i} \sum_{j|i} \mu\left(\frac{i}{j}\right) f_j. \tag{3.2}$$

Now suppose α is an automorphism of the symmetric (v, k, λ)-BIBD $(X, \mathcal{A})$. Then, for all $j \geq 1$, α^j is an automorphism of $(X, \mathcal{A})$, and Lemma 3.12 shows that the permutations of X and $\mathcal{A}$ induced by α^j have the same number of fixed points. Hence, by equation (3.2), the two permutations induced by α have the same cycle type. □

We give an example to illustrate the previous results.

Example 3.15. We refer to the $(7, 3, 1)$-BIBD, $(X, \mathcal{A})$, that was presented in Example 1.23. Let the blocks be named $A_1, A_2, \ldots, A_7$, where

$$A_1 = 123, A_2 = 145, A_3 = 167, A_4 = 246, A_5 = 257, A_6 = 347, A_7 = 356.$$

We showed in Example 1.23 that

$$\alpha = (1)(2)(3)(4\ 5)(6\ 7)$$

is an automorphism of $(X, \mathcal{A})$. The permutation of $\mathcal{A}$ induced by α is the following:

$$(A_1)(A_2)(A_3)(A_4\ A_5)(A_6\ A_7).$$

Hence the two permutations induced by α have the same cycle type; namely $[1, 1, 1, 2, 2]$.

In this example, we have $c_1 = 3$, $c_2 = 2$, $f_1 = 3$, and $f_2 = 7$. Also, $\mu(1) = 1$ and $\mu(2) = -1$. It is easy to verify that equations (3.1) and (3.2) hold for $j = 1, 2$:

$$\begin{aligned} f_1 &= c_1 && = 3, \\ f_2 &= c_1 + 2c_2 && = 7, \\ c_1 &= \mu(1) f_1 && = 3, \quad \text{and} \\ c_2 &= \frac{\mu(2) f_1 + \mu(1) f_2}{2} && = 2. \end{aligned}$$

∎

We are now ready to prove a converse to Theorem 3.8 in the special case where the symmetric BIBD has an automorphism that is a single cycle of length v.

Theorem 3.16. *Suppose $(X, \mathcal{A})$ is a symmetric (v, k, λ)-BIBD having an automorphism α that permutes the points in X in a single cycle of length v. Then there is a (v, k, λ)-difference set in $(\mathbb{Z}_v, +)$.*

Proof. By relabeling the points if necessary, we can assume without loss of generality that $X = \{x_0, \ldots, x_{v-1}\}$ and $\alpha(x_i) = x_{i+1 \bmod v}$ for $0 \le i \le v - 1$, i.e.,

$$\alpha = (x_0 \; x_1 \; \cdots \; x_{v-1}).$$

Choose any block $A \in \mathcal{A}$. Define $A_0 = A$, and for every positive integer j, define

$$A_j = \{\alpha^j(x) : x \in A_0\} = \{x_{i+j \bmod v} : x_i \in A_0\}.$$

Every A_j is a block in $\mathcal{A}$ because $\alpha^j \in \mathsf{Aut}(X, \mathcal{A})$. Also, we have that $\alpha(A_j) = A_{j+1 \bmod v}$ by the way in which the A_j's are defined. Theorem 3.14 establishes that α permutes the blocks in $\mathcal{A}$ in a single cycle of length v. From this, it is seen that $A_0, \ldots, A_{v-1}$ are distinct,

$$\mathcal{A} = \{A_j : 0 \le j \le v - 1\},$$

and α permutes the blocks in $\mathcal{A}$ as follows:

$$\alpha = (A_0 \; A_1 \; \cdots \; A_{v-1}).$$

Now we define

$$D = \{i : x_i \in A_0\}.$$

We will show that D is the desired difference set. Let $g \in \mathbb{Z}_v$, $g \neq 0$. The pair $\{x_0, x_g\}$ occurs in exactly λ blocks in $\mathcal{A}$—say in $A_{i_1}, \ldots, A_{i_\lambda}$. For each occurrence of a pair $\{x_0, x_g\} \subseteq A_{i_j}$, we have a pair with difference g in the set D, namely, $(g - i_j) - (-i_j) \equiv g \pmod{v}$, where

$$\{-i_j \bmod v, g - i_j \bmod v\} \subseteq D.$$

These λ pairs in D are distinct. Thus the difference g occurs λ times in the set D for all nonzero $g \in \mathbb{Z}_v$. All occurrences of g in D are accounted for by this analysis, and hence D is a difference set. □

Theorem 3.16 can be generalized to arbitrary finite groups. Suppose $G \subseteq S_v$ is a permutation group acting on the v-set X. G is *sharply transitive* provided that the following condition holds: for all $x, x' \in X$, there exists a unique permutation $g \in G$ such that $g(x) = x'$. Note that $|G| = v$ if it is sharply transitive.

The following theorem can be proven in a fashion similar to Theorem 3.16.

Theorem 3.17. *Suppose* $(X, \mathcal{A})$ *is a symmetric* (v, k, λ)-BIBD *such that* G *is a sharply transitive subgroup of* $\mathrm{Aut}(X, \mathcal{A})$. *Then there is a* (v, k, λ)*-difference set in the group* $(G, \circ)$.

We present an example to illustrate the application of Theorem 3.17 in the case of a noncyclic group.

Example 3.18. We recall a construction for a symmetric (16,6,2)-BIBD that was mentioned in Exercise 1.13. Write out the integers in the set $X = \{0, \ldots, 15\}$ in a 4×4 array, as follows:

$$M = \begin{array}{cccc} 0 & 1 & 2 & 3 \\ 4 & 5 & 6 & 7 \\ 8 & 9 & 10 & 11 \\ 12 & 13 & 14 & 15 \end{array}$$

For every j, $0 \le j \le 15$, define a block A_j consisting of all the elements in the row and column of M that contains j, excluding j. Then define $\mathcal{A} = \{A_j : 0 \le j \le 15\}$. It is a simple exercise to show that $(X, \mathcal{A})$ is a symmetric (16,6,2)-BIBD.

By the way in which this design is constructed, it is not hard to show that it has many automorphisms. In particular, there is a sharply transitive subgroup, say G, of the automorphism group, such that G is isomorphic to $\mathbb{Z}_4 \times \mathbb{Z}_4$. This is easily seen because a cyclic permutation of the four rows, or the four columns, of the array M leaves the set of blocks unchanged. To be specific, we define two permutations of X:

$$\alpha = (0\,1\,2\,3)(4\,5\,6\,7)(8\,9\,10\,11)(12\,13\,14\,15)$$
$$\beta = (0\,4\,8\,12)(1\,5\,9\,13)(2\,6\,10\,14)(3\,7\,11\,15).$$

It can be shown that $\alpha\beta = \beta\alpha$ and $\alpha^4 = \beta^4 = \mathrm{id}$, where id is the identity permutation. Therefore, α and β generate a subgroup G isomorphic to $\mathbb{Z}_4 \times$

$\mathbb{Z}_4$. It is also easy to see that G is sharply transitive. Therefore Theorem 3.17 asserts that there exists a $(16,6,2)$-difference set in $\mathbb{Z}_4 \times \mathbb{Z}_4$.

Although we did not provide a proof of Theorem 3.17, we can still demonstrate how to construct the difference set using techniques similar to those used in the proof of Theorem 3.16. Suppose we relabel the points in X so that the array M is changed into the following:

$$\begin{array}{cccc} (0,0) & (0,1) & (0,2) & (0,3) \\ (1,0) & (1,1) & (1,2) & (1,3) \\ (2,0) & (2,1) & (2,2) & (2,3) \\ (3,0) & (3,1) & (3,2) & (3,3) \end{array}$$

The reader can verify that the group G (with its points relabeled as described above) is the permutation representation of $\mathbb{Z}_4 \times \mathbb{Z}_4$. Then any block of the design forms the desired difference set. The difference set presented in Example 3.4 is one that can be obtained in this way. ∎

3.2 Quadratic Residue Difference Sets

We introduced the concept of quadratic residues in Section 2.4. We now discuss quadratic residues in a finite field $\mathbb{F}_q$, where q is an odd prime power. The *quadratic residues* of $\mathbb{F}_q$ are the elements in the set

$$\mathsf{QR}(q) = \{z^2 : z \in \mathbb{F}_q, z \neq 0\}.$$

We will also define

$$\mathsf{QNR}(q) = \mathbb{F}_q \backslash (\mathsf{QR}(q) \cup \{0\}).$$

The elements of $\mathsf{QNR}(q)$ are called the *quadratic nonresidues* of $\mathbb{F}_q$.

Using the fact that $z^2 = (-z)^2$, it is not hard to prove that the mapping $z \mapsto z^2$ is a two-to-one mapping if $z \in \mathbb{F}_q \backslash \{0\}$ and q is odd. From this, it can be proven that $\mathsf{QR}(q)$ is a multiplicative subgroup of $\mathbb{F}_q \backslash \{0\}$ having index two, and $\mathsf{QNR}(q)$ is a coset of $\mathsf{QR}(q)$. The following facts can therefore be shown as a consequence:

$$\begin{array}{ll} xy \in \mathsf{QR}(q) & \text{if } x,y \in \mathsf{QR}(q) \\ xy \in \mathsf{QR}(q) & \text{if } x,y \in \mathsf{QNR}(q) \\ xy \in \mathsf{QNR}(q) & \text{if } x \in \mathsf{QR}(q), y \in \mathsf{QNR}(q). \end{array}$$

We will now characterize the quadratic residues and nonresidues in a different way. We make use of the important fact (which we do not prove) that the multiplicative group $(\mathbb{F}_q \backslash \{0\}, \cdot)$ is a cyclic group. A generator of this group, say ω, is called a *primitive element* of the field $\mathbb{F}_q$. Clearly, an element $\omega \in \mathbb{F}_q$ is a primitive element if and only if

$$\{\omega^i : 0 \leq i \leq q-2\} = \mathbb{F}_q \backslash \{0\}.$$

It is obvious that the set

$$\left\{\omega^{2i} : 0 \leq i \leq \frac{q-3}{2}\right\}$$

is a subset of $\mathsf{QR}(q)$. Since

$$\left|\left\{\omega^{2i} : 0 \leq i \leq \frac{q-3}{2}\right\}\right| = \frac{q-1}{2} = |\mathsf{QR}(q)|,$$

we have proven the following result.

Lemma 3.19. *Suppose q is an odd prime power and ω is a primitive element in $\mathbb{F}_q$. Then*

$$\mathsf{QR}(q) = \left\{\omega^{2i} : 0 \leq i \leq \frac{q-3}{2}\right\}.$$

We now state and prove a useful corollary of Lemma 3.19. In the case where q is prime, this result follows from Euler's Criterion. (In fact, Euler's Criterion can be shown to hold in any finite field of odd order.) However, we give a proof using the facts about finite fields that we have discussed above.

Corollary 3.20. *Suppose q is an odd prime power. Then $-1 \in \mathsf{QR}(q)$ if and only if $q \equiv 1 \pmod 4$.*

Proof. Let $\omega \in \mathbb{F}_q$ be a primitive element, and let $\gamma = \omega^{(q-1)/2}$. Now, $\gamma^2 = \omega^{(q-1)} = 1$ and $\gamma \neq 1$, so $\gamma = -1$. The result now follows from Lemma 3.19.

□

It follows that $x \in \mathsf{QR}(q)$ if and only if $-x \in \mathsf{QNR}(q)$ whenever $q \equiv 3 \pmod 4$ is a prime power.

Our next result provides an infinite class of difference sets that are called *quadratic residue difference sets.*

Theorem 3.21 (Quadratic Residue Difference Sets). *Suppose $q \equiv 3 \pmod 4$ is a prime power. Then $\mathsf{QR}(q)$ is a $(q, (q-1)/2, (q-3)/4)$-difference set in $(\mathbb{F}_q, +)$.*

Proof. Denote $D = \mathsf{QR}(q)$. We have already shown that $|D| = (q-1)/2$. Hence, we need only to prove that every nonzero element of $\mathbb{F}_q$ occurs $(q-3)/4$ times as a difference of two elements in D.

For any $d \in \mathbb{F}_q \backslash \{0\}$, define

$$a_d = |\{(x, y) : x, y \in D, x - y = d\}|.$$

Clearly $gx - gy = g(x - y)$ for all g, x, and y, so the number of times any given difference d occurs in D is the same as the number of times the difference gd occurs in gD, where $gD = \{gx : x \in D\}$. Suppose that $g \in \mathsf{QR}(q)$.

Then it is easy to see that $gD = D$, and therefore $a_d = a_{gd}$ for all $g \in \mathsf{QR}(q)$. Hence, there exists a constant λ such that $a_d = \lambda$ for all $d \in \mathsf{QR}(q)$.

Now, suppose that $d \in \mathsf{QNR}(q)$, and let $e = -d$. We have that $-1 \in \mathsf{QNR}(q)$ from Corollary 3.20, and hence $e \in \mathsf{QR}(q)$. Observe that $a_d = a_e$ because $x - y = d$ if and only if $y - x = e$. Therefore it follows that $a_d = \lambda$ for all $d \in \mathbb{F}_q\backslash\{0\}$, and hence D is a $(q, (q-1)/2, \lambda)$-difference set. We can compute λ from the equation $\lambda(v-1) = k(k-1)$, which gives $\lambda = (q-3)/4$, as desired. □

Here is an example to illustrate this.

Example 3.22. An $(11,5,2)$-difference set in $(\mathbb{Z}_{11}, +)$. We compute $1^2 = 1$, $2^2 = 4$, $3^2 = 9$, $4^2 = 5$, and $5^2 = 3$ (where all arithmetic is performed in $\mathbb{Z}_{11}$). Hence, from Theorem 3.21,

$$\mathsf{QR}(11) = \{1,3,4,5,9\}$$

is an $(11,5,2)$-difference set in $(\mathbb{Z}_{11}, +)$. ∎

We mention two related constructions for difference sets that involve quartic residues. For a prime power $q \equiv 1 \pmod 4$, the *quartic residues* in $\mathbb{F}_q$ are the elements of the set $\{z^4 : z \in \mathbb{F}_q, z \neq 0\}$. Equivalently, the quartic residues are $\omega^{4i}, 0 \leq i \leq (q-5)/4$, where ω is a primitive element in $\mathbb{F}_q$.

We state the following two theorems without proof. (The proofs, which are difficult, involve the determination of the so-called "cyclotomic numbers" in the finite fields $\mathbb{F}_q$.)

Theorem 3.23. *Suppose that $p = 4t^2 + 1$ is prime and t is an odd integer. Then the quartic residues in $\mathbb{Z}_p$ form a $(4t^2+1, t^2, (t^2-1)/4)$-difference set in $(\mathbb{Z}_p, +)$.*

Example 3.24. $\{1,7,9,10,12,16,26,33,34\}$ is a $(37,9,2)$-difference set in the group $(\mathbb{Z}_{37}, +)$ that can be constructed using the theorem above. ∎

Theorem 3.25. *Suppose that $p = 4t^2 + 9$ is prime and t is an odd integer. Then the quartic residues in $\mathbb{Z}_p$, together with 0, form a $(4t^2+9, t^2+3, (t^2+3)/4)$-difference set in $(\mathbb{Z}_p, +)$.*

3.3 Singer Difference Sets

In this section we present an infinite class of difference sets, called *Singer difference sets*. These difference sets provide another method of constructing the projective planes of prime power order that we considered in Section 2.3.

Theorem 3.26 (Singer Difference Sets). *Let q be a prime power. Then there exists a $(q^2+q+1, q+1, 1)$-difference set in $(\mathbb{Z}_{q^2+q+1}, +)$.*

Proof. Recall the construction of a symmetric $(q^2+q+1, q+1, 1)$-BIBD that was given in Section 2.3. V is a three-dimensional vector space over $\mathbb{F}_q$; V_1 consists of all the one-dimensional subspaces of V; and the blocks $\mathcal{A}$ correspond to the two-dimensional subspaces of V, which were denoted by V_2.

The finite field $\mathbb{F}_{q^3}$ is a three-dimensional vector space over $\mathbb{F}_q$, so we can take $V = \mathbb{F}_{q^3}$. Let ω be a primitive element of $\mathbb{F}_{q^3}$, and define a mapping $f : V \to V$ by $f(z) = \omega z$. It is easy to see that $f(z+z') = f(z) + f(z')$ for all $z, z' \in V$, and $f(cz) = cf(z)$ for all $z \in V$ and all $c \in \mathbb{F}_q$. It follows that f is an $\mathbb{F}_q$-linear mapping on V, and hence it preserves subspaces of V; i.e., any subspace in V_i is mapped by f to a subspace in V_i, $i = 1, 2$. This implies that f induces an automorphism of the resulting $(q^2+q+1, q+1, 1)$-BIBD.

$\mathbb{F}_q$ is a subfield of $\mathbb{F}_{q^3}$, and it is not hard to see that

$$\mathbb{F}_q = \{\omega^{(q^2+q+1)i} : 0 \le i \le q-2\} \cup \{(0,0,0)\}.$$

For any subspace W of V, it follows that $f^{q^2+q+1}(W) = W$. As a consequence, it can be seen that f permutes the one-dimensional subspaces of $\mathbb{F}_{q^3}$ (i.e., the elements in the set V_1) in a single cycle of length q^2+q+1. Applying Theorem 3.16, we conclude that there exists a $(q^2+q+1, q+1, 1)$-difference set in $(\mathbb{Z}_{q^2+q+1}, +)$. □

We now describe how to carry out the construction of a Singer difference set for a projective plane. We use the same notation as in the proof above. The points of the projective plane can be denoted as C_i $(0 \le i \le q^2+q)$, where

$$C_i = \mathsf{span}(\omega^i),$$

$0 \le i \le q^2+q$. Then $f(C_i) = C_{i+1 \bmod q^2+q}$, $0 \le i \le q^2+q$.

Suppose that the field $\mathbb{F}_{q^3}$ is constructed as $\mathbb{F}_q[x]/(g(x))$, where $g(x) \in \mathbb{F}_q[x]$ is an irreducible cubic polynomial. Then elements of $\mathbb{F}_{q^3}$ can be represented as polynomials in $\mathbb{F}_q[x]$ having degree at most two.

For $j \in \mathbb{F}_q$, define $y_j \in \mathbb{Z}_{q^3-1}$ by the rule $\omega^{y_j} = j + x$ (note that $j + x \in \mathbb{F}_{q^3} \backslash \{0\}$, and hence it can be expressed in this form in a unique way). Then it is easy to see that $\mathsf{span}(1) = C_0$ and

$$\mathsf{span}(j+x) = C_{y_j \bmod q^2+q+1}$$

for all $j \in \mathbb{F}_q$.

Now, let

$$B = \mathsf{span}(1, x) = \{i + jx : i, j \in \mathbb{F}_q\},$$

and consider the block A_B. Then we have that

$$\begin{aligned} A_B &= \{\mathsf{span}(1)\} \cup \{\mathsf{span}(j+x) : j \in \mathbb{F}_q\} \\ &= \{C_0\} \cup \{C_{y_j \bmod q^2+q+1} : j \in \mathbb{F}_q\}. \end{aligned}$$

Then the set

$$D = \{0\} \cup \{y_j \bmod q^2+q+1 : j \in \mathbb{F}_q\}$$

is a $(q^2+q+1, q+1, 1)$-difference set in $(\mathbb{Z}_{q^2+q+1}, +)$.

We do an example to illustrate this.

Example 3.27. Suppose $q = 3$. The field $\mathbb{F}_{27}$ can be constructed as the quotient ring $\mathbb{Z}_3[x]/(x^3+2x^2+1)$ since x^3+2x^2+1 is irreducible in $\mathbb{Z}_3[x]$. It turns out that $\omega = x$ is a primitive element in the resulting field $\mathbb{F}_{27}$. It is possible to compute the powers of ω as follows:

i	ω^i	i	ω^i
0	1	13	2
1	x	14	$2x$
2	x^2	15	$2x^2$
3	x^2+2	16	$2x^2+1$
4	x^2+2x+2	17	$2x^2+x+1$
5	$2x+2$	18	$x+1$
6	$2x^2+2x$	19	x^2+x
7	x^2+1	20	$2x^2+2$
8	x^2+x+2	21	$2x^2+2x+1$
9	$2x^2+2x+2$	22	x^2+x+1
10	x^2+2x+1	23	$2x^2+x+2$
11	$x+2$	24	$2x+1$
12	x^2+2x	25	$2x^2+x$

According to the discussion above, we need only compute the values y_j such that $\omega^{y_j} = j+x$ for $j = 0,1,2$. Referring to the table of values of ω^i constructed above, we see that $y_0 = 1$, $y_1 = 18$, and $y_2 = 11$. Then $D = \{0,1,5,11\}$ is a $(13,4,1)$-difference set in $\mathbb{Z}_{13}$. ∎

Using essentially identical arguments, the following difference sets, corresponding to the projective spaces constructed in Theorem 2.14, can be shown to exist. These are also known as Singer difference sets.

Theorem 3.28 (Singer Difference Sets). *Suppose $q \geq 2$ is a prime power and $d \geq 2$ is an integer. Then there exists a $\left(\frac{q^{d+1}-1}{q-1}, \frac{q^d-1}{q-1}, \frac{q^{d-1}-1}{q-1}\right)$-difference set in $(\mathbb{Z}_{(q^{d+1}-1)/(q-1)}, +)$.*

3.4 The Multiplier Theorem

3.4.1 Multipliers of Difference Sets

In this section, we restrict our attention to Abelian groups. A very useful concept in the study of difference sets in Abelian groups is the idea of a multiplier, which we define now.

Definition 3.29. *Let D be a (v, k, λ)-difference set in an Abelian group $(G, +)$ of order v. For an integer m, define*

$$mD = \{mx : x \in D\},$$

where mx is the sum (computed in G) of m copies of x. Then m is called a multiplier *of D if $mD = D + g$ for some $g \in G$. Also, we say that D is fixed by the multiplier m if $mD = D$.*

Example 3.30. The set $D = \{0, 1, 5, 11\}$ is a $(13, 4, 1)$-difference set in $(\mathbb{Z}_{13}, +)$. It is easy to see that $3D = \{0, 2, 3, 7\} = D + 2$, and hence 3 is a multiplier of D.

$2D = \{0, 2, 9, 10\}$ is a $(13, 4, 1)$-difference set. Suppose that $2D = D + g$ for some $g \in \mathbb{Z}_{13}$. There is a unique occurrence of the difference 1 in D (namely $1 = 1 - 0$) and a unique occurrence in $2D$ (namely $1 = 10 - 9$). Hence $(0, 1) + g = (9, 10)$, which implies $g = 9$. However, $D + 9 = \{3, 7, 9, 10\} \neq 2D$, so 2 is not a multiplier of D. ▌

As another example, any quadratic residue is a multiplier of the difference sets of Theorem 3.21.

We now establish some preliminary results concerning multipliers.

Lemma 3.31. *Suppose that m is a multiplier of a (v, k, λ)-difference set D in an Abelian group $(G, +)$ of order v. Then* $\gcd(m, v) = 1$.

Proof. Suppose that $\gcd(m, v) > 1$, and let p be a prime divisor of m and v. Let $d \in G$ have order p. There must exist $x, y \in D$ such that $x - y = d$. Then $mx - my = md = 0$. Hence, the set mD contains repeated elements, and therefore $mD \neq D + g$ for any g. Therefore m is not a multiplier of D, a contradiction. □

Lemma 3.32. *Suppose that m is a multiplier of a (v, k, λ)-difference set D in an Abelian group $(G, +)$ of order v. Define $\alpha : G \to G$ by the rule $\alpha(x) = mx$. Then $\alpha \in \mathsf{Aut}(G, \mathsf{Dev}(D))$.*

Proof. We have that $mD = D + g$ for some $g \in G$. Now, consider what happens when we apply α to an arbitrary block of the design $(G, \mathsf{Dev}(D))$:

$$\alpha(D + h) = m(D + h) = mD + mh = D + g + mh \in \mathsf{Dev}(D).$$

Therefore α maps any block to a block, as required. □

An important result known as the "Multiplier Theorem" establishes the existence of multipliers in difference sets whose parameters satisfy certain arithmetic conditions. (A proof of this result will be given in Section 3.4.3.)

Theorem 3.33 (Multiplier Theorem). *Suppose there exists a (v, k, λ)-difference set D in an Abelian group $(G, +)$ of order v. Suppose also that the following four conditions are satisfied:*

1. *p is prime,*
2. $\gcd(p, v) = 1$,
3. $k - \lambda \equiv 0 \pmod{p}$, *and*
4. $p > \lambda$.

Then p is a multiplier of D.

Applying the Multiplier Theorem is made easier by the following result.

Theorem 3.34. *Suppose that m is a multiplier of a (v,k,λ)-difference set D in an Abelian group $(G,+)$ of order v. Then there exists a translate of D that is fixed by the multiplier m.*

Proof. Define $\alpha(x) = mx$ for all $x \in G$. We proved in Lemma 3.32 that $\alpha \in \mathrm{Aut}(G, \mathrm{Dev}(D))$. Now, $\alpha(0) = 0$, so α fixes at least one point. By Lemma 3.12, α fixes at least one block of $\mathrm{Dev}(D)$. In other words, there exists a translate of D that is fixed by the multiplier m. □

A more general result can be proven in the case where $\gcd(v,k) = 1$.

Theorem 3.35. *Suppose that $\gcd(k, v) = 1$ and there exists a (v,k,λ)-difference set D in an Abelian group $(G,+)$ of order v. Then there exists a translate of D that is fixed by every multiplier m.*

Proof. Let

$$s = \sum_{x \in D} x.$$

It is easy to see that the following equation holds:

$$\sum_{x \in D+g} x = s + kg. \tag{3.3}$$

Now suppose that $s + kg = s + kh$, where $g, h \in G$ and $g \neq h$. Then $k(g-h) = 0$, so the order of $g - h$ divides k. However, in any finite group, the order of any element divides the order of the group. Hence, the order of $g - h$ divides v. Since $\gcd(k, v) = 1$, it follows that $g - h = 0$, a contradiction.

We have shown that the mapping $g \mapsto s + kg$ is one-to-one. Since this is a mapping from G to G, it must be surjective, and therefore there exists a unique $g \in G$ such that $s + kg = 0$. (In the case where $G = (\mathbb{Z}_v, +)$, it is easy to see that $g = -sk^{-1} \bmod v$.) Hence, from equation (3.3), there is a unique $g \in G$ such that

$$\sum_{x \in D+g} x = 0.$$

Now let m be any multiplier of D. Then m is also a multiplier of the translate $D + g$, and we have

$$\sum_{x \in m(D+g)} x = m \cdot \sum_{x \in D+g} x = 0.$$

Recall that $D + g$ is the unique translate of D whose elements sum to 0. Hence $m(D+g) = D+g$, and the translate $D+g$ is fixed by all multipliers m. □

A difference set (or a translate of a difference set) is said to be *normalized* if the sum of the elements in it is 0. In the proof of Theorem 3.35, we showed that there is a unique normalized translate of any (v,k,λ)-difference set D in an Abelian group of order v when $\gcd(k,v) = 1$, and this unique normalized translate is fixed by all multipliers of D.

Before proceeding to the proof of Theorem 3.33, we give some examples to illustrate the application of the Multiplier Theorem in particular parameter situations.

Example 3.36. We use the Multiplier Theorem to find a $(21,5,1)$-difference set in $(\mathbb{Z}_{21},+)$. Observe that $p = 2$ satisfies the conditions of Theorem 3.33. Hence, 2 is a multiplier of any such difference set. By Theorem 3.34, we can assume that there exists a $(21,5,1)$-difference set in $(\mathbb{Z}_{21},+)$ that is fixed by the multiplier 2. We therefore compute the orbits of $\mathbb{Z}_{21}$ formed by multiplication by 2. (These are in fact the cycles in the disjoint cycle representation of the permutation of $\mathbb{Z}_{21}$ defined by the mapping $x \mapsto 2x \bmod 21$). These cycles (or orbits) are as follows:

(0)
(1 2 4 8 16 11)
(3 6 12)
(5 10 20 19 17 3)
(7 14)
(9 18 15).

The difference set we are looking for must consist of a union of orbits in the list above. Since the difference set has size five, it must be the union of one orbit of length two and one of length three. There are two possible ways to do this, both of which happen to produce difference sets:

$$\{3,6,7,12,14\}$$

and

$$\{7,9,14,15,18\}.$$

■

Example 3.37. We use the Multiplier Theorem to investigate the existence of $(31,10,3)$-difference sets in $(\mathbb{Z}_{31},+)$. It is easily seen that $p = 7$ satisfies the conditions of Theorem 3.33, so 7 will be a multiplier of any such difference set. By Theorem 3.34, we can assume that there exists a $(31,10,3)$-difference set in $(\mathbb{Z}_{31},+)$ that is fixed by the multiplier 7. As in the previous example, we need to consider the orbits of $\mathbb{Z}_{31}$ under multiplication by 7. Of course (0) is one orbit. Let us consider the orbit containing "1". It is as follows:

(1 7 18 2 14 5 4 28 10 8 25 20 16 19 9).

This orbit has length 15, and it is straightforward to verify that there is exactly one other orbit, also having length 15. Clearly there is no way to find a union of orbits having cardinality $k = 10$. We conclude that there is no $(31,10,3)$-difference set in $(\mathbb{Z}_{31},+)$. ∎

Example 3.38. We establish a result about $(n^2+n+1, n+1, 1)$-difference sets. Suppose that $n \equiv 0 \pmod 6$. Then $p = 2$ and $p = 3$ both satisfy the conditions of Theorem 3.33, so they are both multipliers. Using the fact that $n^2+n+1 = n(n+1)+1$, it follows that $\gcd(n^2+n+1, n+1) = 1$. Hence, by Theorem 3.35, we can assume that there exists an $(n^2+n+1, n+1, 1)$-difference set, say D, that is fixed by both of the multipliers 2 and 3. Let $x \in D, x \neq 0$. Then $2x, 3x \in D$. Clearly $x \neq 2x \neq 3x \neq x$. Now, if we compute $2x - x = 3x - 2x = x$, we see that the difference x occurs twice in D. This is not allowed because $\lambda = 1$, and we have a contradiction. We conclude that there is no $(n^2+n+1, n+1, 1)$-difference set when $n \equiv 0 \pmod 6$. ∎

3.4.2 The Group Ring

The proof of the Multiplier Theorem uses an algebraic structure called a group ring. Let $(G,+)$ be an Abelian group. The *group ring* $\mathbb{Z}[G]$ consists of all formal sums of the form

$$\sum_{g \in G} a_g x^g,$$

where $a_g \in \mathbb{Z}$ for all $g \in G$, and x is an indeterminate. Informally, an element of the $\mathbb{Z}[G]$ looks like a polynomial in the indeterminate x having integer coefficients, except that the exponents are elements in the group G rather than nonnegative integers.

If

$$a(x) = \sum_{g \in G} a_g x^g$$

and

$$b(x) = \sum_{g \in G} b_g x^g,$$

then we can define the sum of $a(x)$ and $b(x)$ to be

$$(a+b)(x) = \sum_{g \in G} (a_g + b_g) x^g.$$

The product of $a(x)$ and $b(x)$ is defined to be

$$(a \cdot b)(x) = \sum_{g \in G} \sum_{h \in G} (a_g b_h) x^{g+h}.$$

Thus we compute sums and products of elements of the group ring using the same formulas that are used to compute sums and products of polynomials.

With these operations, it is not hard to see that the group ring $\mathbb{Z}[G]$ is indeed a ring.

Sometimes we will also make use of the group ring $\mathbb{Z}_p[G]$. This is defined in the same way as $\mathbb{Z}[G]$, except that the coefficients are elements of $\mathbb{Z}_p$. Suppose that $a(x), b(x) \in \mathbb{Z}[G]$, $a(x) = \sum a_g x^g$, and $b(x) = b_g x^g$. Then we write $a(x) \equiv b(x) \pmod{p}$ if $a_g \equiv b_g \pmod{p}$ for all $g \in G$.

We need to define some more notation. Recall that, for a positive integer m and any $g \in G$, mg is the m-fold sum of g. For any $a(x) = \sum a_g x^g$, define

$$\begin{aligned} a(x^m) &= \sum_{g \in G} a_g x^{mg}, \\ a(x^{-1}) &= \sum_{g \in G} a_g x^{-g}, \text{and} \\ a(1) &= \sum_{g \in G} a_g. \end{aligned}$$

Finally, define

$$G(x) = \sum_{g \in G} x^g,$$

and for a difference set D in G, define

$$D(x) = \sum_{g \in D} x^g.$$

We now present some easy preliminary lemmas concerning difference sets and the group ring.

Lemma 3.39. *Suppose D is a (v, k, λ)-difference set in an Abelian group G. Then*

$$D(x)D(x^{-1}) = \lambda G(x) + (k - \lambda)x^0.$$

Proof. We have that

$$\begin{aligned} D(x)D(x^{-1}) &= \sum_{g,h \in D} x^{g-h} \\ &= \sum_{d \in G} \alpha_d x^d, \end{aligned}$$

where

$$\alpha_d = |\{(g, h) \in D \times D : g - h = d\}|.$$

Clearly

$$\alpha_d = \begin{cases} k & \text{if } d = 0 \\ \lambda & \text{if } d \neq 0 \end{cases}$$

because D is a difference set. Therefore $D(x)D(x^{-1}) = \lambda G(x) + (k - \lambda)x^0$, as required. □

Lemma 3.40. *Suppose $a(x) \in \mathbb{Z}[G]$. Then*

$$a(x)G(x) = a(1)G(x).$$

Proof. We have that

$$\begin{aligned} a(x)G(x) &= \sum_{g,h\in G} a_g x^{g+h} \\ &= \sum_{i\in G}\left(\sum_{g\in G} a_g\right) x^i, \quad \text{where } g+h=i \\ &= \sum_{i\in G} a(1)x^i \\ &= a(1)G(x), \end{aligned}$$

as desired. □

Lemma 3.41. *Suppose p is prime and $a(x) \in \mathbb{Z}[G]$. Then*

$$(a(x))^p \equiv a(x^p) \pmod{p}. \tag{3.4}$$

Proof. We prove that (3.4) holds by induction on the number of nonzero coefficients in $a(x)$. Suppose that $a(x)$ has no nonzero coefficients; then $a(x) = 0$ and (3.4) is trivially true. If $a(x)$ has one nonzero coefficient, then $a(x) = a_g x^g$ for some $a_g \neq 0$. Then, in $\mathbb{Z}_p[x]$, we have that

$$\begin{aligned} (a(x))^p &= (a_g x^g)^p \\ &= a_g{}^p x^{pg} \\ &= a_g x^{pg} \\ &= a(x^p), \end{aligned}$$

where we use the fact that $a_g{}^p \equiv a_g \pmod{p}$ if $a_g \in \mathbb{Z}_p$.

Now, as an induction assumption, assume that (3.4) holds when $a(x)$ has at most i nonzero coefficients for some integer $i \geq 1$. Suppose that $a(x)$ has exactly $i+1$ nonzero coefficients. We can express $a(x)$ in the form $a(x) = a_i(x) + a_g x^g$, where $a_i(x)$ has exactly i nonzero coefficients and $a_g \neq 0$. Then we compute in $\mathbb{Z}_p[x]$ as follows:

$$\begin{aligned} (a(x))^p &= (a_i(x) + a_g x^g)^p \\ &= (a_i(x))^p + \sum_{j=1}^{p-1}\binom{p}{j}(a_i(x))^j(a_g x^g)^{p-j} + (a_g x^g)^p \\ &= (a_i(x))^p + (a_g x^g)^p \quad \text{because } \binom{p}{j} \equiv 0 \pmod{p} \text{ for } 1 \leq j \leq p-1 \\ &= a_i(x^p) + a_g x^{pg} \quad \text{by induction} \\ &= a(x^p). \end{aligned}$$

By induction, (3.4) holds for all $a(x) \in \mathbb{Z}[G]$. □

If D is a (v,k,λ)-difference set and m is a positive integer such that $\gcd(m,v)=1$, then it is not hard to prove that mD is also a (v,k,λ)-difference set. The next lemma uses this fact and is proven in a fashion similar to Lemma 3.39. We leave the details for the reader.

Lemma 3.42. *Suppose D is a (v,k,λ)-difference set in an Abelian group G. Suppose that m is a positive integer such that $\gcd(m,v)=1$. Then*

$$D(x^m)D(x^{-m}) = \lambda G(x) + (k-\lambda)x^0.$$

3.4.3 Proof of the Multiplier Theorem

In this section, we present the proof of the Multiplier Theorem. For convenience, we restate the theorem now.

Theorem 3.43 (Multiplier Theorem). *Suppose there exists a (v,k,λ)-difference set D in an Abelian group $(G,+)$ of order v. Suppose also that the following four conditions are satisfied:*

1. *p is prime,*
2. $\gcd(p,v)=1$,
3. $k-\lambda \equiv 0 \pmod{p}$, *and*
4. $p>\lambda$.

Then p is a multiplier of D.

Proof. We begin by computing the product $D(x^p)D(x^{-1})$ in $\mathbb{Z}_p[G]$:

$$\begin{aligned}
D(x^p)D(x^{-1}) &= (D(x))^pD(x^{-1}) && \text{by Lemma 3.41}\\
&= (D(x))^{p-1}D(x)D(x^{-1})\\
&= (D(x))^{p-1}(\lambda G(x)+(k-\lambda)x^0) && \text{by Lemma 3.39}\\
&= \lambda k^{p-1}G(x)+(k-\lambda)(D(x))^{p-1} && \text{by Lemma 3.40}\\
&= \lambda k^{p-1}G(x)\\
&= \lambda G(x),
\end{aligned}$$

where we use the facts that $D(1)=k$, $k\equiv\lambda \pmod{p}$, and $\lambda k^{p-1}\equiv\lambda^p\equiv\lambda \pmod{p}$. Define

$$S(x) = D(x^p)D(x^{-1}) - \lambda G(x). \tag{3.5}$$

We have proven that $S(x)\equiv 0 \pmod{p}$; therefore all coefficients of S are divisible by p. Clearly, all coefficients of $D(x^p)D(x^{-1})$ are nonnegative, so it follows that all coefficients of $S(x)$ are greater than or equal to $-\lambda$. We assumed that $p>\lambda$, so it must be the case that all coefficients of $S(x)$ are nonnegative.

We now compute $S(x)S(x^{-1})$:

$$\begin{aligned} S(x)S(x^{-1}) &= (D(x^p)D(x^{-1}) - \lambda G(x))(D(x^{-p})D(x) - \lambda G(x^{-1})) \\ &= (D(x^p)D(x^{-1}) - \lambda G(x))(D(x^{-p})D(x) - \lambda G(x)) \\ &= D(x^p)D(x^{-p})D(x)D(x^{-1}) + \lambda^2(G(x))^2 \\ &\quad -\lambda G(x)(D(x^p)D(x^{-1}) + D(x^{-p})D(x)). \end{aligned}$$

Applying Lemmas 3.39, 3.40, and 3.42 and using the fact that $G(1) = v$, we see that

$$\begin{aligned} D(x^p)D(x^{-p})D(x)D(x^{-1}) &= (\lambda G(x) + (k-\lambda)x^0)^2 \\ &= \lambda^2(G(x))^2 + 2(k-\lambda)\lambda G(x) + (k-\lambda)^2x^0 \\ &= \lambda^2 vG(x) + 2(k-\lambda)\lambda G(x) + (k-\lambda)^2x^0. \end{aligned}$$

Similarly, we have that

$$-\lambda G(x)(D(x^p)D(x^{-1}) + D(x^{-p})D(x)) + \lambda^2(G(x))^2 = -2\lambda k^2 G(x) + \lambda^2 vG(x).$$

Combining everything, we have

$$S(x)S(x^{-1}) = (\lambda^2 v + 2(k-\lambda)\lambda - 2\lambda k^2 + \lambda^2 v)G(x) + (k-\lambda)^2x^0.$$

The coefficient of $G(x)$ can be simplified, as follows:

$$\lambda^2 v + 2(k-\lambda)\lambda - 2\lambda k^2 + \lambda^2 v = 2\lambda(\lambda v + k - \lambda - k^2) = 0.$$

Hence, we have that

$$S(x)S(x^{-1}) = (k-\lambda)^2x^0.$$

Let

$$S(x) = \sum_{g \in G} s_g x^g.$$

We have shown above that $s_g \geq 0$ for all $g \in G$. Suppose that there exist $g, h \in G$, $g \neq h$, such that $s_g > 0$ and $s_h > 0$. Then the coefficient of x^{g-h} in $S(x)S(x^{-1})$ is at least $s_g s_h$, which is greater than 0. This is a contradiction. Hence $S(x) = s_g x^g$ for some $g \in G$. Then

$$S(x)S(x^{-1}) = (s_g x^g)(s_g x^{-g}) = (s_g)^2 x^0.$$

Therefore $(s_g)^2 = (k-\lambda)^2$, and since $s_g \geq 0$, it must be the case that $s_g = k - \lambda$. Hence we have proven that $S(x) = (k-\lambda)x^g$ for some $g \in G$. Substituting into (3.5), we see that

$$D(x^p)D(x^{-1}) = (k-\lambda)x^g + \lambda G(x).$$

Now multiply both sides of this equation by $D(x)$:

$$D(x^p)D(x)D(x^{-1}) = D(x)((k-\lambda)x^g + \lambda G(x)),$$

which can be simplified, using Lemmas 3.39 and 3.40, as follows:

$$\begin{aligned} D(x^p)(\lambda G(x) + (k-\lambda)x^0) &= D(x)(k-\lambda)x^g + \lambda k G(x) \\ \lambda k G(x) + (k-\lambda)D(x^p) &= D(x)(k-\lambda)x^g + \lambda k G(x) \\ (k-\lambda)D(x^p) &= D(x)(k-\lambda)x^g \\ D(x^p) &= x^g D(x). \end{aligned}$$

Therefore $pD = D + g$, and we have shown that p is a multiplier of D, as desired. □

One important conjecture about the Multiplier Theorem concerns the requirement that $p > \lambda$. This assumption certainly is used in the proof of Theorem 3.33. However, there is no known example of a difference set D and a prime p that satisfies the first three conditions of Theorem 3.33, where p is not a multiplier of D. Therefore many people have conjectured that the Multiplier Theorem is true for all primes p satisfying the first three conditions of Theorem 3.33. This conjecture has not been proven, however.

3.5 Difference Families

We begin by generalizing the definition of a difference set to an object called a difference family.

Definition 3.44. *Suppose $(G,+)$ is a finite group of order v in which the identity element is denoted "0". Let k and λ be positive integers such that $2 \le k < v$. A (v,k,λ)-*difference family *in $(G,+)$ is a collection of subsets of G, say $[D_1,\ldots,D_\ell]$, such that the following properties are satisfied:*

1. *$|D_i| = k$ for all i, $1 \le i \le \ell$;*
2. *the multiset union*

$$\bigcup_{i=1}^{\ell} [x - y : x, y \in D_i, x \neq y]$$

contains every element in $G\setminus\{0\}$ exactly λ times.

Example 3.45. A $(13,3,1)$-difference family in $(\mathbb{Z}_{13},+)$:

$$\{\{0,1,4\},\{0,2,8\}\}.$$

The differences obtained from the first block are $1,3,4,9,10$, and 12, and the differences obtained from the second block are $2,5,6,7,8$, and 11. Therefore we obtain every nonzero difference exactly once. ∎

It is not hard to show that $\ell = \lambda(v-1)/(k^2-k)$ if a (v,k,λ)-difference family $[D_1,\ldots,D_\ell]$ exists. Because ℓ is required to be an integer, it must be the case that $\lambda(v-1) \equiv 0 \pmod{k^2-k}$ if a (v,k,λ)-difference family exists.

Given a (v, k, λ)-difference family in $(G, +)$, we define $\mathsf{Dev}(D_1, \ldots, D_\ell)$ to be the collection formed by taking all the blocks in $\mathsf{Dev}(D_i)$, $1 \leq i \leq \ell$. The following result generalizes Theorems 3.8 and 3.10.

Theorem 3.46. *Suppose $D_1, \ldots, D_\ell$ is a (v, k, λ)-difference family in the Abelian group $(G, +)$. Then $(G, \mathsf{Dev}(D_1, \ldots, D_\ell))$ is a (v, k, λ)-BIBD, and $\mathsf{Aut}(G, \mathsf{Dev}(D))$ contains a subgroup $\widehat{G}$ that is isomorphic to G.*

Now we consider the converse of Theorem 3.46. In the case of difference sets, we proved Theorem 3.17, which says that a symmetric BIBD in which the automorphism group has a sharply transitive subgroup implies the existence of a difference set in that subgroup. This theorem does not generalize completely to difference families due to the existence of so-called short orbits. We define and examine these objects now, using some of the concepts and terminology introduced in Section 1.4.1.

Let H be a subgroup of the symmetric group S_v acting on the elements of the v-set X. Let A be a subset of X having cardinality k, and let $\mathsf{orbit}(A)$ be the orbit of A under H. Define

$$\mathsf{stab}(A) = \{\alpha \in H : \alpha(A) = A\};$$

$\mathsf{stab}(A)$ is called the *stabilizer* of A. It is easy to see that $\mathsf{stab}(A)$ is a subgroup of H.

We have the following result.

Lemma 3.47. *Let H be a subgroup of the symmetric group S_v acting on the elements of the v-set X, and suppose $A \subseteq X$. Then $|\mathsf{orbit}(A)| = |H|/|\mathsf{stab}(A)|$. Furthermore, if H is sharply transitive and if $\gcd(|A|, v) = 1$, then $|\mathsf{orbit}(A)| = v$.*

Proof. For every $A' \in \mathsf{orbit}(A)$, define $H_{A'} = \{\alpha \in H : \alpha(A) = A'\}$. Then $H_A = \mathsf{stab}(A)$, and every $H_{A'}$ is a coset of H_A. Since the cosets of the subgroup H_A all have the same size and partition H, it follows that $|\mathsf{orbit}(A)| = |H|/|\mathsf{stab}(A)|$.

Now assume that H is sharply transitive; then $|H| = v$. We will prove that $|A| \equiv 0 \pmod{|\mathsf{stab}(A)|}$. Then, because $v \equiv 0 \pmod{|\mathsf{stab}(A)|}$ and $\gcd(|A|, v) = 1$, it must be the case that $|\mathsf{stab}(A)| = 1$ and hence $|\mathsf{orbit}(A)| = v$.

So, we need to prove that $|A| \equiv 0 \pmod{|\mathsf{stab}(A)|}$. For every $\alpha \in \mathsf{stab}(A)$, define α_A to be the permutation α restricted to the points in A. The set of permutations $\mathsf{stab}(A)_A = \{\alpha_A : \alpha \in \mathsf{stab}(A)\}$ is a permutation group acting on A. Note that $\alpha_A \neq \alpha'_A$ if $\alpha \neq \alpha'$ because H is sharply transitive, and therefore $|\mathsf{stab}(A)_A| = |\mathsf{stab}(A)|$.

We now apply the Cauchy-Frobenius-Burnside Lemma to the group $\mathsf{stab}(A)_A$. We have that $\mathsf{fix}(\mathsf{id}) = |A|$, where id is the identity permutation in $\mathsf{stab}(A)_A$. For any $\alpha_A \in \mathsf{stab}(A)_A$, $\alpha_A \neq \mathsf{id}$, it must be the case that $\mathsf{fix}(\alpha_A) = 0$; this follows from the fact that H is sharply transitive. Using Lemma 1.25, we compute the number of orbits of A under the group

$\text{stab}(A)_A$ to be $|A|/|\text{stab}(A)|$. This number, being the number of orbits of A, must be an integer, so the proof is now complete. □

Suppose that $(G,+)$ is a finite group and $A \subseteq G$, where $|G| = v$ and $|A| = k$. Let $\text{orbit}(A)$ denote the orbit of A under the permutation representation of G. Then it is easy to see that $\text{orbit}(A)$ consists of all the distinct blocks in $\text{Dev}(A)$. Equivalently, $\text{Dev}(A)$ is formed by taking $v/|\text{stab}(A)|$ copies of every block in $\text{orbit}(A)$.

We illustrate the results above in the following example.

Example 3.48. Consider the group $G = (\mathbb{Z}_9,+)$. The permutation representation of G is as follows:

g	$\widehat{g}$
0	(0)(1)(2)(3)(4)(5)(6)(7)(8)
1	(0 1 2 3 4 5 6 7 8)
2	(0 2 4 6 8 1 3 5 7)
3	(0 3 6)(1 4 7)(2 5 8)
4	(0 4 8 3 7 2 6 1 5)
5	(0 5 1 6 2 7 3 8 4)
6	(0 6 3)(1 7 4)(2 8 5)
7	(0 7 5 3 1 8 6 4 2)
8	(0 8 7 6 5 4 3 2 1)

It is straightforward to verify that $\text{stab}(\{0,3,6\}) = \{\widehat{0},\widehat{3},\widehat{6}\}$, or equivalently,

$$\{0,3,6\} = \{0,3,6\} + 3 = \{0,3,6\} + 6.$$

The orbit of the subset $\{0,3,6\}$ has cardinality three:

$$\text{orbit}(\{0,3,6\}) = \{\{0,3,6\},\{1,4,7\},\{2,5,8\}\}.$$

$\text{Dev}(\{0,3,6\})$ consists of three copies of each block in $\text{orbit}(\{0,3,6\})$. ∎

The following is an immediate corollary of Lemma 3.47.

Theorem 3.49. *Suppose that $(G,+)$ is a finite group and $A \subseteq G$. Suppose that $|G| = v$ and $|A| = k$, where $\gcd(k,v) = 1$. Then $\text{Dev}(A) = \text{orbit}(A)$.*

We now state a partial converse to Theorem 3.46. This result can be proven in much the same way as Theorem 3.16 (and the more general Theorem 3.17). Note that Theorem 3.49 ensures that there are no short orbits when $\gcd(k,v) = 1$; in the case of difference sets, we proved a similar result as a consequence of Theorem 3.14 without requiring that k and v be relatively prime.

Theorem 3.50. *Suppose that $\gcd(k,v) = 1$ and $(X,\mathcal{A})$ is a (v,k,λ)-BIBD in which G is a sharply transitive subgroup of $\text{Aut}(X,\mathcal{A})$. Then there is a (v,k,λ)-difference family in the group $(G,\circ)$.*

Theorem 3.50 is not true when $\gcd(k, v) > 1$, as the next example demonstrates.

Example 3.51. It is trivial to see that there is no $(15, 3, 1)$-difference family (such a difference family would consist of $8/3$ blocks, which is not an integer). However, it is not difficult to construct a $(15, 3, 1)$-BIBD in which the automorphism group contains $(\mathbb{Z}_{15}, +)$ as a sharply transitive subgroup. Such a BIBD can be described succinctly by taking the orbits of the three blocks $\{0, 5, 10\}$, $\{0, 1, 4\}$, and $\{0, 2, 8\}$ under the group generated by the permutation $(0\ 1\ \cdots\ 14)$ (this is just the permutation representation of $\mathbb{Z}_{15}$). The 35 blocks in this BIBD consist of two orbits of size 15 and one short orbit of size 5. ∎

3.6 A Construction for Difference Families

In this section, we present a simple yet powerful construction for difference families in finite fields that is due to Wilson. We first define some notation and record a couple of simple preliminary results.

For any two multisets A, B whose elements are from a finite field $\mathbb{F}_q$, define

$$A \circ B = [ab : a \in A, b \in B].$$

For a positive integer r and a multiset A, define

$$rA = \bigcup_{i=1}^{r} A.$$

Also, for any set $A \subseteq \mathbb{F}_q$, define the multiset

$$\Delta(A) = [a - a' : a, a' \in A, a \neq a'].$$

Suppose that q is a prime power and let ω be a primitive element of $\mathbb{F}_q$. For any integer f dividing $q-1$, denote $e = (q-1)/f$ and define

$$H = \{\omega^{ei} : 0 \leq i \leq f-1\}.$$

H is a subgroup of the multiplicative group $(\mathbb{F}_q \backslash \{0\}, \cdot)$ having order f. Denote the cosets of H by $H_0, \ldots, H_{e-1}$, where $H_j = \omega^j H$, $0 \leq j \leq e-1$.

The following lemma, which we state without proof, will be useful.

Lemma 3.52. *For all H as defined above, it holds that*

$$\Delta(H) = [\omega^{ei} - 1 : 1 \leq i \leq f-1] \circ H.$$

Furthermore, if f is odd, it holds that

$$\Delta(H) = [1, -1] \circ [\omega^{ei} - 1 : 1 \leq i \leq (f-1)/2] \circ H.$$

Here is a quite general construction for difference families in finite fields.

Theorem 3.53. *Let $k \geq 2$ and $\lambda \geq 1$ be integers such that k divides 2λ or $k-1$ divides 2λ. Suppose that q is a prime power such that $\lambda(q-1) \equiv 0 \pmod{k^2-k}$. Then there exists a (q,k,λ)-difference family in $(\mathbb{F}_q,+)$.*

Proof. Let ω, e, f, and H be as defined above. Denote the cosets of H by $H_0,\ldots,H_{e-1}$, where $H_j = \omega^j H$, $0 \leq j \leq e-1$.

We now consider four cases separately.

Case 1: $\lambda = k-1$

Let $f = k$. We show that

$$[H_0,\ldots,H_{e-1}]$$

is the desired difference family. Using Lemma 3.52, it is easy to see that

$$\Delta(H_i) = [x-1 : x \in H, x \neq 1] \circ H_i$$

for $0 \leq i \leq e-1$. Then the multiset union of the $\Delta(H_i)$'s is seen to be

$$\begin{aligned}\bigcup_{i=0}^{e-1} \Delta(H_i) &= \bigcup_{i=0}^{e-1} [\omega^{ei} - 1 : 1 \leq i \leq f-1] \circ H_i \\ &= [\omega^{ei} - 1 : 1 \leq i \leq f-1] \circ \left(\bigcup_{i=0}^{e-1} H_i\right) \\ &= [\omega^{ei} - 1 : 1 \leq i \leq f-1] \circ (\mathbb{F}_q \backslash \{0\}) \\ &= (f-1)(\mathbb{F}_q \backslash \{0\}).\end{aligned}$$

Case 2: $\lambda = k$

Let $f = k-1$. We show that

$$[H_0 \cup \{0\},\ldots,H_{e-1} \cup \{0\}]$$

is the desired difference family. First, we have that

$$\Delta(H_i \cup \{0\}) = ([1,-1,] \cup [\omega^{ei} - 1 : 1 \leq i \leq f-1]) \circ H_i,$$

$0 \leq i \leq e-1$. The rest of the proof proceeds as in Case 1; here we have that the multiset union of the sets $\Delta(H_i \cup \{0\})$ is $(f+1)(\mathbb{F}_q \backslash \{0\})$.

Case 3: k is odd and $\lambda = (k-1)/2$

Let $f = k$. Note that $q \equiv 1 \pmod{2k}$, so q is odd. $q-1 = ef$ and f is odd, so e must be even. We show that

$$[H_0,\ldots,H_{\frac{e}{2}-1}]$$

is the desired difference family. Applying Lemma 3.52, we have that

$$\Delta(H_i) = [1, -1] \circ [\omega^{ei} - 1 : 1 \le i \le (f-1)/2] \circ H_i$$

for $0 \le i \le e/2 - 1$. Now, using the facts that $-1 = \omega^{ef/2}$, e is even, and f is odd, it follows that $-1 \in H_{\frac{e}{2}}$. Therefore it holds that

$$[-1] \circ H_i = H_{\frac{e}{2}+i},$$

which implies that

$$\Delta(H_i) = [\omega^{ei} - 1 : 1 \le i \le (f-1)/2] \circ (H_i \cup H_{\frac{e}{2}+i}),$$

$0 \le i \le e/2 - 1$. Now, it is easy to see that the multiset union of the relevant $\Delta(H_i)$'s is

$$\begin{aligned}\bigcup_{i=0}^{\frac{e}{2}-1} \Delta(H_i) &= \bigcup_{i=0}^{\frac{e}{2}-1} [\omega^{ei} - 1 : 1 \le i \le (f-1)/2] \circ (H_i \cup H_{\frac{e}{2}+i}) \\ &= [\omega^{ei} - 1 : 1 \le i \le (f-1)/2] \circ \left(\bigcup_{i=0}^{\frac{e}{2}-1} (H_i \cup H_{\frac{e}{2}+i}) \right) \\ &= \frac{f-1}{2} (\mathbb{F}_q \setminus \{0\}).\end{aligned}$$

Case 4: k is even and $\lambda = k/2$

Let $f = k - 1$; then q is odd and e is even, and

$$[H_0 \cup \{0\}, \ldots, H_{\frac{e}{2}-1} \cup \{0\}]$$

is the desired difference family. (The proof, which uses ideas from Case 2 and Case 3, is omitted.)

The four cases discussed above are sufficient to cover all possibilities. This is seen as follows. First, suppose that k divides 2λ. Then $\lambda = sk/2$, where s is an integer. If k is even, then we can take s copies of the difference family constructed in Case 4. If k is odd, then s must be even, and we can take $s/2$ copies of the difference family constructed in Case 2.

If $k - 1$ divides 2λ, the analysis is similar. Write $\lambda = s(k-1)/2$, where s is an integer. If $k - 1$ is even, then we can take s copies of the difference family constructed in Case 3. If $k - 1$ is odd, then s must be even, and we can take $s/2$ copies of the difference family constructed in Case 1. □

Example 3.54. We construct a $(19, 4, 2)$-difference family using Theorem 3.53. Note that the necessary conditions are satisfied, and we use the construction given in Case 4. We have $k = 4$, $f = 3$, and $e = 6$. $\omega = 2$ is a primitive element in $\mathbb{Z}_{19}$, and the H_i's are as follows:

$$\begin{aligned}
H_0 &= \{1,7,11\}\\
H_1 &= \{2,14,3\}\\
H_2 &= \{4,6,9\}\\
H_3 &= \{8,18,12\}\\
H_4 &= \{16,17,5\}\\
H_5 &= \{13,15,10\}.
\end{aligned}$$

The $(19,4,2)$-difference family is

$$[\{0,1,7,11\},\{0,2,3,14\},\{0,4,6,9\}].$$

Example 3.55. We construct a $(16,3,2)$-difference family using Theorem 3.53. The necessary conditions are satisfied, and we use the construction given in Case 1. We have $k = 3$, $f = 3$, and $e = 5$. $\omega = x$ is a primitive element in $\mathbb{F}_{16} = \mathbb{Z}_2[x]/(x^4+x+1)$, and the H_i's are as follows:

$$\begin{aligned}
H_0 &= \{1, x^2+x, x^2+x+1\}\\
H_1 &= \{x, x^3+x^2, x^3+x^2+x\}\\
H_2 &= \{x^2, x^3+x+1, x^3+x^2+x+1\}\\
H_3 &= \{x^3, x^2+1, x^3+x^2+1\}\\
H_4 &= \{x+1, x^3+x, x^3+1\}.
\end{aligned}$$

The $(16,3,2)$-difference family (written in the additive group $(\mathbb{Z}_2)^4$) is

$$\begin{aligned}
&[\{0001,0110,0111\},\{0010,1100,1110\},\{0100,1011,1111\},\\
&\{1000,0101,1101\},\{0011,1010,1001\}].
\end{aligned}$$

3.7 Notes and References

There is a huge amount of literature on difference sets. The first comprehensive treatise on this topic was the 1971 monograph by Baumert [5]. Good starting points to learn more recent results include the 1992 survey by Jungnickel [65] and Chapter 6 of Beth, Jungnickel, and Lenz [9]. Difference families are discussed thoroughly in Chapter 7 of [9].

Quadratic residue difference sets are also known as Paley difference sets and were first constructed in Paley [83]. Singer difference sets were described in [96].

The concept of a multiplier was introduced by Hall [52]. The Multiplier Theorem is due to Hall and Ryser [55]. Bruck [17] is another important early paper on this toopic.

Theorem 3.53 is due to Wilson [117].

3.8 Exercises

3.1 Give a direct combinatorial proof that the complement of a (v,k,λ)-difference set is a difference set, and determine its parameters.

3.2 Construct the following difference sets.
(a) A $(27,13,6)$-difference set in $(\mathbb{F}_{27},+)$.
Note: $\mathbb{F}_{27}$ can be constructed as $\mathbb{Z}_3[x]/(x^3+2x^2+1)$.
(b) A $(101,25,6)$-difference set in $(\mathbb{Z}_{101},+)$.
(c) A $(109,28,7)$-difference set in $(\mathbb{Z}_{109},+)$.

3.3 Use Singer's Theorem to construct a $(31,6,1)$-difference set in $(\mathbb{Z}_{31},+)$. In order to do this, you need to construct the field $\mathbb{F}_{125}$. $\mathbb{F}_{125} = \mathbb{Z}_5[x]/(x^3+x^2+2)$, and x is a primitive element of $\mathbb{F}_{125}$ in this representation.

3.4 Suppose that m_1 and m_2 are both multipliers of a difference set D. Prove that m_1m_2 is also a multiplier of D.

3.5 Give a complete proof of Lemma 3.42.

3.6 (a) Show that a $(21,5,1)$-difference set in $(\mathbb{Z}_{21},+)$ must have the integer $m=2$ as a multiplier.
(b) Determine all $(21,5,1)$-difference sets in $(\mathbb{Z}_{21},+)$ that are fixed by the multiplier $m=2$.
(c) How many translates of any $(21,5,1)$-difference set in $(\mathbb{Z}_{21},+)$ are fixed by the multiplier $m=2$? Explain briefly.

3.7 Use the Multiplier Theorem to find all $(31,6,1)$ difference sets in $(\mathbb{Z}_{31},+)$ that contain the point "1".

3.8 Prove that there is no $(56,11,2)$ difference set in $(\mathbb{Z}_{56},+)$.
Hint: At some point in the proof, it may be helpful to consider differences in $\mathbb{Z}_{56}$ that are divisible by 7.

3.9 Prove that there do not exist $(n^2+n+1,n+1,1)$ difference sets for $n=10,14$.

3.10 $\{01,02,03,10,20,30\}$ is a $(16,6,2)$-difference set in $(\mathbb{Z}_4\times\mathbb{Z}_4,+)$. How many normalized translates does this difference set have?
Note: This question has a short solution that does not involve checking all the translates.

3.11 (a) Prove there does not exist a $(25,9,3)$-difference set in $(\mathbb{Z}_{25},+)$ having a multiplier $m=2$.
(b) Prove that there does not exist a $(25,9,3)$-difference set in $(\mathbb{Z}_5\times\mathbb{Z}_5,+)$ having a multiplier $m=2$.

3.12 Find all $(15,7,3)$-difference sets in $(\mathbb{Z}_{15},+)$ that are fixed by the multiplier $m=2$.

3.13 Give a complete proof of Lemma 3.52.

3.14 Construct difference families with the following parameters:
(a) $(29,5,5)$.
(b) $(31,5,2)$.
(c) $(41,6,3)$.
(d) $(43,6,5)$.

3.15 Let v be odd. A *difference triple* modulo v is a subset of three integers

$$\{x,y,z\} \subseteq \left\{1,2,\ldots,\frac{v-1}{2}\right\},$$

where $x < y < z$, such that $x + y = z$ or $x + y + z \equiv 0 \pmod{v}$. Suppose $v \equiv 1 \pmod{6}$. A set of $t = \lfloor \frac{v}{6} \rfloor$ difference triples, say $\mathcal{T} = \{T_1,\ldots,T_t\}$, is denoted as an HDP$(v)$ provided that

$$\bigcup_{i=1}^{t} T_i = \left\{1,\ldots,\frac{v-1}{2}\right\}.$$

Remark: HDP is an abbreviation for *Heffter's Difference Problem*.

(a) Suppose that $\mathcal{T} = \{T_1,\ldots,T_t\}$ is an HDP(v). For every $T_i = \{x_i,y_i,z_i\}$, define $D_i = \{0,x_i,x_i+y_i\}$, where $x_i < y_i < z_i$. Prove that $\{D_1,\ldots,D_t\}$ is a $(v,3,1)$-difference family in $(\mathbb{Z}_v,+)$.

(b) By trial and error, construct HDP(v) for $v = 7, 13, 19$, and 25.

4

Hadamard Matrices and Designs

4.1 Hadamard Matrices

Definition 4.1. *A* Hadamard matrix *of* order n *is an $n \times n$ matrix H in which every entry is ± 1 such that $HH^T = nI_n$.*

It is trivial to see that (1) and (-1) are both Hadamard matrices of order 1. In the next examples, we present Hadamard matrices of orders 2 and 4.

Example 4.2. The following matrix is a Hadamard matrix of order 2:

$$\begin{pmatrix} 1 & 1 \\ 1 & -1 \end{pmatrix}.$$

∎

Example 4.3. The following matrix is a Hadamard matrix of order 4:

$$\begin{pmatrix} 1 & 1 & 1 & 1 \\ 1 & -1 & 1 & -1 \\ 1 & 1 & -1 & -1 \\ 1 & -1 & -1 & 1 \end{pmatrix}.$$

∎

Observe that we can multiply all the entries in any row (or column) of a Hadamard matrix by -1 and the result is again a Hadamard matrix. By a sequence of multiplications of this type, we can transform any Hadamard matrix into a Hadamard matrix in which every entry in the first row or column is a "1". Such a Hadamard matrix is called *standardized*.

Let the rows of a Hadamard matrix of order n be denoted r_i, $1 \leq i \leq n$. The (i, j)-entry of HH^T is in fact $r_i \cdot r_j$, where "$\cdot$" denotes the usual inner product of real vectors. Hence, it follows from the definition of a Hadamard matrix that $r_i \cdot r_j = 0$ if $i \neq j$.

We have seen Hadamard matrices of orders 1, 2, and 4. The following result provides a necessary condition for the existence of a Hadamard matrix of order n.

Theorem 4.4. *If there exists a Hadamard matrix of order $n > 2$, then $n \equiv 0 \pmod 4$.*

Proof. Suppose without loss of generality that $H = (h_{i,j})$ is a standardized Hadamard matrix of order $n > 2$. For $1 \le i \le n$, let r_i denote the ith row of H. Since r_1 consists of n "1"s and $r_1 \cdot r_i = 0$ if $i \ge 2$, it follows that any row r_i (where $2 \le i \le n$) contains $n/2$ "1"s and $n/2$ "-1"s. Hence, n is even.

Define

$$\begin{aligned} a &= |\{j : h_{2,j} = h_{3,j} = 1\}|, \\ b &= |\{j : h_{2,j} = 1, h_{3,j} = -1\}|, \\ c &= |\{j : h_{2,j} = -1, h_{3,j} = 1\}|, \quad \text{and} \\ d &= |\{j : h_{2,j} = h_{3,j} = -1\}|. \end{aligned}$$

Then we have the following equations:

$$\begin{aligned} a+b+c+d &= n & & \\ a+b-c-d &= 0 & & \text{since } r_1 \cdot r_2 = 0 \\ a-b+c-d &= 0 & & \text{since } r_1 \cdot r_3 = 0 \\ a-b-c+d &= 0 & & \text{since } r_2 \cdot r_3 = 0. \end{aligned}$$

This system has the unique solution

$$a = b = c = d = \frac{n}{4}.$$

Since a, b, c, and d are integers, it must be the case that $n \equiv 0 \pmod 4$. □

It is a famous open conjecture, first stated by Jacques Hadamard in 1893, that there exists a Hadamard matrix of every order $n \equiv 0 \pmod 4$. In fact, the smallest order $n \equiv 0 \pmod 4$ for which a Hadamard matrix is not currently known to exist is $n = 428$.

4.2 An Equivalence Between Hadamard Matrices and BIBDs

In this section, we show a connection between Hadamard matrices and certain symmetric BIBDs.

Theorem 4.5. *Suppose $m > 1$. Then there exists a Hadamard matrix of order $4m$ if and only if there exists a (symmetric) $(4m-1, 2m-1, m-1)$-BIBD.*

Proof. Suppose H is a standardized Hadamard matrix of order $n = 4m$. Let M be formed by deleting the first row and column of H and then replacing every "-1" entry by "0".

Since every row r_i $(2 \leq i \leq n)$ of H contains $2m$ "1"s (as in the proof of Theorem 4.4), it follows that every row of M contains $2m-1$ "1"s. Further, the inner product of two rows of M is $m-1$ (using the fact, proven in Theorem 4.4, that $a = n/4 = m$). Hence,

$$MM^T = (m-1)J_{4m-1} + mI_{4m-1}.$$

Now, since $HH^T = (4m)I_{4m}$, we have that $H^{-1} = \frac{1}{4m}H^T$, and hence $H^T = 4mH^{-1}$. Then we have that

$$H^T H = (4m)H^{-1}H = (4m)I_{4m},$$

so H^T is a Hadamard matrix. Note that H^T is standardized since H is standardized. Hence, every row of H^T (except the first) contains $2m$ "1"s. This implies that every column of H (except the first) contains $2m$ "1"s, and thus every column of M contains $2m-1$ "1"s. Therefore,

$$\mathbf{u}_{4m-1}M = (2m-1)\mathbf{u}_{4m-1}.$$

Applying Theorem 1.13, we see that M is the incidence matrix of a (symmetric) $(4m-1, 2m-1, m-1)$-BIBD.

Conversely, suppose that M is the incidence matrix of a symmetric $(4m-1, 2m-1, m-1)$-BIBD. Construct H by changing every "0" entry to "-1" and then adjoining a new row and column of "1"s.

Let $1 \leq i \leq 4m$. Then the (i,i)-entry of HH^T is $4m$ since every entry of H is ± 1. Suppose that $1 \leq i < j \leq 4m$. The (i,j)-entry of HH^T is computed to be

$$1 + \lambda - (r-\lambda) - (r-\lambda) + (v - 2r + \lambda) = 1 + (m-1) - m - m + m = 0.$$

Hence, it follows that $HH^T = (4m)I_{4m}$, and therefore H is a Hadamard matrix of order $4m$. □

Example 4.6. We presented a $(7,3,1)$-BIBD in Example 1.3. This BIBD has the following incidence matrix:

$$\begin{pmatrix} 1&1&1&0&0&0&0 \\ 1&0&0&1&1&0&0 \\ 1&0&0&0&0&1&1 \\ 0&1&0&1&0&1&0 \\ 0&1&0&0&1&0&1 \\ 0&0&1&1&0&0&1 \\ 0&0&1&0&1&1&0 \end{pmatrix}$$

If we substitute $0 \to -1$ and add a row and column of "1"s, then we get the following Hadamard matrix of order 8:

$$\begin{pmatrix} 1 & 1 & 1 & 1 & 1 & 1 & 1 & 1 \\ 1 & 1 & 1 & 1 & -1 & -1 & -1 & -1 \\ 1 & 1 & -1 & -1 & 1 & 1 & -1 & -1 \\ 1 & 1 & -1 & -1 & -1 & -1 & 1 & 1 \\ 1 & -1 & 1 & -1 & 1 & -1 & 1 & -1 \\ 1 & -1 & 1 & -1 & -1 & 1 & -1 & 1 \\ 1 & -1 & -1 & 1 & 1 & -1 & -1 & 1 \\ 1 & -1 & -1 & 1 & -1 & 1 & 1 & -1 \end{pmatrix}.$$

∎

The following result is an immediate consequence of Theorems 3.8, 3.21, and 4.5.

Corollary 4.7. *There exists a Hadamard matrix of order* $4m$ *if* $4m - 1$ *is a prime power.*

Given a symmetric BIBD, we can construct residual and derived BIBDs. We therefore have the following immediate consequence of Theorems 4.5 and 2.7.

Theorem 4.8. *Suppose there is a Hadamard matrix of order* $4m$*. If* $m \geq 3$*, then there exists a* $(2m-1, m-1, m-2)$*-BIBD; if* $m \geq 2$*, then there exists a* $(2m, m, m-1)$*-BIBD.*

4.3 Conference Matrices and Hadamard Matrices

In this section, we describe another construction for Hadamard matrices, which will provide a Hadamard matrix of order $2q + 2$ whenever $q \equiv 1 \pmod 4$ is a prime power. We need to define some new concepts before giving the construction.

For an odd prime power q, define the function $\chi_q : \mathbb{F}_q \to \{-1, 0, 1\}$ as follows:

$$\chi_q(x) = \begin{cases} 0 \text{ if } x = 0 \\ 1 \text{ if } x \in \mathsf{QR}(q) \\ -1 \text{ if } x \in \mathsf{QNR}(q). \end{cases}$$

The function χ_q is called the *quadratic character* in the finite field $\mathbb{F}_q$. Observe that Corollary 3.20 states that $\chi_q(-1) = -1$ if $q \equiv 3 \pmod 4$, and $\chi_q(-1) = 1$ if $q \equiv 1 \pmod 4$. We will make use of this fact a bit later. Another important fact about the quadratic character is that it is a multiplicative homomorphism: $\chi_q(x)\chi_q(y) = \chi_q(xy)$ for all $x, y \in \mathbb{F}_q$. This follows easily from results proven in Section 3.2. Additionally, we require the following fundamental properties of the quadratic character.

Lemma 4.9. *Suppose q is an odd prime power. Then the following hold:*

1. $\sum_{x \in \mathbb{F}_q} \chi_q(x) = 0$, *and*
2. $\sum_{x \in \mathbb{F}_q} \chi_q(x)\chi_q(x+y) = -1$ *for all* $y \in \mathbb{Z}_q \backslash \{0\}$.

Proof. Part 1 follows because $|\mathsf{QR}(q)| = |\mathsf{QNR}(q)| = (q-1)/2$.

To prove Part 2, we first observe that

$$\chi_q(x)\chi_q(x+y) = \chi_q(x)\chi_q(x)\chi_q(1+yx^{-1}) = \chi_q(1+yx^{-1})$$

provided that $x \neq 0$. Now, using the fact that $y \neq 0$, it is easily seen that, as x takes on all nonzero values in $\mathbb{F}_q$, the quantity $1 + yx^{-1}$ takes on all values in $\mathbb{F}_q$ except for the value 1. Hence, we have that

$$\begin{aligned}
\sum_{x \in \mathbb{F}_q} \chi_q(x)\chi_q(x+y) &= \sum_{x \in \mathbb{F}_q, x \neq 0} \chi_q(1+yx^{-1}) \\
&= \sum_{x \in \mathbb{F}_q, x \neq 1} \chi_q(x) \\
&= \sum_{x \in \mathbb{F}_q} \chi_q(x) - \chi_q(1) \\
&= 0 - 1 \\
&= -1.
\end{aligned}$$

□

The Hadamard matrix construction also makes use of an auxiliary structure that we define now.

Definition 4.10. *A* conference matrix *of order n is an* $n \times n$ *matrix* $C = (c_{i,j})$ *in which every entry is* 0, 1, *or* -1 *such that* $c_{i,i} = 0$ *for all i and* $CC^T = (n-1)I_n$. *A conference matrix* $C = (c_{i,j})$ *is a* symmetric conference matrix *if* $c_{i,j} = c_{j,i}$ *for all* i, j.

It is easy to see that the only "0" entries in a conference matrix are the entries on the main diagonal. Also, using a counting argument similar to that used in the proof of Theorem 4.4, it can be shown that $n \equiv 2 \pmod 4$ is a necessary condition for a symmetric conference matrix of order n to exist. A further necessary condition can be obtained (via a Bruck-Ryser-Chowla approach), which is stated in the following theorem.

Theorem 4.11. *If a symmetric conference matrix of order n exists, then* $n \equiv 2 \pmod 4$ *and* $n-1$ *is the sum of two integral squares.*

We now give a construction for an infinite class of symmetric conference matrices. Suppose $q \equiv 1 \pmod 4$ is a prime power. Define a matrix $W = (w_{i,j})$, in which the rows and columns are indexed by $\mathbb{F}_q \cup \{\infty\}$, as follows:

$$w_{i,j} = \begin{cases} 0 & \text{if } i = j = \infty \\ 1 & \text{if } i = \infty, j \neq \infty \\ 1 & \text{if } i \neq \infty, j = \infty \\ \chi_q(i-j) & \text{if } i, j \in \mathbb{F}_q. \end{cases}$$

Theorem 4.12. *Suppose $q \equiv 1 \pmod 4$ is a prime power. Then the matrix W defined above is a symmetric conference matrix of order $q+1$.*

Proof. Clearly, the diagonal entries of W are all 0, and every off-diagonal entry is ± 1. This implies that the (i,i) entry of WW^T is q for all $i \in \mathbb{F}_q \cup \{\infty\}$. Furthermore, $\chi_q(-1) = 1$ because $q \equiv 1 \pmod 4$, and therefore it follows that W is symmetric.

It remains to show that, if $i \neq j$, then the (i,j) entry of WW^T is 0. First, suppose that $i, j \in \mathbb{F}_q$, $i \neq j$. Then, using Lemma 4.9, Part 2, the (i,j) entry of WW^T is

$$\begin{aligned} 1 + \sum_{h \in \mathbb{F}_q} \chi_q(i-h)\chi_q(j-h) &= 1 + \sum_{x \in \mathbb{F}_q} \chi_q(x)\chi_q(x+y) \\ &= 1 + (-1) \\ &= 0. \end{aligned}$$

(Note the change of variables $x = i - h, y = j - i$ used in the computation above.) Next, suppose that $i \in \mathbb{F}_q$. The (i,∞) entry (or the (∞,i) entry) of WW^T is

$$\sum_{x \in \mathbb{F}_q} \chi_q(x) = 0$$

from Lemma 4.9, Part 1. This completes the proof. □

Example 4.13. Suppose we take $q = 5$. We have $\mathrm{QR}(5) = \{1,4\}$, so $\chi_q(1) = \chi_q(4) = 1$, $\chi_q(2) = \chi_q(3) = -1$, and $\chi_q(0) = 0$. Then we construct a symmetric conference matrix W as follows:

$$W = \begin{pmatrix} 0 & 1 & 1 & 1 & 1 & 1 \\ 1 & 0 & 1 & -1 & -1 & 1 \\ 1 & 1 & 0 & 1 & -1 & -1 \\ 1 & -1 & 1 & 0 & 1 & -1 \\ 1 & -1 & -1 & 1 & 0 & 1 \\ 1 & 1 & -1 & -1 & 1 & 0 \end{pmatrix}.$$

■

A symmetric conference matrix of order 2 is trivial:

$$\begin{pmatrix} 0 & 1 \\ 1 & 0 \end{pmatrix}.$$

Theorem 4.12 yields symmetric conference matrices of orders 6, 10, 14, 18, 26, 30, 38, 42, etc. Orders 22 and 34 are not possible, by Theorem 4.11. Thus we already are able to determine the existence or nonexistence of symmetric conference matrices of all possible orders less than 46.

We now present a construction of Hadamard matrices from symmetric conference matrices.

Theorem 4.14. *Suppose C is a symmetric conference matrix of order m. Then the matrix*

$$H = \begin{pmatrix} C + I_m & C - I_m \\ C - I_m & -C - I_m \end{pmatrix}$$

is a Hadamard matrix of order 2m.

Proof. Since C is symmetric, we see that $H^T = H$. Also, every entry of H is ± 1. Then we can compute HH^T as follows:

$$\begin{aligned} HH^T &= \begin{pmatrix} C + I_m & C - I_m \\ C - I_m & -C - I_m \end{pmatrix} \begin{pmatrix} C + I_m & C - I_m \\ C - I_m & -C - I_m \end{pmatrix} \\ &= \begin{pmatrix} A_1 & A_2 \\ A_3 & A_4 \end{pmatrix}, \end{aligned}$$

where

$$\begin{aligned} A_1 &= (C + I_m)^2 + (C - I_m)^2, \\ A_2 &= (C + I_m)(C - I_m) + (C - I_m)(-C - I_m), \\ A_3 &= (C - I_m)(C + I_m) + (-C - I_m)(C - I_m), \text{ and} \\ A_4 &= (C - I_m)^2 + (-C - I_m)^2. \end{aligned}$$

It is not hard to verify that A_2 and A_3 are both $m \times m$ matrices of "0"s. Furthermore, we have

$$\begin{aligned} A_1 &= 2C^2 + 2(I_m)^2 \\ &= 2(m-1)I_m + 2(I_m) \\ &= (2m)I_m. \end{aligned}$$

Similarly, $A_4 = (2m)I_m$. Thus, we have

$$HH^T = \begin{pmatrix} (2m)I_m & 0 \\ 0 & (2m)I_m \end{pmatrix} = (2m)I_{2m},$$

as desired. □

Example 4.15. From the conference matrix of order 6 constructed in Example 4.13, we obtain the following Hadamard matrix of order 12:

$$\begin{pmatrix}
1 & 1 & 1 & 1 & 1 & 1 & -1 & 1 & 1 & 1 & 1 & 1 \\
1 & 1 & 1 & -1 & -1 & 1 & 1 & -1 & 1 & -1 & -1 & 1 \\
1 & 1 & 1 & 1 & -1 & -1 & 1 & 1 & -1 & 1 & -1 & -1 \\
1 & -1 & 1 & 1 & 1 & -1 & 1 & -1 & 1 & -1 & 1 & -1 \\
1 & -1 & -1 & 1 & 1 & 1 & 1 & -1 & -1 & 1 & -1 & 1 \\
1 & 1 & -1 & -1 & 1 & 1 & 1 & 1 & -1 & -1 & 1 & -1 \\
-1 & 1 & 1 & 1 & 1 & 1 & -1 & -1 & -1 & -1 & -1 & -1 \\
1 & -1 & 1 & -1 & -1 & 1 & -1 & -1 & -1 & 1 & 1 & -1 \\
1 & 1 & -1 & 1 & -1 & -1 & -1 & -1 & -1 & -1 & 1 & 1 \\
1 & -1 & 1 & -1 & 1 & -1 & -1 & 1 & -1 & -1 & -1 & 1 \\
1 & -1 & -1 & 1 & -1 & 1 & -1 & 1 & 1 & -1 & -1 & -1 \\
1 & 1 & -1 & -1 & 1 & -1 & -1 & -1 & 1 & 1 & -1 & -1
\end{pmatrix}.$$

∎

The following result is an immediate consequence of Theorems 4.12 and 4.14.

Corollary 4.16. *There exists a Hadamard matrix of order* $4m$ *if* $2m-1$ *is a prime power and* m *is odd.*

4.4 A Product Construction

The construction we study in this section is a recursive construction called the *Kronecker Product*. Suppose $H_1 = (h_{i,j})$ is a Hadamard matrix of order n_1 and H_2 is a Hadamard matrix of order n_2. We define the Kronecker Product $H_1 \otimes H_2$ to be the matrix of order $n_1 n_2$ obtained by replacing every entry $h_{i,j}$ of H_1 by the $n_2 \times n_2$ matrix $h_{i,j}H_2$ (where xH_2 denotes the matrix obtained from H_2 by multiplying every entry by x).

Example 4.17. Let H_1 be the Hadamard matrix of order 2 presented in Example 4.2, and let H_2 be the Hadamard matrix of order 4 presented in Example 4.3. Then $H_1 \otimes H_2$ is the following matrix of order 8:

$$\begin{pmatrix}
1 & 1 & 1 & 1 & 1 & 1 & 1 & 1 \\
1 & -1 & 1 & -1 & 1 & -1 & 1 & -1 \\
1 & 1 & -1 & -1 & 1 & 1 & -1 & -1 \\
1 & -1 & -1 & 1 & 1 & -1 & -1 & 1 \\
1 & 1 & 1 & 1 & -1 & -1 & -1 & -1 \\
1 & -1 & 1 & -1 & -1 & 1 & -1 & 1 \\
1 & 1 & -1 & -1 & -1 & -1 & 1 & 1 \\
1 & -1 & -1 & 1 & -1 & 1 & 1 & -1
\end{pmatrix}.$$

∎

Theorem 4.18 (Kronecker Product). *If H_1 is a Hadamard matrix of order n_1 and H_2 is a Hadamard matrix of order n_2, then $H_1 \otimes H_2$ is a Hadamard matrix of order $n_1 n_2$.*

Proof. Suppose the rows of $H_1 \otimes H_2$ are indexed by $\{1, \ldots, n_1\} \times \{1, \ldots, n_2\}$, so that row (i, j) of $H_1 \otimes H_2$ is in fact row j within the $n_2 \times (n_1 n_2)$ submatrix

$$(h_{i,1}H_2 \ \ldots \ h_{i,n_1}H_2).$$

We need to compute the inner product of two rows of $H_1 \otimes H_2$, say rows (i, j) and (k, ℓ). We have the following:

$$\begin{aligned}
\text{row } (i,j) \cdot \text{row } (k,\ell) &= \sum_{a=1}^{n_1} h_{ia}(\text{row } j \text{ of } H_2) \cdot h_{ka}(\text{row } \ell \text{ of } H_2) \\
&= ((\text{row } i \text{ of } H_1) \cdot (\text{row } k \text{ of } H_1)) \\
&\quad \times ((\text{row } j \text{ of } H_2) \cdot (\text{row } \ell \text{ of } H_2)) \\
&= \begin{cases} n_1 n_2 & \text{if } (i,j) = (k,\ell) \\ 0 & \text{otherwise.} \end{cases}
\end{aligned}$$

Hence, $H_1 \otimes H_2$ is a Hadamard matrix. □

Ths following corollary of Theorem 4.18 is obtained by letting $n_1 = 2$, $n_2 = n$.

Corollary 4.19. *If there exists a Hadamard matrix of order n, then there exists a Hadamard matrix of order $2n$.*

4.5 Williamson's Method

The constructions described to this point allow us to construct Hadamard matrices of all possible orders $n \leq 88$. A Hadamard matrix of order 92 was first constructed using a method suggested by Williamson, which we describe in this section.

The basis for the construction is the following matrix identity, which is essentially the same as the one stated as Lemma 2.18: If a, b, c, and d are integers (or, indeed, elements of any commutative ring), and

$$H = \begin{pmatrix} -a & b & c & d \\ b & a & d & -c \\ c & -d & a & b \\ d & c & -b & a \end{pmatrix},$$

then $HH^T = (a^2 + b^2 + c^2 + d^2)I_4$. The Hadamard matrix construction is obtained by replacing a, b, c, and d by matrices that satisfy certain properties. The proof of the following result is straightforward.

Theorem 4.20. *Suppose that $A, B, C,$ and D are $n \times n$ matrices that satisfy the following properties:*

1. *$A, B, C,$ and D are symmetric matrices having entries ± 1;*
2. *the matrices $A, B, C,$ and D commute.*

Define the matrix

$$H = \begin{pmatrix} -A & B & C & D \\ B & A & D & -C \\ C & -D & A & B \\ D & C & -B & A \end{pmatrix},$$

and denote $A^2 + B^2 + C^2 + D^2 = M$. Then

$$HH^T = \begin{pmatrix} M & 0 & 0 & 0 \\ 0 & M & 0 & 0 \\ 0 & 0 & M & 0 \\ 0 & 0 & 0 & M \end{pmatrix},$$

where the "0" entries denote $n \times n$ blocks of "0"s.

Corollary 4.21. *Suppose there exist $n \times n$ matrices, $A, B, C,$ and D, that satisfy the following properties:*

1. *$A, B, C,$ and D are symmetric matrices having entries ± 1;*
2. *the matrices $A, B, C,$ and D commute;*
3. *$A^2 + B^2 + C^2 + D^2 = 4nI_n$.*

Then there is a Hadamard matrix of order $4n$.

Example 4.22. Let

$$A = \begin{pmatrix} 1 & 1 & 1 \\ 1 & 1 & 1 \\ 1 & 1 & 1 \end{pmatrix}$$

and let

$$B = C = D = \begin{pmatrix} 1 & -1 & -1 \\ -1 & 1 & -1 \\ -1 & -1 & 1 \end{pmatrix}.$$

The conditions of Corollary 4.21 are easily verified. In particular, $A^2 = 3J_3$ and $B^2 = C^2 = D^2 = 4I_3 - J_3$, so $A^2 + B^2 + C^2 + D^2 = 12I_3$. Hence there exists a Hadamard matrix of order 12. ∎

An $n \times n$ matrix, say $A = (a_{i,j})$, is said to be a *circulant matrix* provided that $a_{i+1 \bmod n, j+1 \bmod n} = a_{i,j}$ for all i, j. In other words, the entries on any (circulant) diagonal are constant. In practice, it is convenient to take $A, B, C,$ and D to be circulant matrices, as was done in Example 4.22.

Fix a positive integer n, and let $U = (u_{i,j})$ be the matrix where

$$u_{i,j} = \begin{cases} 1 \text{ if } j - i \equiv 1 \pmod{n} \\ 0 \text{ otherwise.} \end{cases}$$

Now, it is easy to see that any matrix of the form $\sum_{i=0}^{n-1} a_i U^i$ is a circulant matrix. In fact, any circulant matrix can be expressed in this way in a unique fashion; this is clear because

$$\sum_{i=0}^{n-1} a_i U^i = \begin{pmatrix} a_0 & a_1 & \cdots & a_{n-2} & a_{n-1} \\ a_{n-1} & a_0 & \cdots & a_{n-3} & a_{n-2} \\ \vdots & \vdots & \ddots & \vdots & \vdots \\ a_1 & a_2 & \cdots & a_{n-1} & a_0 \end{pmatrix}.$$

The sequence $(a_0, a_1, \ldots, a_{n-2}, a_{n-1})$ is just the first row of the matrix A.

Now suppose we stipulate that A, B, C, and D are circulant matrices as follows:

$$A = \sum_{i=0}^{n-1} a_i U^i,$$

$$B = \sum_{i=0}^{n-1} b_i U^i,$$

$$C = \sum_{i=0}^{n-1} c_i U^i, \quad \text{and}$$

$$D = \sum_{i=0}^{n-1} d_i U^i.$$

Let us consider the conditions of Theorem 4.20. Since the four matrices A, B, C, and D are all expressed as polynomials in the matrix U, it is clear that they commute. If $a_i, b_i, c_i, d_i = \pm 1$ for all i, then A, B, C, and D will all have entries ± 1. The condition that the matrix A is symmetric is that $a_i = a_{n-i}$ for $0 \leq i \leq n-1$. Similar conditions will ensure that B, C, and D are symmetric. There still remains the condition that $A^2 + B^2 + C^2 + D^2 = 4nI_n$, which is, in general, quite difficult to satisfy. In fact, most applications of Corollary 4.21 have required computer searches to find suitable input matrices.

Example 4.23. A Hadamard matrix of order 92 was discovered in 1962 by Baumert, Golomb, and Hall using the method described above. The first rows of the matrices A, B, C, and D are as follows, where we encode "1" as "+" and "−1" as "−":

$$A: + + - - - + - - - + - + + - + - - - + - - - +$$
$$B: + - + + - + + - - + + + + + + - - + + - + + -$$
$$C: + + + - - - + + - + - + + - + - + + - - - + +$$
$$D: + + + - + + + - + - - - - - - + - + + + - + +$$

∎

4.6 Existence Results for Hadamard Matrices of Small Orders

The constructions we have presented thus far allow us to obtain Hadamard matrices of all possible orders $n \leq 100$. We summarize the details in Table 4.1.

order	equation	authority
2		Example 4.2
4	2×2	Theorem 4.19
8	2×4	Theorem 4.19
12	$11 + 1$	Corollary 4.7
16	2×8	Theorem 4.19
20	$19 + 1$	Corollary 4.7
24	2×12	Theorem 4.19
28	$27 + 1$	Corollary 4.7
32	2×16	Theorem 4.19
36	$2 \times 17 + 2$	Corollary 4.16
40	2×20	Theorem 4.19
44	$43 + 1$	Corollary 4.7
48	2×24	Theorem 4.19
52	$2 \times 25 + 2$	Corollary 4.16
56	2×28	Theorem 4.19
60	$59 + 1$	Corollary 4.7
64	2×32	Theorem 4.19
68	$67 + 1$	Corollary 4.7
72	2×36	Theorem 4.19
76	$2 \times 37 + 2$	Corollary 4.16
80	2×40	Theorem 4.19
84	$83 + 1$	Corollary 4.7
88	2×44	Theorem 4.19
92		Example 4.23
96	2×48	Theorem 4.19
100	$2 \times 49 + 2$	Corollary 4.16

Table 4.1. Constructions of Hadamard Matrices of all Orders $n \leq 100$

4.7 Regular Hadamard Matrices

A *regular Hadamard matrix* is one in which every row and every column contains the same number of "1"s. Regular Hadamard matrices are interesting for several reasons. First, they turn out to be equivalent to certain symmetric BIBDs. In addition, they have the maximum number of "1" entries (among all possible Hadamard matrices of a given order). We pursue these topics in the rest of this section.

We begin with a small example.

Example 4.24. The following matrix is a regular Hadamard matrix of order 4:

$$\begin{pmatrix} -1 & 1 & 1 & 1 \\ 1 & -1 & 1 & 1 \\ 1 & 1 & -1 & 1 \\ 1 & 1 & 1 & -1 \end{pmatrix}.$$

∎

A Hadamard matrix in which every row has the same number of "1"s is called a *row-regular Hadamard matrix;* one in which every column has the same number of "1"s is called a *column-regular Hadamard matrix*. We begin by investigating necessary conditions for the existence of row-regular Hadamard matrices. Suppose that $H = (h_{i,j})$ is a Hadamard matrix of order $n > 1$ in which every row contains exactly ℓ entries equal to 1. For $1 \leq i \leq n$, let r_i denote the ith row of H. Define

$$\begin{aligned} a &= |\{j : h_{1,j} = h_{2,j} = 1\}|, \\ b &= |\{j : h_{1,j} = 1, h_{2,j} = -1\}|, \\ c &= |\{j : h_{1,j} = -1, h_{2,j} = 1\}|, \quad \text{and} \\ d &= |\{j : h_{1,j} = h_{2,j} = -1\}|. \end{aligned}$$

Then we have the following equations:

$$\begin{aligned} a+b+c+d &= n & & \\ a+b &= \ell & & \text{since } r_1 \text{ contains } \ell \text{ "1"s} \\ a+c &= \ell & & \text{since } r_2 \text{ contains } \ell \text{ "1"s} \\ a-b-c+d &= 0 & & \text{since } r_1 \cdot r_2 = 0. \end{aligned}$$

This system has the following unique solution:

$$\begin{aligned} a &= \ell - \frac{n}{4} \\ b &= \frac{n}{4} \\ c &= \frac{n}{4} \\ d &= \frac{3n}{4} - \ell. \end{aligned}$$

Now suppose we change every "-1" entry of H to "0". The resulting $0-1$ matrix M satisfies the equation $MM^T = \lambda J_n + (\ell - \lambda) I_n$, where $\lambda = a = \ell - n/4$. Therefore, by Theorem 1.15, M is the incidence matrix of a pairwise balanced design having n points and n blocks in which every point occurs in ℓ blocks and every pair of points occurs in λ blocks. Theorem 2.3 tells us that this PBD is in fact a BIBD, and therefore M is the incidence matrix of a symmetric $(n, \ell, \ell - \frac{n}{4})$-BIBD. This in turn implies that

$$\ell(\ell - 1) = \left(\ell - \frac{n}{4}\right)(n - 1).$$

For any fixed value of n, the equation above is a quadratic equation in ℓ. Therefore we can solve for ℓ as a function of n using the quadratic formula. We obtain the following:

$$\ell = \frac{n \pm \sqrt{n}}{2}.$$

This implies that n must be a perfect square. It is also the case that $n \equiv 0 \pmod 4$. (All Hadamard matrices have orders $n \equiv 0 \pmod 4$ except for matrices of orders $n = 1$ and 2. We are assuming that $n > 1$. Furthermore, $n \neq 2$ because 2 is not a perfect square.) Therefore we can write $n = (2u)^2$, where u is a positive integer, and it follows that our symmetric BIBD has parameters $(4u^2, 2u^2 \pm u, u^2 \pm u)$.

Conversely, if we begin with the incidence matrix of a $(4u^2, 2u^2 \pm u, u^2 \pm u)$-BIBD and replace every "0" by "-1", then it is not difficult to show that the result is a regular Hadamard matrix of order $4u^2$.

Summarizing the discussion above, we have the following theorem.

Theorem 4.25. *A row-regular Hadamard matrix, say H, of order $n > 4$ exists only if $n = 4u^2$ for some integer $u \geq 2$ and every row of H contains ℓ "1"s, where $\ell = 2u^2 \pm u$. Furthermore, such a Hadamard matrix is equivalent to a (symmetric) $(4u^2, 2u^2 \pm u, u^2 \pm u)$-BIBD.*

We have also proven the following result.

Theorem 4.26. *The following are equivalent:*

- *H is a row-regular Hadamard matrix of order n;*
- *H is a column-regular Hadamard matrix of order n;*
- *H is a regular Hadamard matrix of order n.*

Example 4.27. We constructed a $(16, 6, 2)$-difference set in $(\mathbb{Z}_4 \times \mathbb{Z}_4, +)$ in Example 3.4 and a $(36, 16, 5)$-difference set in $(\mathbb{Z}_6 \times \mathbb{Z}_6, +)$ in Example 3.6. Therefore there exists a $(16, 6, 2)$-BIBD and a $(36, 16, 5)$-BIBD. Applying Theorem 4.25, there exist regular Hadamard matrices of orders 16 and 36. ∎

It is not difficult to show that the Kronecker Product of two regular Hadamard matrices is a regular Hadamard matrix. Therefore we can also construct a regular Hadamard matrix of order 16 as the Kronecker Product of regular Hadamard matrices of order 4. More generally, we can easily obtain infinite classes of regular Hadamard matrices as follows.

Theorem 4.28. *Suppose that $n = 4^a 9^b$, where a and b are nonnegative integers such that $a \geq b$. Then there is a regular Hadamard matrix of order n.*

Proof. If $a = b = 0$, then $n = 1$, and there exists a regular Hadamard matrix of order 1, namely (1). Therefore, we can assume that $a + b \geq 1$. Write $n = 4^{a-b}36^b$. Then a regular Hadamard matrix of order n can be constructed by taking the Kronecker Product of $a - b$ regular Hadamard matrices of order 4 and b regular Hadamard matrices of order 36. □

4.7.1 Excess of Hadamard Matrices

Let $H = (h_{i,j})$ be a Hadamard matrix of order n. Define the *excess* of H to be

$$\text{excess}(H) = \sum_{i=1}^{n}\sum_{j=1}^{n} h_{i,j}.$$

Clearly excess(H) is the amount by which the number of "1"s in H exceeds the number of "-1"s. For an integer n such that a Hadamard matrix of order n exists, define

$$\sigma(n) = \max\{\text{excess}(H) : H \text{ is a Hadamard matrix of order } n\}.$$

Lemma 4.29. $\sigma(n) \leq n^{3/2}$.

Proof. Let H be a Hadamard matrix of order n. For $1 \leq k \leq n$, define

$$s_k = \sum_{i=1}^{n} h_{i,k}.$$

The quantity s_k is the sum of the entries in column k of H, so it is obvious that

$$\text{excess}(H) = \sum_{k=1}^{n} s_k.$$

Let $r_1, \ldots, r_n$ denote the rows of H. We compute the quantity

$$\sum_{i=1}^{n}\sum_{j=1}^{n} r_i \cdot r_j$$

in two ways. It is clear that $r_i \cdot r_j = n$ if $i = j$ and $r_i \cdot r_j = 0$ if $i \neq j$. It therefore follows that

$$\sum_{i=1}^{n}\sum_{j=1}^{n} r_i \cdot r_j = n^2.$$

On the other hand, we have that

$$\begin{aligned}\sum_{i=1}^{n}\sum_{j=1}^{n} r_i \cdot r_j &= \sum_{i=1}^{n}\sum_{j=1}^{n}\sum_{k=1}^{n} h_{i,k}h_{j,k} \\ &= \sum_{k=1}^{n}\left(\sum_{i=1}^{n} h_{i,k}\right)\left(\sum_{j=1}^{n} h_{j,k}\right) \\ &= \sum_{k=1}^{n} {s_k}^2.\end{aligned}$$

Hence,

$$\sum_{k=1}^{n} s_k{}^2 = n^2. \tag{4.1}$$

Now, the classical *Cauchy-Schwartz Inequality* asserts that

$$\left(\sum_{k=1}^{n} x_k y_k\right)^2 \leq \left(\sum_{k=1}^{n} x_k{}^2\right) \times \left(\sum_{k=1}^{n} y_k{}^2\right) \tag{4.2}$$

for arbitrary real numbers $x_1, \ldots, x_n, y_1, \ldots, y_n$. Setting $x_k = 1$ and $y_k = s_k$ for $1 \leq k \leq n$, it follows immediately that

$$\left(\sum_{k=1}^{n} s_k\right)^2 \leq n \sum_{k=1}^{n} s_k{}^2. \tag{4.3}$$

Combining (4.1) and (4.3), we have that

$$n^2 = \sum_{k=1}^{n} s_k{}^2 \geq \frac{\left(\sum_{k=1}^{n} s_k\right)^2}{n}, \tag{4.4}$$

and hence

$$\text{excess}(H) = \sum_{j=1}^{n} s_j \leq n^{3/2}.$$

□

We now show that Hadamard matrices having the maximum possible excess are equivalent to regular Hadamard matrices.

Theorem 4.30. $\sigma(n) = n^{3/2}$ *if and only if there exists a regular Hadamard matrix of order* n.

Proof. Suppose that H is a regular Hadamard matrix of order n. We proved earlier that H has exactly ℓ "1"s in every row and column, where

$$\ell = \frac{n \pm \sqrt{n}}{2}.$$

If $\ell = (n - \sqrt{n})/2$, then multiply every entry of H by -1. The result is a regular Hadamard matrix in which every row and column contains exactly $(n + \sqrt{n})/2$ "1"s. This Hadamard matrix has excess equal to

$$n\left(\frac{n+\sqrt{n}}{2} - \frac{n-\sqrt{n}}{2}\right) = n^{3/2}.$$

Conversely, suppose that H is a Hadamard matrix of order n such that $\text{excess}(H) = n^{3/2}$. Then, in the proof of Lemma 4.29, it must be the case that (4.4) is in fact an equality:

$$\left(\sum_{k=1}^{n} s_k\right)^2 = n\sum_{k=1}^{n} {s_k}^2.$$

It is well-known that equality occurs in the Cauchy-Schwartz Inequality (4.2) if and only if

$$\frac{y_1}{x_1} = \frac{y_2}{x_2} = \cdots = \frac{y_n}{x_n}.$$

Hence, equality occurs in (4.4) if and only if $s_1 = s_2 = \cdots = s_n$, which implies that H is column-regular. However, Theorem 4.26 asserts that a column-regular Hadamard matrix is regular. This completes the proof. □

4.8 Bent Functions

Suppose $n \geq 1$ is an integer. A function $f : (\mathbb{Z}_2)^n \rightarrow \mathbb{Z}_2$ is called a *Boolean function* of n variables. Define $\mathcal{B}_n$ to be the set of all Boolean functions of n variables.

Suppose $f \in \mathcal{B}_n$. We can list the values $f(\mathbf{x})$, for all $\mathbf{x} \in (\mathbb{Z}_2)^n$, in a vector of length 2^n. Denote this vector by $\phi(f)$, where $\phi(f)_{\mathbf{x}} = f(\mathbf{x})$ for all $\mathbf{x} \in (\mathbb{Z}_2)^n$. For the sake of consistency, we will index the coordinates of $\phi(f)$ in lexicographic order.

Note that $\phi(f) \in (\mathbb{Z}_2)^{2^n}$ and therefore $|\mathcal{B}_n| = 2^{2^n}$ (i.e., there are 2^{2^n} Boolean functions of n variables). For any $f \in \mathcal{B}_n$, define $(-1)^f$ to be the function $(-1)^f : (\mathbb{Z}_2)^n \rightarrow \{-1, 1\}$ such that $((-1)^f)(\mathbf{x}) = (-1)^{f(\mathbf{x})}$ for all $\mathbf{x} \in (\mathbb{Z}_2)^n$. In other words, $(-1)^f$ is formed from f by replacing every output equal to "0" by "1" and every output equal to "1" by "−1". (We already performed a similar operation when we constructed a Hadamard matrix of order $4n$ from a symmetric $(4n-1, 2n-1, n-1)$-BIBD.)

Define the *inner product* of two vectors $\mathbf{x}, \mathbf{y} \in (\mathbb{Z}_2)^n$ as follows:

$$\mathbf{x} \cdot \mathbf{y} = \sum_{i=1}^{n} x_i y_i \bmod 2,$$

where $\mathbf{x} = (x_1, \ldots, x_n)$ and $\mathbf{y} = (y_1, \ldots, y_n)$. Let F be any real-valued function defined on $(\mathbb{Z}_2)^n$. The *Fourier transform* of F is the function $\widehat{F} : (\mathbb{Z}_2)^n \rightarrow \mathbb{R}$ defined by the following formula:

$$\widehat{F}(\mathbf{x}) = \sum_{\mathbf{y} \in (\mathbb{Z}_2)^n} (-1)^{\mathbf{x}\cdot\mathbf{y}} F(\mathbf{y})$$

for all $\mathbf{x} \in (\mathbb{Z}_2)^n$.

For any two vectors $\mathbf{x}, \mathbf{y} \in (\mathbb{Z}_2)^n$, define

$$\delta_{\mathbf{x},\mathbf{y}} = \begin{cases} 1 & \text{if } \mathbf{x} = \mathbf{y} \\ 0 & \text{if } \mathbf{x} \neq \mathbf{y}. \end{cases}$$

Lemma 4.31. *For any* $\mathbf{y} \in (\mathbb{Z}_2)^n$*, it holds that*

$$\sum_{\mathbf{x} \in (\mathbb{Z}_2)^n} (-1)^{\mathbf{x} \cdot \mathbf{y}} = 2^n \, \delta_{\mathbf{y},(0,\ldots,0)}.$$

Proof. If $\mathbf{y} = (0,\ldots,0)$, then every term in the sum equals 1, and the result follows. If $\mathbf{y} \neq (0,\ldots,0)$, then there are the same number of terms equal to 1 as there are terms equal to -1, so the sum is 0. □

Let $S_n = (s_{\mathbf{x},\mathbf{y}})$ be the $2^n \times 2^n$ matrix in which the rows and columns are indexed by $(\mathbb{Z}_2)^n$ (in lexicographic order) and $s_{\mathbf{x},\mathbf{y}} = (-1)^{\mathbf{x} \cdot \mathbf{y}}$ for all $\mathbf{x}, \mathbf{y} \in (\mathbb{Z}_2)^n$. S_n is called the *Sylvester matrix* of order 2^n.

Lemma 4.32. S_n *is a Hadamard matrix.*

Proof. Let $\mathbf{x}, \mathbf{y} \in (\mathbb{Z}_2)^n$. Then, applying Lemma 4.31, we have that

$$\begin{aligned}
\sum_{\mathbf{z} \in (\mathbb{Z}_2)^n} s_{\mathbf{x},\mathbf{z}} s_{\mathbf{y},\mathbf{z}} &= \sum_{\mathbf{z} \in (\mathbb{Z}_2)^n} (-1)^{\mathbf{x} \cdot \mathbf{z} + \mathbf{y} \cdot \mathbf{z}} \\
&= \sum_{\mathbf{z} \in (\mathbb{Z}_2)^n} (-1)^{(\mathbf{x}+\mathbf{y}) \cdot \mathbf{z}} \\
&= 2^n \, \delta_{\mathbf{x}+\mathbf{y},(0,\ldots,0)} \\
&= 2^n \, \delta_{\mathbf{x},\mathbf{y}}.
\end{aligned}$$

□

For any function $F : \{0,1\}^n \to \mathbb{R}$, define $\phi(F)$ in the same way that $\phi(f)$ was defined from f, i.e., $\phi(F)$ is the vector of values $F(\mathbf{x})$. Then we have the following result, which follows immediately from the definition of $\widehat{F}$.

Lemma 4.33. *Suppose that* $F : \{0,1\}^n \to \mathbb{R}$*. Then* $\phi(\widehat{F}) = \phi(F) S_n$.

The following corollary will be useful.

Corollary 4.34. *Suppose that* $F : \{0,1\}^n \to \mathbb{R}$*. Then* $\widehat{\widehat{F}} = 2^n F$.

Proof. We have that $\phi(\widehat{F}) = \phi(F) S_n$. Multiplying on the right by S_n and using the fact that $(S_n)^2 = 2^n I_{2^n}$ (which holds because S_n is a Hadamard matrix and $S_n = (S_n)^T$), we have that $\phi(\widehat{\widehat{F}}) = 2^n \phi(F)$. Hence, $\widehat{\widehat{F}} = 2^n F$. □

Example 4.35. Suppose that $n = 2$, $f(x_1, x_2) = x_1 x_2$, and $F = (-1)^f$, where $x_1, x_2 \in \mathbb{Z}_2$. Then $\phi(f) = (0,0,0,1)$, where the coordinates of $\phi(f)$ are in lexicographic order; i.e., $\phi(f) = (f(0,0), f(0,1), f(1,0), f(1,1))$.

Then $\phi(F) = (1,1,1,-1)$, and

$$S_2 = \begin{pmatrix} 1 & 1 & 1 & 1 \\ 1 & -1 & 1 & -1 \\ 1 & 1 & -1 & -1 \\ 1 & -1 & -1 & 1 \end{pmatrix},$$

so

$$\phi(\widehat{F}) = (1,1,1,-1)\begin{pmatrix} 1 & 1 & 1 & 1 \\ 1 & -1 & 1 & -1 \\ 1 & 1 & -1 & -1 \\ 1 & -1 & -1 & 1 \end{pmatrix} = (2,2,2,-2).$$

Theorem 4.36. *Suppose that $f \in \mathcal{B}_n$ and $F = (-1)^f$. Let $\mathbf{y} \in (\mathbb{Z}_2)^n$. Then it holds that*

$$\sum_{\mathbf{x}\in(\mathbb{Z}_2)^n} \widehat{F}(\mathbf{x})\widehat{F}(\mathbf{x}+\mathbf{y}) = \begin{cases} 2^{2n} & \text{if } \mathbf{y} = (0,\ldots,0) \\ 0 & \text{if } \mathbf{y} \neq (0,\ldots,0). \end{cases}$$

Proof.

$$\begin{aligned}
&\sum_{\mathbf{x}\in(\mathbb{Z}_2)^n} \widehat{F}(\mathbf{x})\widehat{F}(\mathbf{x}+\mathbf{y}) \\
&= \sum_{\mathbf{x}\in(\mathbb{Z}_2)^n} \sum_{\mathbf{u}\in(\mathbb{Z}_2)^n} (-1)^{\mathbf{x}\cdot\mathbf{u}} F(\mathbf{u}) \sum_{\mathbf{v}\in(\mathbb{Z}_2)^n} (-1)^{(\mathbf{x}+\mathbf{y})\cdot\mathbf{v}} F(\mathbf{v}) \\
&= \sum_{\mathbf{x}\in(\mathbb{Z}_2)^n} \sum_{\mathbf{u}\in(\mathbb{Z}_2)^n} \sum_{\mathbf{v}\in(\mathbb{Z}_2)^n} (-1)^{\mathbf{x}\cdot\mathbf{u}+(\mathbf{x}+\mathbf{y})\cdot\mathbf{v}} F(\mathbf{u})F(\mathbf{v}) \\
&= \sum_{\mathbf{x}\in(\mathbb{Z}_2)^n} \sum_{\mathbf{u}\in(\mathbb{Z}_2)^n} \sum_{\mathbf{v}\in(\mathbb{Z}_2)^n} (-1)^{\mathbf{x}\cdot(\mathbf{u}+\mathbf{v})+\mathbf{y}\cdot\mathbf{v}} F(\mathbf{u})F(\mathbf{v}) \\
&= \sum_{\mathbf{u}\in(\mathbb{Z}_2)^n} \sum_{\mathbf{v}\in(\mathbb{Z}_2)^n} (-1)^{\mathbf{y}\cdot\mathbf{v}} F(\mathbf{u})F(\mathbf{v}) \sum_{\mathbf{x}\in(\mathbb{Z}_2)^n} (-1)^{\mathbf{x}\cdot(\mathbf{u}+\mathbf{v})} \\
&= \sum_{\mathbf{u}\in(\mathbb{Z}_2)^n} \sum_{\mathbf{v}\in(\mathbb{Z}_2)^n} (-1)^{\mathbf{y}\cdot\mathbf{v}} F(\mathbf{u})F(\mathbf{v}) 2^n \,\delta_{\mathbf{u},\mathbf{v}} \quad \text{from Lemma 4.31} \\
&= 2^n \sum_{\mathbf{u}\in(\mathbb{Z}_2)^n} (-1)^{\mathbf{y}\cdot\mathbf{u}} (F(\mathbf{u}))^2 \\
&= 2^n \sum_{\mathbf{u}\in(\mathbb{Z}_2)^n} (-1)^{\mathbf{y}\cdot\mathbf{u}} \quad \text{because } F(\mathbf{u}) = \pm 1 \\
&= 2^{2n}\,\delta_{\mathbf{y},(0,\ldots,0)} \quad \text{from Lemma 4.31,}
\end{aligned}$$

as required. □

We state two corollaries. The first corollary is just the case $\mathbf{y} = (0,\ldots,0)$ in the previous theorem.

Corollary 4.37 (Parseval's Equation). *Suppose that $f \in \mathcal{B}_n$ and $F = (-1)^f$. Then it holds that*

$$\sum_{\mathbf{x}\in(\mathbb{Z}_2)^n} (\widehat{F}(\mathbf{x}))^2 = 2^{2n}.$$

The second corollary follows from the proof of Theorem 4.36 by noting that the first part of the proof (all but the last two lines of the displayed equations, in fact) applies to any real-valued function.

Theorem 4.38. *Suppose that $F : (\mathbb{Z}_2)^n \to \mathbb{R}$ and let $\mathbf{y} \in (\mathbb{Z}_2)^n$. Then it holds that*

$$\sum_{\mathbf{x}\in(\mathbb{Z}_2)^n} \widehat{F}(\mathbf{x})\widehat{F}(\mathbf{x}+\mathbf{y}) = 2^n \sum_{\mathbf{u}\in(\mathbb{Z}_2)^n} (-1)^{\mathbf{y}\cdot\mathbf{u}}(F(\mathbf{u}))^2.$$

It turns out that the Fourier coefficients $\widehat{F}(\mathbf{x})$ provide a measure of the nonlinearity of Boolean functions. Suppose that $f, g \in \mathcal{B}_n$. We define the *distance* between f and g to be the quantity

$$\mathsf{d}(f,g) = |\{\mathbf{x} \in (\mathbb{Z}_2)^n : f(\mathbf{x}) \neq g(\mathbf{x})\}|.$$

Equivalently, $\mathsf{d}(f,g)$ is the Hamming distance between the vectors $\phi(f)$ and $\phi(g)$.

A function $f \in \mathcal{B}_n$ is a *linear function* if f has the form

$$f(\mathbf{x}) = \mathbf{a}\cdot\mathbf{x},$$

where $\mathbf{a} \in (\mathbb{Z}_2)^n$. Clearly there are 2^n linear functions in $\mathcal{B}_n$. For brevity, we will denote the function $\mathbf{a}\cdot\mathbf{x}$ by $L_{\mathbf{a}}$. By $L_{\mathbf{a}}+1$ we mean the function taking on the value $L_{\mathbf{a}}(\mathbf{x})+1 \bmod 2$ for all $\mathbf{x}$. A function $f \in \mathcal{B}_n$ is an *affine function* if $f = L_{\mathbf{a}}$ or $f = L_{\mathbf{a}}+1$ for some $\mathbf{a} \in (\mathbb{Z}_2)^n$. Note that there are 2^{n+1} affine functions in $\mathcal{B}_n$.

The following formula relates the distance between a function f and a linear or affine function to the Fourier transform of f.

Theorem 4.39. *Suppose that $f \in \mathcal{B}_n$ and $F = (-1)^f$. Let $\mathbf{a} \in (\mathbb{Z}_2)^n$. Then*

$$\mathsf{d}(f, L_{\mathbf{a}}) = 2^{n-1} - \frac{1}{2}\widehat{F}(\mathbf{a})$$

and

$$\mathsf{d}(f, L_{\mathbf{a}}+1) = 2^{n-1} + \frac{1}{2}\widehat{F}(\mathbf{a}).$$

Proof.

$$\begin{aligned}\widehat{F}(\mathbf{a}) &= |\{\mathbf{y} \in (\mathbb{Z}_2)^n : \mathbf{a}\cdot\mathbf{y} = f(\mathbf{y})\}| - |\{\mathbf{y} \in (\mathbb{Z}_2)^n : \mathbf{a}\cdot\mathbf{y} \neq f(\mathbf{y})\}| \\ &= 2^n - 2|\{\mathbf{y} \in (\mathbb{Z}_2)^n : \mathbf{a}\cdot\mathbf{y} \neq f(\mathbf{y})\}| \\ &= 2^n - 2\mathsf{d}(f, L_{\mathbf{a}}).\end{aligned}$$

From this it follows immediately that

$$\mathsf{d}(f, L_{\mathbf{a}}) = 2^{n-1} - \frac{1}{2}\widehat{F}(\mathbf{a}).$$

The second formula is obtained by observing that

$$\mathsf{d}(f, L_{\mathbf{a}}+1) = 2^n - \mathsf{d}(f, L_{\mathbf{a}}).$$

□

We illustrate the concepts described above by continuing Example 4.35.

Example 4.40. Suppose that $n = 2$ and $f(x_1, x_2) = x_1 x_2$, where $x_1, x_2 \in \mathbb{Z}_2$. We observed in Example 4.35 that $\phi(f) = (0,0,0,1)$ and $\phi(\widehat{F}) = (2,2,2,-2)$. The affine functions of two Boolean variables (denoted by g in the following table) and their distances to f are as follows:

a	g	$\phi(g)$	$\widehat{F}(\mathbf{a})$	$\mathsf{d}(f,g)$
(0,0)	$L_{(0,0)}$	(0,0,0,0)	2	1
(0,0)	$L_{(0,0)}+1$	(1,1,1,1)	2	3
(0,1)	$L_{(0,1)}$	(0,1,0,1)	2	1
(0,1)	$L_{(0.1)}+1$	(1,0,1,0)	2	3
(1,0)	$L_{(1,0)}$	(0,0,1,1)	2	1
(1,0)	$L_{(1,0)}+1$	(1,1,0,0)	2	3
(1,1)	$L_{(1,1)}$	(0,1,1,0)	−2	3
(1,1)	$L_{(1,1)}+1$	(1,0,0,1)	−2	1

It can be verified that $\mathsf{d}(f,g) = 1$ or 3 for all affine functions g and, moreover, $\mathsf{d}(f,g)$ is given by the formula proven in Theorem 4.39. ∎

The *nonlinearity* of f, denoted N_f, is defined as follows:

$$N_f = \min\{\mathsf{d}(f, L_{\mathbf{a}}), \mathsf{d}(f, L_{\mathbf{a}}+1) : \mathbf{a} \in (\mathbb{Z}_2)^n\}.$$

In view of Theorem 4.39, we have that

$$N_f = 2^{n-1} - \frac{1}{2}\max\{|\widehat{F}(\mathbf{a})| : \mathbf{a} \in (\mathbb{Z}_2)^n\}. \tag{4.5}$$

A function $f \in \mathcal{B}_n$ is a *bent function* if $|\widehat{F}(\mathbf{x})| = 2^{n/2}$ for all $\mathbf{x} \in (\mathbb{Z}_2)^n$, where $F = (-1)^f$. Note that the function f in Example 4.35 is bent. A bent function can exist in $\mathcal{B}_n$ only when n is even because $\widehat{F}(\mathbf{x})$ is an integer for all $\mathbf{x} \in (\mathbb{Z}_2)^n$ when $f \in \mathcal{B}_n$, and $2^{n/2}$ is not an integer if n is odd.

We prove in the next theorem that bent functions have maximum possible nonlinearity (this is the reason for the terminology "bent").

Theorem 4.41. *For any $f \in \mathcal{B}_n$, it holds that $N_f \leq 2^{n-1} - 2^{n/2-1}$. Furthermore, equality holds if and only if f is a bent function.*

Proof. Denote

$$M = \max\{|\widehat{F}(\mathbf{a})| : \mathbf{a} \in (\mathbb{Z}_2)^n\};$$

then $N_f = 2^{n-1} - M/2$. Applying Parseval's Equation (Corollary 4.37), we have that

$$2^n M^2 \geq \sum_{\mathbf{x} \in (\mathbb{Z}_2)^n} (\widehat{F}(\mathbf{x}))^2 = 2^{2n},$$

so $M \geq 2^{n/2}$. Furthermore, $M = 2^{n/2}$ if and only if $|\widehat{F}(\mathbf{x})| = 2^{n/2}$ for all $\mathbf{x} \in (\mathbb{Z}_2)^n$. In other words, $N_f \leq 2^{n-1} - 2^{n/2-1}$, and equality occurs if and only if f is bent. □

Here is an interesting way to characterize bent functions in terms of Hadamard matrices.

Theorem 4.42. *Suppose that* $f \in \mathcal{B}_n$ *and* $F = (-1)^f$. *Define the matrix* $H_f = (h_{\mathbf{x},\mathbf{y}})$, *where* $h_{\mathbf{x},\mathbf{y}} = F(\mathbf{x}+\mathbf{y})$ *for all* $\mathbf{x},\mathbf{y} \in (\mathbb{Z}_2)^n$. *Then* f *is a bent function if and only if* H_f *is a Hadamard matrix.*

Proof. Suppose that f is a bent function. Define the function

$$G = \frac{1}{2^{n/2}}\widehat{F}.$$

Then $G = (-1)^g$ for some Boolean function $g \in \mathcal{B}_n$. Applying Corollary 4.34, it follows that $\widehat{G} = 2^{n/2}F$. Now, to verify that H_f is a Hadamard matrix, we must show that the equation

$$\sum_{\mathbf{z}\in(\mathbb{Z}_2)^n} h_{\mathbf{x},\mathbf{z}}h_{\mathbf{y},\mathbf{z}} = 2^n\,\delta_{\mathbf{x},\mathbf{y}} \tag{4.6}$$

holds for all $\mathbf{x},\mathbf{y} \in (\mathbb{Z}_2)^n$. This is done as follows:

$$\begin{aligned}
\sum_{\mathbf{z}\in(\mathbb{Z}_2)^n} h_{\mathbf{x},\mathbf{z}}h_{\mathbf{y},\mathbf{z}} &= \sum_{\mathbf{z}\in(\mathbb{Z}_2)^n} F(\mathbf{x}+\mathbf{z})F(\mathbf{y}+\mathbf{z})\\
&= \sum_{\mathbf{w}\in(\mathbb{Z}_2)^n} F(\mathbf{w})F(\mathbf{x}+\mathbf{y}+\mathbf{w})\\
&= \frac{1}{2^n}\sum_{\mathbf{w}\in(\mathbb{Z}_2)^n} \widehat{G}(\mathbf{w})\widehat{G}(\mathbf{x}+\mathbf{y}+\mathbf{w})\\
&= \frac{1}{2^n}\times 2^{2n}\,\delta_{\mathbf{x}+\mathbf{y},(0,\ldots,0)}\\
&= 2^n\,\delta_{\mathbf{x},\mathbf{y}},
\end{aligned}$$

where we apply Theorem 4.36 to the Boolean function g.

Conversely, suppose that (4.6) holds for all $\mathbf{x},\mathbf{y} \in (\mathbb{Z}_2)^n$. Define the real-valued function

$$G = \frac{1}{2^{n/2}}\widehat{F}.$$

Setting $\mathbf{y} = (0,\ldots,0)$ in (4.6), we obtain the following:

$$\begin{aligned}
2^n\,\delta_{\mathbf{x},(0,\ldots,0)} &= \sum_{\mathbf{z}\in(\mathbb{Z}_2)^n} h_{\mathbf{x},\mathbf{z}}h_{(0,\ldots,0),\mathbf{z}}\\
&= \sum_{\mathbf{z}\in(\mathbb{Z}_2)^n} F(\mathbf{x}+\mathbf{z})F(\mathbf{z})\\
&= \frac{1}{2^n}\sum_{\mathbf{z}\in(\mathbb{Z}_2)^n} \widehat{G}(\mathbf{z})\widehat{G}(\mathbf{x}+\mathbf{z})\\
&= \sum_{\mathbf{z}\in(\mathbb{Z}_2)^n} (-1)^{\mathbf{x}\cdot\mathbf{z}}(G(\mathbf{z}))^2\\
&= \frac{1}{2^n}\sum_{\mathbf{z}\in(\mathbb{Z}_2)^n} (-1)^{\mathbf{x}\cdot\mathbf{z}}(\widehat{F}(\mathbf{z}))^2.
\end{aligned}$$

Therefore we have that

$$2^{2n}(1,0,\ldots,0) = \phi((\widehat{F}(z))^2)S_n.$$

Multiplying on the right by S_n, we obtain

$$2^{2n}(1,0,\ldots,0)S_n = 2^n\phi((\widehat{F}(z))^2),$$

which simplifies to give

$$2^n(1,1,\ldots,1) = \phi((\widehat{F}(z))^2).$$

Therefore $|\widehat{F}(\mathbf{x})| = 2^{n/2}$ for all $\mathbf{x} \in (\mathbb{Z}_2)^n$, and f is bent. □

Example 4.43. Again, suppose that $n = 2$ and $f(x_1, x_2) = x_1x_2$, where $x_1, x_2 \in \mathbb{Z}_2$. We have that $\phi(f) = (0,0,0,1)$ and $\phi(F) = (1,1,1,-1)$. The matrix H_f is

$$H_f = \begin{pmatrix} 1 & 1 & 1 & -1 \\ 1 & 1 & -1 & 1 \\ 1 & -1 & 1 & 1 \\ -1 & 1 & 1 & 1 \end{pmatrix},$$

which is easily seen to be a Hadamard matrix of order 4. ∎

Our next theorem ties together all the results we have presented so far in this section. This theorem proves an equivalence between bent functions and certain difference sets.

Theorem 4.44 (Dillon). *There exists a bent function $f : (\mathbb{Z}_2)^n \to \mathbb{Z}_2$ if and only if there exists a $(2^n, 2^{n-1} \pm 2^{(n-2)/2}, 2^{n-2} \pm 2^{(n-2)/2})$-difference set in $(\mathbb{Z}_2)^n$.*

Proof. Suppose that $f \in \mathcal{B}_n$ is a bent function. Then, the matrix $H_f = (h_{\mathbf{x},\mathbf{y}})$ constructed in Theorem 4.42 is a Hadamard matrix. It is also easy to see that H_f is regular; this is because every row and column of H_f is a permutation of the list of values $F(\mathbf{x})$, $\mathbf{x} \in (\mathbb{Z}_2)^n$.

We next show that $(\mathbb{Z}_2)^n$ is a sharply transitive automorphism group of this Hadamard matrix. For any $\mathbf{u} \in (\mathbb{Z}_2)^n$, define $t_{\mathbf{u}} : (\mathbb{Z}_2)^n \to (\mathbb{Z}_2)^n$ as follows: $t_{\mathbf{u}}(\mathbf{x}) = \mathbf{x} + \mathbf{u}$ for all $\mathbf{x} \in (\mathbb{Z}_2)^n$. It is clear that $t_{\mathbf{u}}$ is a permutation (i.e., a bijection) of $(\mathbb{Z}_2)^n$, and $\{t_{\mathbf{u}} : \mathbf{u} \in (\mathbb{Z}_2)^n\}$ is a sharply transitive set of permutations. Furthermore, every $t_{\mathbf{u}}$ is an automorphism of H_f because

$$h_{t_{\mathbf{u}}(\mathbf{x}),t_{\mathbf{u}}(\mathbf{y})} = h_{\mathbf{u}+\mathbf{x},\mathbf{u}+\mathbf{y}} = F(\mathbf{u}+\mathbf{x}+\mathbf{u}+\mathbf{y}) = F(\mathbf{x}+\mathbf{y}) = h_{\mathbf{x},\mathbf{y}}$$

for all $\mathbf{x}, \mathbf{y} \in (\mathbb{Z}_2)^n$.

This implies that the symmetric $(2^n, 2^{n-1} \pm 2^{(n-2)/2}, 2^{n-2} \pm 2^{(n-2)/2})$-BIBD, whose incidence matrix is constructed from H_f by replacing every entry -1 by 0, has $(\mathbb{Z}_2)^n$ as a sharply transitive automorphism group (apply Theorem 4.25 with $u = 2^{n-2}$). Therefore, by Theorem 3.17, the desired difference set in $(\mathbb{Z}_2)^n$ exists.

Conversely, suppose that the stated difference set exists. From this difference set, we can construct a symmetric $(2^n, 2^{n-1} \pm 2^{(n-2)/2}, 2^{n-2} \pm 2^{(n-2)/2})$-BIBD having $(\mathbb{Z}_2)^n$ as a sharply transitive automorphism group. Then, replacing every entry 0 in the incidence matrix of this BIBD by -1, we obtain a regular Hadamard matrix of order 2^n having $(\mathbb{Z}_2)^n$ as a sharply transitive automorphism group. The fact that the Hadamard matrix has this automorphism group means that

$$h_{\mathbf{u}+\mathbf{x},\mathbf{u}+\mathbf{y}} = h_{\mathbf{x},\mathbf{y}}$$

for all $\mathbf{u}, \mathbf{x}, \mathbf{y} \in (\mathbb{Z}_2)^n$. Suppose we define a function $f \in \mathcal{B}_n$ as follows:

$$f(\mathbf{x}) = \begin{cases} 0 & \text{if } h_{\mathbf{x},(0,\ldots,0)} = 1 \\ 1 & \text{if } h_{\mathbf{x},(0,\ldots,0)} = -1. \end{cases}$$

Then, we have that

$$h_{\mathbf{x},\mathbf{y}} = h_{\mathbf{x}+\mathbf{y},(0,\ldots,0)} = (-1)^{f(\mathbf{x}+\mathbf{y})}$$

for all $\mathbf{x}, \mathbf{y} \in (\mathbb{Z}_2)^n$. Therefore Theorem 4.42 establishes that f is a bent function. □

The proof of Theorem 4.44 involved several steps to show that a bent function can be transformed into the relevant difference set and vice versa. However, if we examine the sequence of operations performed, we can easily describe a direct transformation between these objects. We state the following result, which is primarily a consequence of the proof of Theorem 4.44.

Corollary 4.45. *Suppose that $f \in \mathcal{B}_n$ is a bent function. Let $i = 0$ or 1 and define*

$$D_i = \{\mathbf{x} \in (\mathbb{Z}_2)^n : f(\mathbf{x}) = i\}.$$

Then D_i is a $(2^n, 2^{n-1} \pm 2^{(n-2)/2}, 2^{n-2} \pm 2^{(n-2)/2})$-difference set in $(\mathbb{Z}_2)^n$. Conversely, suppose that $D \subseteq (\mathbb{Z}_2)^n$ is a $(2^n, 2^{n-1} \pm 2^{(n-2)/2}, 2^{n-2} \pm 2^{(n-2)/2})$-difference set. Define $f \in \mathcal{B}_n$ by $f(\mathbf{x}) = 0$ if and only if $\mathbf{x} \in D$. Then f is a bent function.

So far, we have seen one example of a bent function, namely the function $x_1x_2 \in \mathcal{B}_2$ that was introduced in Example 4.35. We will prove for all even integers $n \geq 2$ that there exists a bent function in $\mathcal{B}_n$. First, we will state and prove an easy result concerning the Fourier coefficients of the sum of two Boolean functions on disjoint sets of input variables. Suppose that $f_1 \in \mathcal{B}_{n_1}$ and $f_2 \in \mathcal{B}_{n_2}$. Define the function $f = f_1 \oplus f_2 \in \mathcal{B}_{n_1+n_2}$ as follows:

$$f(x_1, \ldots, x_{n_1+n_2}) = f_1(x_1, \ldots, x_{n_1}) + f_2(x_{n_1+1}, \ldots, x_{n_1+n_2}) \bmod 2.$$

We have the following.

Lemma 4.46. *Suppose that $f_1 \in \mathcal{B}_{n_1}$, $f_2 \in \mathcal{B}_{n_2}$, and $f = f_1 \oplus f_2$. Let $F = (-1)^f$, $F_1 = (-1)^{f_1}$, and $F_2 = (-1)^{f_2}$. Then*

$$\widehat{F}(\mathbf{x}) = \widehat{F_1}(\mathbf{x_1})\widehat{F_2}(\mathbf{x_2}),$$

where $\mathbf{x} = (x_1, \ldots, x_{n_1+n_2})$, $\mathbf{x_1} = (x_1, \ldots, x_{n_1})$, and $\mathbf{x_2} = (x_{n_1+1}, \ldots, x_{n_1+n_2})$.

Proof.

$$\begin{aligned}
\widehat{F}(\mathbf{x}) &= \sum_{\mathbf{y} \in (\mathbb{Z}_2)^{n_1+n_2}} (-1)^{\mathbf{x}\cdot\mathbf{y}} F(\mathbf{y}) \\
&= \sum_{\mathbf{y_1} \in (\mathbb{Z}_2)^{n_1}} \sum_{\mathbf{y_2} \in (\mathbb{Z}_2)^{n_2}} (-1)^{\mathbf{x_1}\cdot\mathbf{y_1} + \mathbf{x_2}\cdot\mathbf{y_2}} F_1(\mathbf{y_1}) F_2(\mathbf{y_2}) \\
&= \left(\sum_{\mathbf{y_1} \in (\mathbb{Z}_2)^{n_1}} (-1)^{\mathbf{x_1}\cdot\mathbf{y_1}} F_1(\mathbf{y_1}) \right) \times \left(\sum_{\mathbf{y_2} \in (\mathbb{Z}_2)^{n_2}} (-1)^{\mathbf{x_2}\cdot\mathbf{y_2}} F_2(\mathbf{y_2}) \right) \\
&= \widehat{F_1}(\mathbf{x_1})\widehat{F_2}(\mathbf{x_2}).
\end{aligned}$$

□

We now apply the lemma above to bent functions. The following corollary is immediate.

Corollary 4.47. *Suppose that f_1 and f_2 are both bent functions. Then $f = f_1 \oplus f_2$ is a bent function.*

We now state an existence result for bent functions, which follows from the previous results by induction.

Theorem 4.48. *Suppose that $n = 2m$. Then the function*

$$x_1x_2 + x_3x_4 + \cdots + x_{2m-1}x_{2m} \bmod 2$$

is a bent function.

Proof. The proof is by induction on m. For $m = 1$, the function x_1x_2 was shown to be bent by the computations performed in Example 4.35.

As an induction hypothesis, assume that the function

$$x_1x_2 + x_3x_4 + \cdots + x_{2m-3}x_{2m-2} \bmod 2$$

is bent. Using the fact that $x_{2m-1}x_{2m}$ is bent, we can apply Corollary 4.47 to establish that the function

$$x_1x_2 + x_3x_4 + \cdots + x_{2m-1}x_{2m} \bmod 2$$

is a bent function.

Therefore, by induction, the proof is complete. □

We close this section by examining the bent function $x_1x_2 + x_3x_4 \bmod 2$ and the difference set equivalent to it.

Example 4.49. Suppose that $n = 4$ and $f(x_1, x_2, x_3, x_4) = x_1x_2 + x_3x_4 \bmod 2$. Then

$$\phi(f) = (0,0,0,1,0,0,0,1,0,0,0,1,1,1,1,0),$$

where the coordinates of $\phi(f)$ are in lexicographic order.

We construct a difference set from the function f by recording the values $\mathbf{x}$ where $f(\mathbf{x}) = 1$. We obtain the set

$$D = \{(0,0,1,1),(0,1,1,1),(1,0,1,1),(1,1,0,0),(1,1,0,1),(1,1,1,0)\}.$$

D is a $(16,6,2)$-difference set in the group $((\mathbb{Z}_2)^4, +)$. ∎

4.9 Notes and References

Seberry and Yamada [92] is a thorough survey on Hadamard matrices and related concepts. Craigen and Wallis [36] is more tightly focussed on Hadamard matrices and contains some interesting historical information; both surveys are useful references. Up-to-date general asymptotic existence results for Hadamard matrices are found in Craigen [34].

Theorem 4.5 is due to Todd [109], and Corollary 4.16 is due to Paley [83].

Conference matrices were introduced in 1950 by Belevitch. Conference matrices and related objects such as weighing matrices have been studied extensively since then. Recent results on these topics can be found in Koukouvinos and Seberry [68].

Williamson's method is presented in [116]. The discovery, in 1962, of a Hadamard matrix of order 92 using this techinque is reported in [6].

Theorem 4.25 is well-known, but its origin seems not to be known. For some relatively recent results on regular Hadamard matrices, see Craigen and Kharaghani [35]. The concept of excess of a Hadamard matrix is due to Best [8]; Theorem 4.30 is also proven in [8].

Bent functions were introduced by Rothaus in [88]. They have been an active area of research in recent years, in part due to their applications in coding theory and cryptography. Theorem 4.44 was first proven by Dillon in his Ph.D. thesis [40].

4.10 Exercises

4.1 Construct Hadamard matrices of orders 12, 16, and 20.

4.2 Construct symmetric conference matrices of orders 10, 14, and 18.

4.3 (a) Prove that a $W(n, n-1)$ (as defined in Exercise 2.10) exists if and only if a conference matrix of order n exists.

(b) Deduce from Exercise 2.10 that a conference matrix of order $n \equiv 2 \pmod 4$ exists only if $n-1$ is the sum of two squares.

4.4 (a) A conference matrix is *standardized* if every entry in the first row or column is equal to "1". Let $C = (c_{i,j})$ be a symmetric conference matrix of order n. For $2 \leq i \leq n$, suppose we multiply every entry in row i of C by $c_{i,1}$. Then, for $2 \leq j \leq n$, suppose we multiply every entry in column j of C by $c_{1,j}$. Prove that the resulting matrix is a standardized symmetric conference matrix of order n.

(b) Let $C = (c_{i,j})$ be a standardized symmetric conference matrix of order n. Define

$$a = |\{j : 4 \leq j \leq n, c_{2,j} = c_{3,j} = 1\}|,$$
$$b = |\{j : 4 \leq j \leq n, c_{2,j} = 1, c_{3,j} = -1\}|,$$
$$c = |\{j : 4 \leq j \leq n, c_{2,j} = -1, c_{3,j} = 1\}|, \quad \text{and}$$
$$d = |\{j : 4 \leq j \leq n, c_{2,j} = c_{3,j} = -1\}|.$$

Determine the values of a, b, c, and d (note that there are two cases to consider, depending on whether $c_{2,3} = c_{3,2} = 1$ or $c_{2,3} = c_{3,2} = -1$).

(c) Prove that a symmetric conference matrix of order n exists only if $n \equiv 2 \pmod 4$.

4.5 Suppose that $C = (c_{i,j})$ is a standardized conference matrix of order $n \equiv 2 \pmod 4$. Prove that C is symmetric by using a counting argument similar to that used in Exercise 4.4.

4.6 Extend Table 4.1 considering orders $n \leq 200$. To be specific, show that Hadamard matrices of all possible orders in the range $100 < n \leq 200$ can be constructed using the methods described in this chapter, except for $n = 116, 156, 172$, and 188.

4.7 **Note:** This exercise requires some knowledge of linear algebra pertaining to eigenvalues and eigenvectors.

Suppose we want to apply Williamson's construction. Thus we are looking for $n \times n$ matrices, A, B, C, and D, that satisfy the following properties:

- A, B, C, and D are symmetric matrices having entries ± 1,
- $A^2 + B^2 + C^2 + D^2 = 4nI_n$, and
- A, B, C, and D are circulant matrices.

Let s_A, s_B, s_C, and s_D denote the sum of the entries of any row of A, B, C, and D, respectively, and let $\mathbf{u} = (1, \cdots, 1)$.

(a) Prove that $\mathbf{u}$ is an eigenvector of A, B, C, and D, and prove that the corresponding eigenvalues are s_A, s_B, s_C, and s_D, respectively.

(b) Prove that $\mathbf{u}$ is an eigenvector of A^2, B^2, C^2, and D^2, and prove that the corresponding eigenvalues are $(s_A)^2, (s_B)^2, (s_C)^2$, and $(s_D)^2$, respectively.

(c) Prove that $(s_A)^2 + (s_B)^2 + (s_C)^2 + (s_D)^2 = 4n$.

(d) Suppose that n is odd. When applying Williamson's construction, prove that we can assume without loss of generality that s_A, s_B, s_C, and s_D are all odd, nonnegative integers.
Hint: Replace A by $-A$ if necessary, etc.

(e) For $n = 5$, find the unique solution (up to permutation) in odd nonnegative integers to the equation $(s_A)^2 + (s_B)^2 + (s_C)^2 + (s_D)^2 = 4n$.

(f) For $n = 5$, find circulant matrices A, B, C, and D that satisfy the conditions for Williamson's construction. Verify that all the conditions are satisfied.
Hint: Make use of the fact that s_A, s_B, s_C, and s_D are determined, as well as the fact that the matrices A, B, C, and D must be symmetric, in order to reduce the number of cases that need to be considered.

4.8 (a) Prove that the Kronecker Product of two regular Hadamard matrices is a regular Hadamard matrix.
(b) Construct a regular Hadamard matrix of order 16 using the Kronecker Product.
(c) Use this regular Hadamard matrix to construct a $(16, 6, 2)$-BIBD.

4.9 (a) Let H_1 and H_2 be Hadamard matrices, and define $H = H_1 \otimes H_2$. Prove that $\text{excess}(H) = \text{excess}(H_1) \times \text{excess}(H_2)$.
(b) Prove that $\sigma(8) \geq 16$.

4.10 Define the function $f \in \mathcal{B}_4$ to be

$$f(x_1, x_2, x_3, x_4) = x_1x_2 + x_2x_3 + x_3x_4 \bmod 2.$$

(a) Compute $\phi(f)$, $\phi(F)$, and $\phi(\widehat{F})$.
(b) Compute N_f using equation (4.5) and observe that f is a bent function.
(c) Construct a $(16, 6, 2)$-difference set in the group $((\mathbb{Z}_2)^4, +)$ from the function f by using the technique described in Corollary 4.45.

5

Resolvable BIBDs

5.1 Introduction

Definition 5.1. *Suppose* $(X, \mathcal{A})$ *is a* (v, k, λ)-BIBD. *A* parallel class *in* $(X, \mathcal{A})$ *is a subset of disjoint blocks from* $\mathcal{A}$ *whose union is* X. *A partition of* $\mathcal{A}$ *into r parallel classes is called a* resolution, *and* $(X, \mathcal{A})$ *is said to be a* resolvable BIBD *if* $\mathcal{A}$ *has at least one resolution.*

Observe that a parallel class contains v/k blocks, and therefore a BIBD can have a parallel class only if $v \equiv 0 \bmod k$.

We begin by constructing resolvable $(v, 2, 1)$-BIBDs for all even v. (Note that a $(v, 2, 1)$-BIBD consists of all 2-subsets of a v-set, so it exists trivially. The interesting thing is to show that it is resolvable.)

Theorem 5.2. *A resolvable* $(v, 2, 1)$-BIBD *exists if and only if* v *is an even integer and* $v \geq 4$.

Proof. Clearly it is necessary that v is even and $v \geq 4$. We construct a resolvable $(v, 2, 1)$-BIBD for all such v as follows: Take the set of points to be $\mathbb{Z}_{v-1} \cup \{\infty\}$. For $j \in \mathbb{Z}_{v-1}$, define

$$\Pi_j = \{\{\infty, j\}\} \cup \{\{i + j \bmod (v-1), j - i \bmod (v-1)\} : 1 \leq i \leq (v-2)/2\}.$$

It is not difficult to see that each Π_j is a parallel class, and each pair of points occurs in exactly one Π_j. Hence, we have a resolvable BIBD, as required. □

Example 5.3. A resolvable $(6, 2, 1)$-BIBD. The parallel classes are as follows:

$$\begin{aligned}
\Pi_0 &= \{\{\infty, 0\}, \{1, 4\}, \{2, 3\}\} \\
\Pi_1 &= \{\{\infty, 1\}, \{2, 0\}, \{3, 4\}\} \\
\Pi_2 &= \{\{\infty, 2\}, \{3, 1\}, \{4, 0\}\} \\
\Pi_3 &= \{\{\infty, 3\}, \{4, 2\}, \{0, 1\}\} \\
\Pi_4 &= \{\{\infty, 4\}, \{0, 3\}, \{1, 2\}\}.
\end{aligned}$$

∎

5.2 Affine Planes and Geometries

Recall from Section 2.3 that an affine plane of order $n \geq 2$ is an $(n^2, n, 1)$-BIBD. An affine plane of order n has $r = n + 1$ and $b = n^2 + n$. Theorem 2.13 asserts that affine planes exist for all prime power orders because they are residual BIBDs of projective planes. Affine planes of prime power order are also easy to construct directly; we prove the following theorem.

Theorem 5.4. *For any prime power q, there exists an affine plane of order q (i.e., a $(q^2, q, 1)$-BIBD).*

Proof. Define $P = \mathbb{F}_q \times \mathbb{F}_q$. For any $a, b \in \mathbb{F}_q$, define a block

$$L_{a,b} = \{(x, y) \in P : y = ax + b\}.$$

For any $c \in \mathbb{F}_q$, define

$$L_{\infty,c} = \{(c, y) : y \in \mathbb{F}_q\}.$$

Finally, define

$$\mathcal{L} = \{L_{a,b} : a, b \in \mathbb{F}_q\} \cup \{L_{\infty,c} : c \in \mathbb{F}_q\}.$$

We will show that $(P, \mathcal{L})$ is a $(q^2, q, 1)$-BIBD.

Clearly, there are q^2 points in P, and every block contains exactly q points. Hence, we need only show that every pair of points is contained in a unique block. Let $(x_1, y_1), (x_2, y_2) \in P$. We consider two cases:

1. If $x_1 = x_2$, then the unique block containing the pair $\{(x_1, y_1), (x_2, y_2)\}$ is L_{∞,x_1}.
2. If $x_1 \neq x_2$, consider the system of equations in $\mathbb{F}_q$:

$$y_1 = ax_1 + b$$
$$y_2 = ax_2 + b.$$

We will show that this system of equations has a unique solution for a and b. Subtracting the second equation from the first, we obtain

$$y_1 - y_2 = a(x_1 - x_2).$$

Since $x_1 \neq x_2$, there is a unique multiplicative inverse $(x_1 - x_2)^{-1} \in \mathbb{F}_q$. Multiply both sides of the previous equation by $(x_1 - x_2)^{-1}$, obtaining

$$a = (x_1 - x_2)^{-1}(y_1 - y_2).$$

Having determined a, it is a simple matter to determine b by back-substitution:

$$b = y_1 - ax_1 = y_1 - (x_1 - x_2)^{-1}(y_1 - y_2)x_1.$$

Therefore, the unique block containing the pair $\{(x_1, y_1), (x_2, y_2)\}$ is $L_{a,b}$, where a and b are computed from the formulas above.

Summarizing, we have shown that $(P, \mathcal{L})$ is a $(q^2, q, 1)$-BIBD. □

Example 5.5. We use Theorem 5.4 to construct an affine plane of order 3. The set of points is $\mathbb{Z}_3 \times \mathbb{Z}_3$, and the blocks are as follows:

$$\begin{aligned}
L_{0,0} &= \{(0,0),(1,0),(2,0)\}\\
L_{0,1} &= \{(0,1),(1,1),(2,1)\}\\
L_{0,2} &= \{(0,2),(1,2),(2,2)\}\\
L_{1,0} &= \{(0,0),(1,1),(2,2)\}\\
L_{1,1} &= \{(0,1),(1,2),(2,0)\}\\
L_{1,2} &= \{(0,2),(1,0),(2,1)\}\\
L_{2,0} &= \{(0,0),(1,2),(2,1)\}\\
L_{2,1} &= \{(0,1),(1,0),(2,2)\}\\
L_{2,2} &= \{(0,2),(1,1),(2,0)\}\\
L_{\infty,0} &= \{(0,0),(0,1),(0,2)\}\\
L_{\infty,1} &= \{(1,0),(1,1),(1,2)\}\\
L_{\infty,2} &= \{(2,0),(2,1),(2,2)\}.
\end{aligned}$$

▮

At this point, we have two constructions for affine planes of prime power order: the direct construction given in Theorem 5.4 and forming the residual BIBD of the projective plane $\mathrm{PG}_2(q)$ constructed in Theorem 2.10. With a bit of work, we can show that these two constructions of affine planes of order q (q a prime power) yield isomorphic BIBDs.

First, it is not difficult to show that all affine planes constructed as residual designs of the projective plane $\mathrm{PG}_2(q)$ are isomorphic. In other words, it does not matter which block in $\mathrm{PG}_2(q)$ we use to construct the residual design. Therefore, we can suppose without loss of generality that we choose the block A_{B_0} corresponding to the two-dimensional subspace

$$B_0 = \{(x_1,x_2,x_3) \in (\mathbb{F}_q)^3 : (0,0,1)\cdot(x_1,x_2,x_3) = 0\}$$

of $(\mathbb{F}_q)^3$ (i.e., the subspace $B_0 = \mathsf{span}((1,0,0),(0,1,0))$). The points in the block A_{B_0} are the following one-dimensional subspaces of $(\mathbb{F}_q)^3$:

$$\begin{aligned}
&\mathsf{span}((1,i,0)), i \in \mathbb{F}_q, \quad \text{and}\\
&\mathsf{span}((0,1,0)).
\end{aligned}$$

The q^2 points not in A_{B_0} are

$$\mathsf{span}((x,y,1)), x, y \in \mathbb{F}_q.$$

Let $(P,\mathcal{L}) = (\mathbb{F}_q \times \mathbb{F}_q, \{L_{a,b} : a,b \in \mathbb{F}_q\} \cup \{L_{\infty,c} : c \in \mathbb{F}_q\})$ be the affine plane of order q constructed in Theorem 5.4. We will show that the bijection α, defined by

$$(x,y) \mapsto \mathsf{span}((x,y,1))$$

for all $(x,y) \in P$, yields an isomorphism of the affine plane $(P,\mathcal{L})$ and the residual BIBD of $\mathsf{PG}_2(q)$ through the block A_{B_0}.

We must demonstrate that blocks are mapped to blocks under the bijection α. The q^2+q blocks (other than A_0) in $\mathsf{PG}_2(q)$ are obtained from the following two-dimensional subspaces:

$$B_{a,b} = \{(x_1,x_2,x_3) \in (\mathbb{F}_q)^3 : (a,-1,b)\cdot(x_1,x_2,x_3) = 0\}, a,b \in \mathbb{F}_q, \quad \text{and}$$
$$B_c = \{(x_1,x_2,x_3) \in (\mathbb{F}_q)^3 : (1,0,-c)\cdot(x_1,x_2,x_3) = 0\}, c \in \mathbb{F}_q.$$

(To see this, observe that these q^2+q subspaces are distinct, and different from B_0.)

Let $a,b \in \mathbb{F}_q$. The $q+1$ points in the block $A_{B_{a,b}}$ of $\mathsf{PG}_2(q)$ are

$$\mathsf{span}((x,ax+b,1)), x \in \mathbb{F}_q, \quad \text{and}$$
$$\mathsf{span}((1,a,0)).$$

The point $\mathsf{span}((1,a,0))$ is deleted from $A_{B_{a,b}}$ when the residual design is constructed, and $\alpha(x,ax+b) = \mathsf{span}((x,ax+b,1))$ for all $x \in \mathbb{F}_q$. Thus the block $L_{a,b}$ is mapped by α to the block

$$A_{B_{a,b}} \backslash \{\mathsf{span}((1,a,0))\}.$$

Finally, let's consider a block A_{B_c}, where $c \in \mathbb{F}_q$. The $q+1$ points in this block are

$$\mathsf{span}((c,y,1)), y \in \mathbb{F}_q, \quad \text{and}$$
$$\mathsf{span}((0,1,0)).$$

The point $\mathsf{span}((0,1,0))$ is deleted from A_{B_c} when the residual design is constructed, and $\alpha(c,y) = \mathsf{span}((c,y,1))$ for all $y \in \mathbb{F}_q$. Thus the block $L_{\infty,c}$ is mapped by α to the block

$$A_{B_c} \backslash \{\mathsf{span}((0,1,0))\}.$$

We have therefore shown that the two designs are isomorphic.

5.2.1 Resolvability of Affine Planes

Affine planes provide interesting examples of resolvable BIBDs because any affine plane can be shown to be resolvable. The main steps in proving this are as follows. First, the following lemma is proved by a simple counting argument.

Lemma 5.6. *Suppose $(P,\mathcal{L})$ is an affine plane of order n. Suppose $L \in \mathcal{L}$, $x \in P$, and $x \notin L$. Then there is exactly one block $M \in \mathcal{L}$ such that $x \in M$ and $L \cap M = \emptyset$.*

Proof. $(P, \mathcal{L})$ is a BIBD with $k = n$ and $\lambda = 1$. Hence, for every point $y \in L$, there is a unique block L_y such that $x \in L_y$ and $L \cap L_y = \{y\}$. This accounts for n blocks containing the point x. Since $r = n + 1$, there is one further block containing x, say M, and $L \cap M = \emptyset$. □

Now, suppose that $(P, \mathcal{L})$ is an affine plane of order n, and define a binary relation $\sim$ on the set of blocks, $\mathcal{L}$, as follows:

$$L \sim M \text{ if } L = M \text{ or } L \cap M = \emptyset.$$

The following can now be proved.

Lemma 5.7. *Suppose $(X, \mathcal{L})$ is an affine plane of order n. Then the relation $\sim$, as defined above, is an equivalence relation.*

Proof. We need to show that $\sim$ is reflexive, symmetric, and transitive. First, $L \sim L$ for every $L \in \mathcal{L}$ by definition. Second, it follows easily from the definition that $L \sim M$ if and only if $M \sim L$. Third, suppose that $L \sim M$ and $M \sim N$. There are four cases that arise:

1. If $L = M$ and $M = N$, then $L = N$ and hence $L \sim N$.
2. If $L = M$ and $M \cap N = \emptyset$, then $L \cap N = \emptyset$ and hence $L \sim N$.
3. If $L \cap M = \emptyset$ and $M = N$, then $L \cap N = \emptyset$ and hence $L \sim N$.
4. Suppose $L \cap M = \emptyset$ and $M \cap N = \emptyset$. If $L = N$, then $L \sim N$, so suppose $L \neq N$. In this case, we want to prove that $L \cap N = \emptyset$. If it does not, then there is a unique point $x \in L \cap N$. Now, L and N are two blocks that contain the point x and are both disjoint from M. This contradicts Lemma 5.6, so we conclude that $L \cap N = \emptyset$ and hence $L \sim N$.

We have proved that $\sim$ is reflexive, symmetric, and transitive, and hence it is an equivalence relation. □

The next step is to prove the following.

Lemma 5.8. *Suppose $(X, \mathcal{L})$ is an affine plane of order n. Then each equivalence class of $\sim$ is a parallel class in $(X, \mathcal{L})$.*

Proof. Let Π be an equivalence class of $\sim$ and let $L \in \Pi$. Then,

$$\Pi = \{M \in \mathcal{L} : L \sim M\}.$$

Clearly, all the blocks in Π are disjoint. Furthermore, for any point x, Lemma 5.6 tells us that there exists a block $M \in \Pi$ such that $x \in M$. It follows that each equivalence class of $\sim$ is a partition of X. □

Using this lemma, it is easy to see that $(X, \mathcal{L})$ is resolvable, as follows.

Theorem 5.9. *Any affine plane is resolvable.*

Proof. By Lemma 5.8, each equivalence class of $\sim$ is a parallel class of the BIBD. Also, every block of the BIBD is in exactly one equivalence class of $\sim$. Therefore, the equivalence classes of $\sim$ form a resolution of the affine plane. □

In the case of the affine planes of prime order that we constructed in Theorem 5.4, it is easy to determine the parallel classes. For any $a \in \mathbb{F}_q$,

$$\{L_{a,b} : b \in \mathbb{F}_q\}$$

is a parallel class. Furthermore,

$$\{L_{\infty,c} : c \in \mathbb{F}_q\}$$

is a parallel class. These $q+1$ parallel classes form a resolution of the BIBD. Observe that each of these parallel classes consists of all "lines" having a given "slope". In this fashion, the finite affine planes can be thought of as finite analogs of the classical real Euclidean plane.

5.2.2 Projective and Affine Planes

Recall that a projective plane of order n is an $(n^2+n+1, n+1, 1)$-BIBD. The next theorem establishes a close connection between affine and projective planes.

Theorem 5.10. *There exists an affine plane of order n if and only if there exists a projective plane of order n.*

Proof. First, the residual BIBD of a projective plane of order n is an affine plane of order n by Theorem 2.7. Conversely, given any affine plane of order n, we will show how to embed it into a projective plane of order n. Let $(X, \mathcal{L})$ be an affine plane of order n. By Theorem 5.9, $(X, \mathcal{L})$ is resolvable; let $\Pi_1, \ldots, \Pi_{n+1}$ be the $n+1$ parallel classes. Let $\infty_1, \ldots, \infty_{n+1} \notin X$, define $\Omega = \{\infty_1, \ldots, \infty_{n+1}\}$, and define $X' = X \cup \Omega$. For every $L \in \mathcal{L}$, define $L' = L \cup \{\infty_i\}$, where $L \in \Pi_i$ (in other words, adjoin the point ∞_i to every block in the ith parallel class, $1 \leq i \leq n+1$). Finally, define $\mathcal{L}' = \{L' : L \in \mathcal{L}\} \cup \{\Omega\}$.

We show that $(X', \mathcal{L}')$ is a projective plane of order n. There are n^2+n+1 points, and every block contains exactly $n+1$ points. Thus we need only to show that every pair of points $x, y \in X'$ $(x \neq y)$ occurs in a unique block. If $x, y \in X$, then x and y occur in a unique block $L \in \mathcal{L}$, and hence x and y occur in a unique block in $\mathcal{L}'$, namely L'. If $x \in X$ and $y \in \Omega$, say $y = \infty_i$, then $\{x, y\} \subseteq L'$, where L is the unique block in Π_i that contains x. Finally, if $x = \infty_i$ and $y = \infty_j$, then $\{x, y\} \subseteq \Omega$. □

Example 5.11. The affine plane of order 3 constructed in Example 5.5 can be embedded into a projective plane of order 3 consisting of the following blocks:

$$
\begin{aligned}
L'_{0,0} &= \{(0,0),(1,0),(2,0),\infty_1\}\\
L'_{0,1} &= \{(0,1),(1,1),(2,1),\infty_1\}\\
L'_{0,2} &= \{(0,2),(1,2),(2,2),\infty_1\}\\
L'_{1,0} &= \{(0,0),(1,1),(2,2),\infty_2\}\\
L'_{1,1} &= \{(0,1),(1,2),(2,0),\infty_2\}\\
L'_{1,2} &= \{(0,2),(1,0),(2,1),\infty_2\}\\
L'_{2,0} &= \{(0,0),(1,2),(2,1),\infty_3\}\\
L'_{2,1} &= \{(0,1),(1,0),(2,2),\infty_3\}\\
L'_{2,2} &= \{(0,2),(1,1),(2,0),\infty_3\}\\
L'_{\infty,0} &= \{(0,0),(0,1),(0,2),\infty_4\}\\
L'_{\infty,1} &= \{(1,0),(1,1),(1,2),\infty_4\}\\
L'_{\infty,2} &= \{(2,0),(2,1),(2,2),\infty_4\}\\
\Omega &= \{\infty_1,\infty_2,\infty_3,\infty_4\}.
\end{aligned}
$$

∎

5.2.3 Affine Geometries

In this section, we generalize the construction of affine planes to higher dimensional affine geometries. We use a slightly different presentation. Let q be a prime power, let $m \geq 2$, and let $X = (\mathbb{F}_q)^m$. Let $1 \leq d \leq m-1$. A *d-flat* in X is a subspace of X having dimension d or an additive coset of such a subspace. Note that X itself is a vector space of dimension m over $\mathbb{F}_q$.

A d-dimensional subspace is the same thing as the solution set to a system of $m-d$ linearly independent homogeneous linear equations in m variables $x_1, \ldots, x_m \in \mathbb{F}_q$. A d-flat is the solution set to a system of $m-d$ independent linear equations, which can be homogeneous or nonhomogeneous.

The set of points X and the set of all d-flats of X (for $1 \leq d \leq m-1$) comprise the m-dimensional *affine geometry* over $\mathbb{F}_q$, which will be denoted $\mathsf{AG}_m(q)$.

For $0 \leq d \leq m$, define the *Gaussian coefficient* $\begin{bmatrix} m \\ d \end{bmatrix}_q$ as follows:

$$
\begin{bmatrix} m \\ d \end{bmatrix}_q = \begin{cases} \frac{(q^m-1)(q^{m-1}-1)\cdots(q^{m-d+1}-1)}{(q^d-1)(q^{d-1}-1)\cdots(q-1)} & \text{if } d \neq 0 \\ 1 & \text{if } d = 0. \end{cases}
$$

The geometry $\mathsf{AG}_m(q)$ gives rise to various resolvable BIBDs, as shown in the following theorem.

Theorem 5.12. *Let q be a prime power, let $m \geq 2$, and let $1 \leq d \leq m-1$. Let X denote the set of points in $\mathsf{AG}_m(q)$ and let $\mathcal{A}$ denote the set of all d-flats in $\mathsf{AG}_m(q)$. Then $(X, \mathcal{A})$ is a resolvable $(q^m, b, r, q^d, \lambda)$-BIBD, where $b = q^{m-d}\begin{bmatrix} m \\ d \end{bmatrix}_q$, $r = \begin{bmatrix} m \\ d \end{bmatrix}_q$, and $\lambda = \begin{bmatrix} m-1 \\ d-1 \end{bmatrix}_q$.*

Proof. The fact that this set system is resolvable is easy to see because any subspace together with all of its cosets forms a parallel class. Therefore, we just need to prove that the design is a BIBD.

First, we show that every pair of points occurs in λ blocks, where $\lambda = \begin{bmatrix} m-1 \\ d-1 \end{bmatrix}_q$. Suppose that $\mathbf{x} = (x_1, \ldots, x_m)$ and $\mathbf{y} = (y_1, \ldots, y_m)$ are any two distinct points. The number of d-flats that contain $\mathbf{x}$ and $\mathbf{y}$ is the same as the number of d-dimensional subspaces that contain the two points $(0, \ldots, 0)$ and $\mathbf{z} = \mathbf{x} - \mathbf{y}$. A subspace of dimension d that contains $\mathbf{z}$ is determined by choosing $d-1$ vectors, say $\mathbf{z}^2, \ldots, \mathbf{z}^d$, such that $\mathbf{z}, \mathbf{z}^2, \ldots, \mathbf{z}^d$ are d linearly independent vectors. Denote $\mathbf{z}^1 = \mathbf{z}$; then the d-tuple $(\mathbf{z}^1, \ldots, \mathbf{z}^d)$ is an ordered basis for a subspace containing $\mathbf{z}$. The number of ordered bases of this type is easily seen to be

$$(q^m - q)(q^m - q^2) \cdots (q^m - q^{d-1}).$$

The terms in the product above are determined as follows: there are $q^m - q$ vectors in $(\mathbb{Z}_q)^m$ that are not scalar multiples of $\mathbf{z}^1$; there are $q^m - q^2$ vectors that are not in $\mathsf{span}(\mathbf{z}^1, \mathbf{z}^2)$; etc.

Now, a similar argument shows that every subspace containing $\mathbf{z}$ is generated by a constant number of ordered bases of this form, namely

$$(q^d - q)(q^d - q^2) \cdots (q^d - q^{d-1}).$$

The total number of subspaces containing $\mathbf{z}$ is therefore equal to

$$\begin{aligned} \frac{(q^m - q)(q^m - q^2) \cdots (q^m - q^{d-1})}{(q^d - q)(q^d - q^2) \cdots (q^d - q^{d-1})} &= \frac{(q^{m-1} - 1)(q^{m-2} - 1) \cdots (q^{m-d+1} - 1)}{(q^{d-1} - 1)(q^{d-2} - 1) \cdots (q-1)} \\ &= \begin{bmatrix} m-1 \\ d-1 \end{bmatrix}_q \\ &= \lambda. \end{aligned}$$

Now, it is easy to see that every block has size q^d. Given that k and λ are constants, it follows that we have a BIBD, and the parameters b and r can be determined by straightforward algebra. □

The construction above includes affine planes in the special case $m = 2$, $d = 1$. A line in the affine plane is the same thing as a 1-flat in $\mathsf{AG}_2(q)$. Here is an example of this construction with $d > 1$.

Example 5.13. Suppose we take $q = 3$, $m = 3$, and $d = 2$ in Theorem 5.12. The resulting BIBD is a resolvable (27, 39, 13, 9, 4)-BIBD. There are $r = 13$ parallel classes, each of which contains one two-dimensional subspace of $(\mathbb{Z}_3)^3$ and its two cosets.

The thirteen two-dimensional subspaces of $(\mathbb{Z}_3)^3$ are the solutions to homogeneous linear equations over $\mathbb{Z}_3$ in three variables. These are tabulated as follows:

equation	subspace
$x_1 = 0$	{000, 001, 002, 010, 011, 012, 020, 021, 022}
$x_2 = 0$	{000, 001, 002, 100, 101, 102, 200, 201, 202}
$x_3 = 0$	{000, 010, 020, 100, 110, 120, 200, 210, 220}
$x_1 + x_2 = 0$	{000, 001, 002, 120, 121, 122, 210, 211, 212}
$x_1 + 2x_2 = 0$	{000, 001, 002, 110, 111, 112, 220, 221, 222}
$x_1 + x_3 = 0$	{000, 010, 020, 102, 112, 122, 201, 211, 221}
$x_1 + 2x_3 = 0$	{000, 010, 020, 101, 111, 121, 202, 212, 222}
$x_2 + x_3 = 0$	{000, 100, 200, 012, 112, 212, 021, 121, 221}
$x_2 + 2x_3 = 0$	{000, 100, 200, 011, 111, 211, 022, 122, 222}
$x_1 + x_2 + x_3 = 0$	{000, 111, 222, 012, 120, 201, 021, 102, 210}
$x_1 + x_2 + 2x_3 = 0$	{000, 112, 221, 011, 120, 202, 022, 101, 210}
$x_1 + 2x_2 + x_3 = 0$	{000, 121, 212, 011, 102, 220, 022, 110, 201}
$x_1 + 2x_2 + 2x_3 = 0$	{000, 211, 122, 101, 012, 220, 202, 110, 021}

∎

Recall that we showed in Theorem 5.9 that any affine plane is resolvable. However, this result does not carry over to all designs having parameters as given in Theorem 5.12. It turns out that, if $d > 1$, there are BIBDs having parameters of the given form that are not resolvable. For example, there exist (8, 4, 3)-BIBDs that are not resolvable.

5.3 Bose's Inequality and Affine Resolvable BIBDs

The following inequality of Bose provides a necessary condition for the existence of a resolvable BIBD.

Theorem 5.14 (Bose's Inequality). *If there exists a resolvable (v, b, r, k, λ)-BIBD, then $b \geq v + r - 1$.*

Proof. We again use the technique of Theorem 1.33 and Theorem 2.2. In the proof of Theorem 1.33, Equation (1.5) showed that each basis vector $\mathbf{e}_i \in \mathbb{R}^v$ can be expressed as a linear combination of the vectors in $S = \{\mathbf{s}_1, \dots, \mathbf{s}_b\}$.

We are now given a resolvable BIBD, $(X, \mathcal{A})$. For $1 \leq i \leq r$, define

$$m_i = \frac{(i-1)v}{k} + 1 \quad \text{and} \quad n_i = \frac{iv}{k}.$$

Suppose that the blocks are labeled so that the r parallel classes are

$$\Pi_i = \{A_j : m_i \leq j \leq n_i\},$$

$1 \leq i \leq r$. Since each Π_i is a parallel class, we have that

$$\sum_{j=m_i}^{n_i} \mathbf{s}_j = (1, \ldots, 1)$$

for $1 \leq i \leq r$. From this, it follows that

$$\mathbf{s}_{m_i} = \sum_{j=m_1}^{n_1} \mathbf{s}_j - \sum_{j=m_i+1}^{n_i} \mathbf{s}_j \tag{5.1}$$

for $2 \leq i \leq r$. In other words, the $r-1$ vectors in the set

$$S' = \{\mathbf{s}_{m_2}, \ldots, \mathbf{s}_{m_r}\}$$

can be expressed as linear combinations of the $b - r + 1$ vectors in $S \backslash S'$.

Now, since the b vectors in S span $\mathbb{R}^v$, it follows that the $b - r + 1$ vectors in $S \backslash S'$ span $\mathbb{R}^v$. Since $\mathbb{R}^v$ has dimension v and is spanned by a set of $b - r + 1$ vectors, it must be the case that $b \geq v - r + 1$. □

Recall that Fisher's Inequality (Theorem 1.33) says that $b \geq v$ in any BIBD. Bose's Inequality strengthens Fisher's Inequality whenever the BIBD is resolvable.

The following lemma provides an alternate way of stating Bose's Inequality.

Lemma 5.15. *Suppose (v, b, r, k, λ) are the parameters of a BIBD. Then $b \geq v + r - 1$ if and only if $r \geq k + \lambda$.*

Proof. Suppose that $b \geq v + r - 1$. This implies that $b > v$ and hence $r > k$. Then we have the following:

$$\begin{aligned}
& & \frac{vr}{k} &\geq v + r - 1 \\
&\Longleftrightarrow & \frac{v(r-k)}{k} &\geq r - 1 \\
&\Longleftrightarrow & v &\geq \frac{k(r-1)}{r-k} \\
&\Longleftrightarrow & \frac{r(k-1)+\lambda}{\lambda} &\geq \frac{k(r-1)}{r-k} \\
&\Longleftrightarrow & r(k-1)(r-k) + \lambda(r-k) &\geq \lambda k(r-1) \\
&\Longleftrightarrow & r(k-1)(r-k) &\geq \lambda r(k-1) \\
&\Longleftrightarrow & r - k &\geq \lambda.
\end{aligned}$$

□

Corollary 5.16. *If there exists a resolvable (v, b, r, k, λ)-BIBD, then $r \geq k + \lambda$.*

The results above motivate the following definition.

Definition 5.17. *A resolvable BIBD with $b = v + r - 1$ (or, equivalently, if $r = k + \lambda$) is said to be an* affine resolvable BIBD.

Affine planes are affine resolvable because $r = n + 1 = k + \lambda$. More generally, we obtain an affine resolvable BIBD from Theorem 5.12 whenever $d = m - 1$. This follows by verifying that

$$\frac{q^m - 1}{q - 1} = \begin{bmatrix} m \\ m-1 \end{bmatrix}_q = q^{m-1} + \begin{bmatrix} m-1 \\ m-2 \end{bmatrix}_q = q^{m-1} + \frac{q^{m-1} - 1}{q - 1},$$

which can be done by simple algebra. Thus we have the following result.

Corollary 5.18. *Let q be a prime power and let $m \geq 2$. Then there is an affine resolvable (q^m, q^{m-1}, λ)-BIBD, where $\lambda = (q^{m-1} - 1)/(q - 1)$.*

Observe that affine resolvable BIBDs are quasiresidual. In Corollary 2.15, we already constructed residual BIBDs having the same parameters as those from Corollary 5.18. It can be shown that the BIBDs obtained from these two corollaries are, in fact, isomorphic.

There are not many other known constructions for affine resolvable BIBDs. One such infinite class of affine resolvable BIBDs is derived from Hadamard matrices. We show how to construct this class of designs now.

We know from Theorem 4.5 that a Hadamard matrix of order $4m$ is equivalent to a (symmetric) $(4m - 1, 2m - 1, m - 1)$-BIBD, say $(X, \mathcal{A})$. Applying Theorem 1.32, the block complement of $(X, \mathcal{A})$ is a $(4m - 1, 2m, m)$-BIBD, say $(X, \mathcal{B})$. Let $\infty \notin X$, and define $X' = X \cup \{\infty\}$. For every $A \in \mathcal{A}$, define $A' = A \cup \{\infty\}$, and define $\mathcal{A}' = \{A' : A \in \mathcal{A}\}$. Then it is not hard to prove that $(X', \mathcal{A}' \cup \mathcal{B})$ is an affine resolvable $(4m, 8m - 2, 4m - 1, 2m, 2m - 1)$-BIBD, where each parallel class consists of two blocks. Thus we have the following.

Theorem 5.19. *If there exists a Hadamard matrix of order $4m$, then there exists an affine resolvable $(4m, 2m, 2m - 1)$-BIBD.*

Example 5.20. $\{1, 2, 4\}$ is a $(7, 3, 1)$-difference set in $\mathbb{Z}_7$ which generates a $(7, 3, 1)$-BIBD. The affine resolvable $(8, 4, 3)$-BIBD produced by the construction preceding Theorem 5.19 has the following blocks:

$$\begin{array}{ll}
\{\infty,1,2,4\} & \{0,3,5,6\} \\
\{\infty,2,3,5\} & \{1,4,6,0\} \\
\{\infty,3,4,6\} & \{2,5,0,1\} \\
\{\infty,4,5,0\} & \{3,6,1,2\} \\
\{\infty,5,6,1\} & \{4,0,2,3\} \\
\{\infty,6,0,2\} & \{5,1,3,4\} \\
\{\infty,0,1,3\} & \{6,2,4,5\}.
\end{array}$$

∎

Recall that Theorem 2.2 states that any two distinct blocks in a symmetric BIBD intersect in exactly λ points. There is a similar result for affine resolvable BIBDs.

Theorem 5.21. *Any two blocks from different parallel classes of an affine resolvable* (v, k, λ)-BIBD *intersect in exactly* k^2/v *points.*

Proof. We will show that $|A_1 \cap A_j| = k^2/v$ for $m_2 \le j \le b$. We start by setting $h = 1$ in Equation (2.1):

$$(r-\lambda)\mathbf{s}_1 + \sum_{j=1}^{b} \frac{\lambda k}{r}\mathbf{s}_j = \sum_{j=1}^{b} |A_1 \cap A_j| \mathbf{s}_j. \tag{5.2}$$

Using Equation (1.2), which states that $\sum \mathbf{s}_j = (r, \ldots, r)$, and the fact that Π_1 is a parallel class, we can rewrite Equation (5.2) as follows:

$$(r-\lambda)\mathbf{s}_1 + \sum_{j=m_1}^{n_1} \lambda k \mathbf{s}_j = \sum_{j=1}^{b} |A_1 \cap A_j| \mathbf{s}_j. \tag{5.3}$$

In the proof of Theorem 5.14, we showed that the $b - r + 1$ vectors in $S \backslash S'$ span $\mathbb{R}^v$. Since we are now assuming that $b = v - r + 1$, it must be the case that $S \backslash S'$ is a basis for $\mathbb{R}^v$.

Equation (5.3) can be rewritten in terms of the basis $S \backslash S'$. This can be done by using Equation (5.1) to eliminate the vectors in S' from the right side of Equation (5.3). (Note that none of the vectors in S' appear on the left side of Equation (5.3).) Denote $I_1 = \{m_1, \ldots, n_1\}$, $I_2 = \{m_i : 2 \le i \le r\}$, and $I_3 = \{1, \ldots, b\} \backslash (I_1 \cup I_2)$.

We obtain the following:

$$\begin{aligned}
&(r-\lambda)\mathbf{s}_1 + \sum_{j \in I_1} \lambda k \mathbf{s}_j \\
&= \sum_{j \in I_1} |A_1 \cap A_j| \mathbf{s}_j + \sum_{j \in I_2} |A_1 \cap A_j| \mathbf{s}_j + \sum_{j \in I_3} |A_1 \cap A_j| \mathbf{s}_j \\
&= \sum_{j \in I_1} |A_1 \cap A_j| \mathbf{s}_j + \sum_{i=2}^{r} |A_1 \cap A_{m_i}| \mathbf{s}_{m_i} + \sum_{j \in I_3} |A_1 \cap A_j| \mathbf{s}_j \\
&= \sum_{j \in I_1} |A_1 \cap A_j| \mathbf{s}_j + \sum_{i=2}^{r} |A_1 \cap A_{m_i}| \left(\sum_{j \in I_1} \mathbf{s}_j - \sum_{j=m_i+1}^{n_i} \mathbf{s}_j \right) \\
&\qquad + \sum_{j \in I_3} |A_1 \cap A_j| \mathbf{s}_j.
\end{aligned}$$

Now, consider the coefficient of a vector $\mathbf{s}_j$, $j \in I_3$. For any such j, we have $m_i + 1 \le j \le n_i$ for some i, $2 \le i \le r$. The coefficient of $\mathbf{s}_j$ on the left side of the equation above is 0, and the coefficient on the right side is

$$|A_1 \cap A_j| - |A_1 \cap A_{m_i}|.$$

Since $S \backslash S'$ is a basis, it must be the case that $|A_1 \cap A_j| - |A_1 \cap A_{m_i}| = 0$, so $|A_1 \cap A_j| = |A_1 \cap A_{m_i}|$.

It follows that there exists a constant μ such that $|A_1 \cap A_j| = \mu$ for all j, $m_i \le j \le n_i$. Since Π is a parallel class consisting of v/k blocks, we have

$$k = |A_1| = \sum_{j=m_i}^{n_i} |A_1 \cap A_j| = \frac{\mu v}{k},$$

so $\mu = k^2/v$. This completes the proof. □

We present an example to illustrate how Theorem 5.21 can be used to show that certain resolvable BIBDs do not exist.

Example 5.22. A resolvable (28,7,2)-BIBD would have $r = 9$ and $b = 63$. Since $9 = 7 + 2$ (i.e., $r = k + \lambda$), a resolvable (28,7,2)-BIBD would be affine resolvable. By Theorem 5.21, any two blocks from different parallel classes would intersect in $k^2/v = 7/4$ points. Since $7/4$ is not an integer, there does not exist a resolvable (28,7,2)-BIBD. (We note, however, that there do exist (28,7,2)-BIBDs that are not resolvable.) ▮

Finally, we describe a convenient way to parameterize affine resolvable BIBDs. In an affine resolvable BIBD, we write $\mu = k^2/v$. As above, μ must be an integer. Now, the number of blocks in a parallel class is

$$\frac{v}{k} = \frac{k}{\mu},$$

so it must be the case that $k \equiv 0 \bmod \mu$. If we write $n = k/\mu$, then we have

$$v = \frac{k^2}{\mu} = n^2\mu.$$

Now, let us proceed to express λ in terms of n and μ. Since $\lambda(v-1) = r(k-1)$ and $r = k + \lambda$, we have

$$\lambda(v-1) = (k+\lambda)(k-1)$$

and hence

$$\lambda(v-k) = k(k-1).$$

Thus

$$\lambda = \frac{k(k-1)}{v-k} = \frac{n\mu(n\mu-1)}{n^2\mu - n\mu} = \frac{n\mu - 1}{n-1}.$$

Any affine resolvable BIBD must have parameters of the form

$$\left(n^2\mu, n\mu, \frac{n\mu - 1}{n - 1}\right),$$

and, conversely, any resolvable BIBD having parameters of this form is affine resolvable. We will denote such a BIBD as an (n, μ)-ARBIBD. The designs constructed in Theorem 5.19 are $(2, m)$-ARBIBDs, and those obtained from Corollary 5.18 are (q, q^{m-1})-ARBIBDs. For example, we constructed an affine resolvable $(27, 9, 4)$-BIBD in Example 5.13. This is denoted as a $(3, 3)$-ARBIBD.

5.3.1 Symmetric BIBDs from Affine Resolvable BIBDs

In this section, we present a construction of certain symmetric BIBDs from affine resolvable BIBDs. Suppose that there is an affine resolvable (v, b, r, k, λ)-BIBD, say $(X, \mathcal{A})$, having parallel classes $\Pi_1, \ldots, \Pi_r$. Let $X = \{x_i : 1 \leq i \leq v\}$. We define several $v \times v$ matrices, denoted $M_1, \ldots, M_r$, as follows. Let $1 \leq h \leq r$. Then $M_h = (m^h_{i,j})$, where

$$m^h_{i,j} = \begin{cases} 1 & \text{if there exists } A \in \Pi_h \text{ such that } x_i, x_j \in A \\ 0 & \text{otherwise.} \end{cases}$$

Let M_0 be a $v \times v$ matrix of zeroes, and define M to be the following $(r+1)v \times (r+1)v$ matrix:

$$M = \begin{pmatrix} M_0 & M_1 & M_2 & \cdots & M_r \\ M_1 & M_2 & M_3 & \cdots & M_0 \\ M_2 & M_3 & M_4 & \cdots & M_1 \\ \vdots & \vdots & \vdots & & \vdots \\ M_r & M_0 & M_1 & \cdots & M_{r-1} \end{pmatrix}.$$

The matrix M, as described above, can be shown to be the incidence matrix of a symmetric BIBD. Therefore, we have the following result.

Theorem 5.23. *Suppose there exists an affine resolvable* (v, b, r, k, λ)-BIBD. *Then there exists a (symmetric)* $((r+1)v, kr, k\lambda)$-BIBD.

The following corollary is obtained by using affine planes of prime power order in Theorem 5.23.

Corollary 5.24. *Suppose that q is a prime power. Then there exists a (symmetric)* $(q^2(q+2), q(q+1), q)$-BIBD.

Example 5.25. From an affine plane of order 3, we can construct a (symmetric) $(45, 12, 3)$-BIBD. Suppose we start with the $(9, 3, 1)$-BIBD presented in Example 1.4. The four parallel classes are easily seen to be the following:

$$\Pi_1 = \{123, 456, 789\},$$
$$\Pi_2 = \{147, 258, 369\},$$
$$\Pi_3 = \{159, 267, 348\}, \quad \text{and}$$
$$\Pi_4 = \{168, 249, 357\}.$$

The matrices M_1 and M_2 are as follows:

$$M_1 = \begin{pmatrix} 1&1&1&0&0&0&0&0&0 \\ 1&1&1&0&0&0&0&0&0 \\ 1&1&1&0&0&0&0&0&0 \\ 0&0&0&1&1&1&0&0&0 \\ 0&0&0&1&1&1&0&0&0 \\ 0&0&0&1&1&1&0&0&0 \\ 0&0&0&0&0&0&1&1&1 \\ 0&0&0&0&0&0&1&1&1 \\ 0&0&0&0&0&0&1&1&1 \end{pmatrix}$$

and

$$M_2 = \begin{pmatrix} 1&0&0&1&0&0&1&0&0 \\ 0&1&0&0&1&0&0&1&0 \\ 0&0&1&0&0&1&0&0&1 \\ 1&0&0&1&0&0&1&0&0 \\ 0&1&0&0&1&0&0&1&0 \\ 0&0&1&0&0&1&0&0&1 \\ 1&0&0&1&0&0&1&0&0 \\ 0&1&0&0&1&0&0&1&0 \\ 0&0&1&0&0&1&0&0&1 \end{pmatrix}$$

The matrices M_3 and M_4 are constructed in a similar fashion, and then the matrix

$$M = \begin{pmatrix} M_0 & M_1 & M_2 & M_3 & M_4 \\ M_1 & M_2 & M_3 & M_4 & M_0 \\ M_2 & M_3 & M_4 & M_0 & M_1 \\ M_3 & M_4 & M_0 & M_1 & M_2 \\ M_4 & M_0 & M_1 & M_2 & M_3 \end{pmatrix}$$

is the incidence matrix of a (45, 12, 3)-BIBD. ∎

5.4 Orthogonal Resolutions

Suppose $(X, \mathcal{A})$ is a (v, k, λ)-BIBD. Suppose that $\Pi_1, \ldots, \Pi_r$ are the parallel classes in a resolution of $(X, \mathcal{A})$, and suppose that $\Pi'_1, \ldots, \Pi'_r$ are the parallel classes in a second resolution of $(X, \mathcal{A})$. These two resolutions of $(X, \mathcal{A})$ are *orthogonal resolutions* if $|\Pi_j \cap \Pi'_h| \leq 1$ for all $1 \leq j, h \leq r$. (In other words, no two parallel classes, one from each resolution, contain more than one block in common.)

Closely related to the notion of orthogonal resolutions is an object called a "generalized Room square", which we define now.

Definition 5.26. *Suppose that v, k, and λ are integers with $2 \le k < v$ and $\lambda \ge 1$. A* generalized Room square $\mathsf{GRS}(v,k,\lambda)$ *is an r by r array (where $r = \lambda(v-1)/(k-1)$), say R, that satisfies the following properties:*

1. *each cell of R either is empty or contains a k-subset of a set X of v points;*
2. *every point appears in exactly one cell in each row (or column) of R;*
3. *$(X, \mathcal{A})$ is a (v,k,λ)-BIBD, where the set of blocks $\mathcal{A}$ is obtained from the nonempty cells of R.*

Theorem 5.27. *There exists a $\mathsf{GRS}(v,k,\lambda)$ if and only if there exists a (v,k,λ)-BIBD having orthogonal resolutions.*

Proof. It is clear that the nonempty cells in each row (column, respectively) of a $\mathsf{GRS}(v,k,\lambda)$ yield a parallel class of the (v,k,λ)-BIBD. The set of all parallel classes formed from the rows (columns, resp.) of the GRS comprise a resolution of the (v,k,λ)-BIBD. Furthermore, these two resolutions are orthogonal: two parallel classes (one from each resolution) contain one common block if the cell that is the intersection of the corresponding row and column in the GRS is nonempty; and they have no blocks in common otherwise.

Conversely, suppose we have two orthogonal resolutions of a (v,k,λ)-BIBD, say Π_i $(1 \le i \le r)$ and Π'_j $(1 \le j \le r)$, where, as usual, r is the replication number of the BIBD. Construct an r by r array, R, in which $R(i,j) = \Pi_i \cap \Pi'_j$ for $1 \le i,j \le r$. It is easy to see that the array R is a $\mathsf{GRS}(v,k,\lambda)$. □

Example 5.28. We exhibit a $\mathsf{GRS}(8,2,1)$:

∞0			64		32	51
62	∞1			05		43
54	03	∞2			16	
	65	14	∞3			20
31		06	25	∞4		
	42		10	36	∞5	
		53		21	40	∞6

This generalized Room square is equivalent to two orthogonal resolutions of an $(8,2,1)$-BIBD, denoted by Π_i $(0 \le i \le 6)$ and Π'_i $(0 \le i \le 6)$, respectively, which are depicted in Figure 5.1. ∎

It is trivial to show that a $\mathsf{GRS}(4,2,1)$ does not exist. It is also true, but not so easy to prove, that a $\mathsf{GRS}(6,2,1)$ does not exist. Therefore, the $\mathsf{GRS}(8,2,1)$ presented in Example 5.28 is the smallest $\mathsf{GRS}(v,2,1)$ that exists.

$$\begin{array}{ll}
\Pi_0 = \{\infty 0, 64, 32, 51\} & \Pi'_0 = \{\infty 0, 62, 31, 54\} \\
\Pi_1 = \{\infty 1, 05, 43, 62\} & \Pi'_1 = \{\infty 1, 03, 42, 65\} \\
\Pi_2 = \{\infty 2, 16, 54, 03\} & \Pi'_2 = \{\infty 2, 14, 53, 06\} \\
\Pi_3 = \{\infty 3, 20, 65, 14\} & \Pi'_3 = \{\infty 3, 25, 64, 10\} \\
\Pi_4 = \{\infty 4, 31, 06, 25\} & \Pi'_4 = \{\infty 4, 36, 05, 21\} \\
\Pi_5 = \{\infty 5, 42, 10, 36\} & \Pi'_5 = \{\infty 5, 40, 16, 32\} \\
\Pi_6 = \{\infty 6, 53, 21, 40\} & \Pi'_6 = \{\infty 6, 51, 20, 43\}.
\end{array}$$

Fig. 5.1. Orthogonal Resolutions of an (8, 2, 1)-BIBD

We next describe a technique whereby infinite classes of GRS$(v, 2, 1)$ can be constructed. First, we require a definition. Suppose that G is an additive Abelian group of order n, where n is odd. A *strong starter* in G is a set of $(n-1)/2$ unordered pairs $\{\{s_i, t_i\} : 1 \leq i \leq (n-1)/2\}$ such that the following properties are satisfied:

1. $\{s_i, t_i : 1 \leq i \leq (n-1)/2\} = G\backslash\{0\}$;
2. $\{\pm(s_i - t_i) : 1 \leq i \leq (n-1)/2\} = G\backslash\{0\}$;
3. $s_i + t_i \neq 0$ for all i; and $s_i + t_i \neq s_j + t_j$ if $i \neq j$.

As an example, it is easy to verify that $\{\{3, 2\}, \{6, 4\}, \{5, 1\}\}$ is a strong starter in the group $(\mathbb{Z}_7, +)$.

We have the following construction method for GRS$(v, 2, 1)$ using strong starters.

Theorem 5.29. *Suppose that G is an additive Abelian group of order n, where n is odd, and suppose that there exists a strong starter in G. Then there is a* GRS$(n+1, 2, 1)$.

Proof. Let S be a strong starter in G. We construct an n by n array, denoted R, the rows and columns of which are indexed by the elements of G. The points in R will be the elements in $G \cup \{\infty\}$, where $\infty \notin G$. Here is how R is constructed: For all $g \in G$, place the pair $\{\infty, g\}$ in $R(g, g)$; and for all $g \in G$ and all $\{s, t\} \in S$, place the pair $\{s+g, t+g\}$ in $R(g, s+t+g)$.

It is not hard to see that every cell of R contains an unordered pair of points or is empty (this follows from property 3 of a strong starter). It is also easy to see that every unordered pair of points occurs in exactly one cell of R (this follows from property 2 of a strong starter). The fact that row 0 of R contains each point follows from property 1 of a strong starter. From this, it is easy to see that every row of R contains each point.

Thus it remains only to show that each column of R contains each point. Consider column 0. It is not hard to see that the pairs occurring in column 0 are $\{\infty, 0\}$ and $\{-s, -t\}$ for all $\{s, t\} \in S$. Property 1 of a strong starter then can be used to show that every point occurs in column 0 of R. From this, it is easy to see that every column of R contains each point. □

Strong starters in many finite fields can be constructed by the following method due to Mullin and Nemeth.

Theorem 5.30 (Mullin-Nemeth Strong Starters). *Suppose that $q = 2^a b + 1$ is an odd prime power, where a is a positive integer and $b > 1$ is odd. Then there exists a strong starter in $(\mathbb{F}_q, +)$.*

Proof. We use notation introduced in Section 3.6. Let ω be a primitive element of $\mathbb{F}_q$, and define

$$H = \{\omega^{2^a i} : 0 \leq i \leq b-1\}.$$

H is a subgroup of the multiplicative group $(\mathbb{F}_q \backslash \{0\}, \cdot)$ having order b. Denote the cosets of H by $H_0, \ldots, H_{2^a-1}$, where $H_j = \omega^j H, 0 \leq j \leq 2^a - 1$.

Now define

$$H^* = \bigcup_{j=0}^{2^{a-1}-1} H_j$$

and

$$S = \{\{x, \omega^{2^{a-1}} x\} : x \in H^*\}.$$

We will show that S is the desired strong starter.

First, we observe that $\omega^{2^{a-1}} \in H_{2^{a-1}}$, which implies that

$$[1, \omega^{2^{a-1}}] \circ H^* = G \backslash \{0\}.$$

This implies that $\{s_i, t_i\} = G \backslash \{0\}$.

Next, we observe that $-1 = \omega^{2^{a-1}b}$. Using the fact that b is odd, it is easy to show that $2^{a-1}b \bmod 2^a = 2^{a-1}$, and hence $-1 \in H_{2^{a-1}}$. This implies that

$$[1, -1] \circ H^* = G \backslash \{0\}.$$

Also, $\omega^{2^{a-1}} \neq 1$. It follows from these observations that

$$\begin{aligned}\{\pm(s_i - t_i)\} &= [1, -1] \circ [\omega^{2^{a-1}} - 1] \circ H^* \\ &= [\omega^{2^{a-1}} - 1] \circ G \backslash \{0\} \\ &= G \backslash \{0\}.\end{aligned}$$

It is also true that $\omega^{2^{a-1}} \neq -1$ because $b > 1$. Then, in a similar fashion, we see that

$$\begin{aligned}\{\pm(s_i + t_i)\} &= [1, -1] \circ [\omega^{2^{a-1}} + 1] \circ H^* \\ &= [\omega^{2^{a-1}} + 1] \circ G \backslash \{0\} \\ &= G \backslash \{0\}.\end{aligned}$$

This implies that $s_i + t_i \neq 0$ for all i, and $s_i + t_i \neq s_j + t_j$ if $i \neq j$. This completes the proof that S is a strong starter. □

Example 5.31. We construct a strong starter in $(\mathbb{Z}_{13}, +)$ using Theorem 5.30. $13 = 2^2 3 + 1$, so $a = 2$ and $b = 3$. $\omega = 2$ is a primitive element of $\mathbb{Z}_{13}$ and $H = \{3, 9, 1\}$. The cosets of H are

$$\begin{aligned} H_0 &= \{3,9,1\} \\ H_1 &= \{6,5,2\} \\ H_2 &= \{12,10,4\} \\ H_3 &= \{11,7,8\}, \end{aligned}$$

and therefore

$$H^* = \{3,9,1,6,5,2\}.$$

$\omega^{2^{a-1}} = 4$, and the strong starter is

$$\{\{x, 4x\} : x \in H^*\} = \{\{3,12\}, \{9,10\}, \{1,4\}, \{6,11\}, \{5,7\}, \{2,8\}\}.$$

∎

Let us examine the hypotheses of Theorem 5.30. It is clear that any odd prime power q can be written in the form $q = 2^a b + 1$, where a and b are positive integers and b is odd. Theorem 5.30 can be applied unless $b = 1$; i.e., the only bad cases are when $q = 2^a + 1$. It is possible to show that the only prime powers q of the form $2^a + 1$ are the following:

- q is a Fermat prime. (For an integer $n \geq 0$, the mth Fermat number is defined to be $F_m = 2^{2^m} + 1$. If F_m is prime, it is called a Fermat prime. The only known Fermat primes are F_0, F_1, F_2, F_3, and F_4.)
- $q = 9$.

There is a bit more that can be said. By a different construction, it can be shown that there exists a strong starter in $(\mathbb{Z}_n, +)$ whenever n is a Fermat number F_m with $m \geq 2$. It is known that there is no strong starter in the groups $(\mathbb{Z}_3, +)$, $(\mathbb{Z}_5, +)$, $(\mathbb{Z}_9, +)$, or $(\mathbb{Z}_3 \times \mathbb{Z}_3, +)$, and it has been conjectured that these are the only finite Abelian groups of odd order exceeding 1 that do not contain a strong starter.

5.5 Notes and References

Furino, Miao, and Yin [46] is a monograph that is devoted to resolvable BIBDs and related designs.

Bose's Inequality was proven in [14]. Shrikhande [94] is a survey on affine resolvable designs. Theorem 5.23 is due to Wallis [114].

Theorem 5.30 is proven in [82]. A GRS$(v, 2, 1)$ is often called a *Room square*; for a survey of these objects, see Dinitz and Stinson [42].

5.6 Exercises

5.1 Construct an affine plane of order 4 using the finite field $\mathbb{F}_4$, where $\mathbb{F}_4 = \mathbb{Z}_2[x]/(x^2+x+1)$.

5.2 Use a $(21,5,1)$-difference set to construct a projective plane of order 4. Then, construct an affine plane of order 4 from this projective plane, and write out the parallel classes in this affine plane.

5.3 Prove that there does not exist a resolvable $(n, \frac{n}{2}, \frac{n}{2}-1)$-BIBD if $n \equiv 2 \pmod 4$.

5.4 Prove that there exists a resolvable $(n, \frac{n}{2}, \frac{n}{2}-1)$-BIBD if $n \equiv 0 \pmod 4$ and a Hadamard matrix of order n exists.

5.5 Let G be the permutation group of order 7 on the set $X = \{0,\ldots,7\}$ that is generated by the permutation $\alpha = (0\,1\,2\,3\,4\,5\,6)(7)$.
  (a) Show that the two orbits containing the blocks $\{1,2,4,7\}$ and $\{0,1,2,4\}$ yield a $(7,4,3)$-BIBD.
  (b) Prove that this BIBD is not resolvable.

5.6 (a) Suppose there exists a (symmetric) $\left(\frac{n^3\mu-1}{n-1}, \frac{n^2\mu-1}{n-1}, \frac{n\mu-1}{n-1}\right)$-BIBD, say $(X, \mathcal{A})$, where $n > 1$ and $\mu > 1$ are integers. Suppose also that the residual BIBD of $(X, \mathcal{A})$ is affine resolvable. Prove that there exists a (symmetric) $\left(\frac{n^2\mu-1}{n-1}, \frac{n\mu-1}{n-1}, \frac{\mu-1}{n-1}\right)$-BIBD.
  Hint: The derived BIBD of $(X, \mathcal{A})$ is a $\left(\frac{n^2\mu-1}{n-1}, \frac{n\mu-1}{n-1}, \frac{n(\mu-1)}{n-1}\right)$-BIBD in which every block is repeated n times.
  (b) Suppose that there is an (n,μ)-ARBIBD and a $\left(\frac{n^2\mu-1}{n-1}, \frac{n\mu-1}{n-1}, \frac{\mu-1}{n-1}\right)$-BIBD, where $n > 1$ and $\mu > 1$ are integers. Prove that there is a $\left(\frac{n^3\mu-1}{n-1}, \frac{n^2\mu-1}{n-1}, \frac{n\mu-1}{n-1}\right)$-BIBD.

5.7 This exercise provides a proof of Theorem 5.23.
  (a) Prove that $(r-1)\mu = k\lambda$ in an affine resolvable BIBD.
  (b) Prove the following regarding the matrices $M_1, \ldots, M_r$.
    i. $M_i {M_i}^T = kM_i$ for $1 \le i \le r$.
    ii. $M_i {M_j}^T = \mu J_v$ for $1 \le i, j \le r, i \ne j$.
    iii. $M_1 + \cdots + M_r = \lambda J_v + (r-\lambda) I_v$.
  (c) Prove that $MM^T = k\lambda J_{v(r+1)} + k(r-\lambda) I_{v(r+1)}$, and hence M is the incidence matrix of a symmetric BIBD.

5.8 A strong starter $S = \{\{s_i, t_i\} : 1 \le i \le (n-1)/2\}$ in an additive group G of odd order n is *skew* provided that

$$\{\pm(s_i + t_i) : 1 \le i \le (n-1)/2\} = G \backslash \{0\}.$$

Prove that the Mullin-Nemeth strong starters are skew.

5.9 A *starter* in an additive group G of odd order n is a set of $(n-1)/2$ unordered pairs $\{\{s_i, t_i\} : 1 \le i \le (n-1)/2\}$ such that the following properties are satisfied:
  (a) $\{s_i, t_i : 1 \le i \le (n-1)/2\} = G \backslash \{0\}$; and

(b) $\{\pm(s_i - t_i) : 1 \leq i \leq (n-1)/2\} = G\backslash\{0\}$.

Suppose that $S = \{\{s_i, t_i\} : 1 \leq i \leq (n-1)/2\}$ and $U = \{\{u_i, v_i\} : 1 \leq i \leq (n-1)/2\}$ are both starters in G. Without loss of generality, suppose that $s_i - t_i = u_i - v_i$, $1 \leq i \leq (n-1)/2$, and denote $a_i = s_i - u_i$, $1 \leq i \leq (n-1)/2$. We say that S and U are *orthogonal starters* provided that $a_i \neq 0$ for all i; and $a_i \neq a_j$ if $i \neq j$.

(a) Prove that the existence of a strong starter in G implies the existence of orthogonal starters in G.

(b) Prove that orthogonal starters in G can be used to construct a $\mathsf{GRS}(n+1,2,1)$.

(c) Find orthogonal starters in $(\mathbb{Z}_9,+)$ and use them to construct a $\mathsf{GRS}(10,2,1)$.

5.10 Suppose $(X, \mathcal{A})$ is a (v,k,λ)-BIBD. A *near parallel class* in $(X, \mathcal{A})$ is a subset of disjoint blocks from $\mathcal{A}$ whose union is $X\backslash\{x\}$ for some point $x \in X$, which is called the *deficient point* of the near parallel class. A partition of $\mathcal{A}$ into near parallel classes is called a *near resolution*, and $(X, \mathcal{A})$ is said to be a *near resolvable BIBD* if $\mathcal{A}$ has at least one near resolution.

(a) Suppose that $(X, \mathcal{A})$ has a near resolution. Prove that every point $x \in X$ is the deficient point of exactly $r/(v-1)$ near parallel classes.

(b) Then prove that $\lambda = \alpha(k-1)$, where α is a positive integer.

6

Latin Squares

6.1 Latin Squares and Quasigroups

We begin with a definition.

Definition 6.1. *A* Latin square *of* order n *with entries from an n-set X is an $n \times n$ array L in which every cell contains an element of X such that every row of L is a permutation of X and every column of L is a permutation of X.*

It is easy to construct a Latin square of any order $n \geq 1$. For example, we could take the first row to be

1	2	$\cdots$	n

and then shift this row cyclically to the right by $1, 2, \ldots, n-1$ positions to construct the remaining $n-1$ rows.

Example 6.2. A Latin square of order 4.

1	2	3	4
4	1	2	3
3	4	1	2
2	3	4	1

.

∎

Closely related to Latin squares are algebraic objects called quasigroups, which we define now.

Definition 6.3. *Let X be a finite set of cardinality n, and let $\circ$ be a binary operation defined on X (i.e., $\circ : X \times X \to X$). We say that the pair $(X, \circ)$ is a* quasigroup *of* order n *provided that the following two properties are satisfied:*

1. *For every $x, y \in X$, the equation $x \circ z = y$ has a unique solution for $z \in X$.*
2. *For every $x, y \in X$, the equation $z \circ x = y$ has a unique solution for $z \in X$.*

The *operation table* of a binary operation $\circ$ defined on X is the $|X| \times |X|$ array $A = (a_{x,y})$, where $a_{x,y} = x \circ y$. The following simple observation relates quasigroups to Latin squares.

Theorem 6.4. *Suppose $\circ$ is a binary operation defined on a finite set X of cardinality n. Then $(X, \circ)$ is a quasigroup if and only if its operation table is a Latin square of order n.*

It should be clear that the notions of quasigroups and Latin squares provide two different ways of looking at the same thing. We will use both points of view at various times.

We begin by investigating quasigroups (or Latin squares) that satisfy two special properties that we define now.

Definition 6.5. *Suppose $(X, \circ)$ is a quasigroup. We say that $(X, \circ)$ is an* idempotent quasigroup *if $x \circ x = x$ for all $x \in X$, and we say that $(X, \circ)$ is a* symmetric quasigroup *if $x \circ y = y \circ x$ for all $x, y \in X$.*

These concepts can also be defined for Latin squares in the obvious way: A *symmetric Latin square* $L = (\ell_{x,y})$ is one in which $\ell_{x,y} = \ell_{y,x}$ for all x, y, and an *idempotent Latin square* is one in which $\ell_{x,x} = x$ for all x.

Example 6.6. Let $X = \{1, 2\}$. There are exactly two Latin squares defined on X, as follows:

$$\begin{array}{|c|c|}\hline 1 & 2 \\ \hline 2 & 1 \\ \hline \end{array} \quad \begin{array}{|c|c|}\hline 2 & 1 \\ \hline 1 & 2 \\ \hline \end{array}$$

Both of these Latin squares are symmetric, but neither of them is idempotent. ∎

Example 6.7. Let $X = \{1, 2, 3\}$. There are exactly twelve Latin squares defined on X, as follows:

$$L_1 = \begin{array}{|c|c|c|}\hline 1 & 2 & 3 \\ \hline 2 & 3 & 1 \\ \hline 3 & 1 & 2 \\ \hline \end{array} \quad L_2 = \begin{array}{|c|c|c|}\hline 1 & 2 & 3 \\ \hline 3 & 1 & 2 \\ \hline 2 & 3 & 1 \\ \hline \end{array} \quad L_3 = \begin{array}{|c|c|c|}\hline 1 & 3 & 2 \\ \hline 2 & 1 & 3 \\ \hline 3 & 2 & 1 \\ \hline \end{array} \quad L_4 = \begin{array}{|c|c|c|}\hline 1 & 3 & 2 \\ \hline 3 & 2 & 1 \\ \hline 2 & 1 & 3 \\ \hline \end{array}$$

$$L_5 = \begin{array}{|c|c|c|}\hline 2 & 1 & 3 \\ \hline 1 & 3 & 2 \\ \hline 3 & 2 & 1 \\ \hline \end{array} \quad L_6 = \begin{array}{|c|c|c|}\hline 2 & 1 & 3 \\ \hline 3 & 2 & 1 \\ \hline 1 & 2 & 3 \\ \hline \end{array} \quad L_7 = \begin{array}{|c|c|c|}\hline 2 & 3 & 1 \\ \hline 1 & 2 & 3 \\ \hline 3 & 1 & 2 \\ \hline \end{array} \quad L_8 = \begin{array}{|c|c|c|}\hline 2 & 3 & 1 \\ \hline 3 & 1 & 2 \\ \hline 1 & 2 & 3 \\ \hline \end{array}$$

$$L_9 = \begin{array}{|c|c|c|}\hline 3 & 1 & 2 \\ \hline 1 & 2 & 3 \\ \hline 2 & 3 & 1 \\ \hline \end{array} \quad L_{10} = \begin{array}{|c|c|c|}\hline 3 & 1 & 2 \\ \hline 2 & 3 & 1 \\ \hline 1 & 2 & 3 \\ \hline \end{array} \quad L_{11} = \begin{array}{|c|c|c|}\hline 3 & 2 & 1 \\ \hline 1 & 3 & 2 \\ \hline 2 & 1 & 3 \\ \hline \end{array} \quad L_{12} = \begin{array}{|c|c|c|}\hline 3 & 2 & 1 \\ \hline 2 & 1 & 3 \\ \hline 1 & 3 & 2 \\ \hline \end{array}$$

The only idempotent square in the list above is L_4; the squares L_1, L_4, L_5, L_8, L_9, and L_{12} are symmetric. ∎

It is not difficult to construct idempotent Latin squares (or quasigroups) of all orders $n > 2$ and symmetric Latin squares (or quasigroups) of all orders $n \geq 1$. In the rest of this section, we discuss the problem of constructing quasigroups that are both symmetric and idempotent. These will have applications to the construction of $(v,3,1)$-BIBDs; see Section 6.2.

We begin by establishing a simple necessary condition for the existence of a symmetric idempotent quasigroup of order n.

Lemma 6.8. *If there exists a symmetric idempotent quasigroup of order n, then n is odd.*

Proof. Suppose that $\circ : X \times X \to X$ is a symmetric quasigroup. Let $z \in X$, and define $S = \{(x,y) : x \circ y = z\}$. Since $\circ$ is idempotent, it follows that $(x,x) \in S$ if and only if $x = z$. Since $\circ$ is symmetric, it follows that $(x,y) \in S$ if and only if $(y,x) \in S$. Hence, $\{\{x,y\} : x \neq y, x \circ y = z\}$ is a partition of $X \backslash \{z\}$ into sets of size two. Therefore $|X| - 1$ is even, and hence $|X|$ is odd. □

We now construct the desired quasigroups for every odd order. Suppose n is odd, and consider the group $(\mathbb{Z}_n, +)$. Because $(\mathbb{Z}_n, +)$ is a group, it is automatically a quasigroup. It is also symmetric because addition modulo n is commutative.

This quasigroup is not idempotent; however, we will be able to modify it so it is. When n is odd, the list of values on the main diagonal of the operation table of $(\mathbb{Z}_n, +)$ is (in order)

$$(x + x \bmod n : x \in \mathbb{Z}_n) = (0, 2, 4, \ldots, n-1, 1, 3, \ldots, n-3).$$

This is a permutation of $\mathbb{Z}_n$. Therefore the operation table of $(\mathbb{Z}_n, +)$ has all the elements of $\mathbb{Z}_n$ on its main diagonal but not in the correct order. However, we can rectify this by permuting (i.e., relabeling) the symbols so that the diagonal elements are $0, 1, \ldots, n-1$ (in this order). We therefore define a permutation π to be $\pi(0) = 0$, $\pi(2) = 1$, ..., $\pi(n-1) = (n-1)/2$, $\pi(1) = (n+1)/2$, $\pi(3) = (n+3)/2$, ..., $\pi(n-3) = n-1$. In fact, the permutation π can be described by the formula

$$\pi(x) = 2^{-1}x \bmod n = \left(\frac{n+1}{2}\right) x \bmod n$$

since the multiplicative inverse $2^{-1} \bmod n = (n+1)/2$ whenever n is odd. Hence, one binary operation $\circ$, defined on $\{0, \ldots, n-1\}$, that yields a symmetric idempotent quasigroup, is as follows:

$$x \circ y = \left(\frac{n+1}{2}\right)(x+y) \bmod n.$$

Example 6.9. Suppose $n = 5$. The binary operation

$$x \circ y = 3(x + y) \bmod 5$$

defines a symmetric idempotent quasigroup on the set $\{0,1,2,3,4\}$. The corresponding Latin square is as follows:

0	3	1	4	2
3	1	4	2	0
1	4	2	0	3
4	2	0	3	1
2	0	3	1	4

■

The discussion above, together with Lemma 6.8, establishes the following.

Theorem 6.10. *There exists a symmetric idempotent quasigroup of order n if and only if n is odd.*

6.2 Steiner Triple Systems

A *Steiner triple system* of *order* v, or $\mathsf{STS}(v)$, is a $(v,3,1)$-BIBD. Since BIBDs with $k = 2$ are trivial, Steiner triple systems are the simplest type of "interesting" BIBDs. We have already seen examples of Steiner triple systems: an $\mathsf{STS}(7)$ was constructed in Example 1.3 and an $\mathsf{STS}(9)$ was constructed in Example 1.4. Steiner triple systems are, by far, the most-studied type of BIBD. In this section, we will determine necessary and sufficient conditions for existence of an $\mathsf{STS}(v)$.

We begin by deriving necessary conditions for existence of an $\mathsf{STS}(v)$.

Lemma 6.11. *There exists an* $\mathsf{STS}(v)$ *only if* $v \equiv 1,3 \pmod 6$, $v \geq 7$.

Proof. Since $k = 3$ and $\lambda = 1$, we have $r = \lambda(v-1)/(k-1) = (v-1)/2$. Hence $v = 2r+1$; i.e., v is odd. Now we can compute $b = vr/k = v(v-1)/6$. Since b is an integer, it must be the case that $v(v-1) \equiv 0 \pmod 6$. This congruence is satisfied if and only if $v \equiv 0,1,3,4 \pmod 6$. However, since v is odd, we see that $v \equiv 1,3 \pmod 6$. Finally, since $v > k$ in a BIBD, an $\mathsf{STS}(v)$ can exist only if $v \geq 7$. □

In the next two subsections, we will show that these necessary conditions are sufficient by constructing an $\mathsf{STS}(v)$ for every v allowed by Lemma 6.11.

6.2.1 The Bose Construction

We now present a construction, due to Bose, that uses symmetric idempotent quasigroups to construct Steiner triple systems of all orders $v \equiv 3 \pmod 6$. (A modified construction due to Skolem, which we present a bit later, will handle the cases $v \equiv 1 \pmod 6$.)

Let $v = 6t + 3$, $t \geq 1$. Suppose $(X, \circ)$ is a symmetric idempotent quasigroup of (odd) order $2t + 1$, which exists by Theorem 6.1. Let "$<$" be any total ordering defined on X. Define $Y = X \times \mathbb{Z}_3$. ($Y$ will be the set of points in the STS(v) that we construct.) For every $x \in X$, define a block

$$A_x = \{(x, 0), (x, 1), (x, 2)\}.$$

Then for every $x, y \in X$, $x < y$, and for every $i \in \mathbb{Z}_3$, define a block

$$B_{x,y,i} = \{(x, i), (y, i), (x \circ y, (i + 1) \bmod 3)\}.$$

Then define

$$\mathcal{B} = \{A_x : x \in X\} \cup \{B_{x,y,i} : x, y \in X, x < y, i \in \mathbb{Z}_3\}.$$

In Figure 6.1, we show pictorially how three blocks are constructed from one entry of $(X, \circ)$, say $x \circ y = z$.

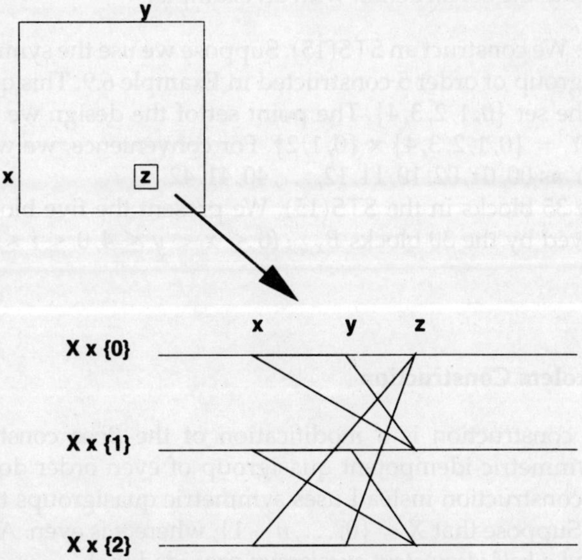

Fig. 6.1. The Bose Construction

We will show that $(Y, \mathcal{B})$ is an STS(v). Clearly there are v points in Y, and every block in $\mathcal{B}$ contains three points. Hence, it suffices to show that every pair of points occurs in exactly one block.

Consider the pair of points (α, j), (β, k). If $\alpha = \beta$, then $j \neq k$, and this pair occurs in the block A_α and in no other block. Hence we can assume that $\alpha \neq \beta$. Without loss of generality, suppose that $\alpha < \beta$.

We consider three cases:

1. If $k = j$, then this pair occurs in the block $B_{\alpha,\beta,j}$ and in no other block.
2. If $k = (j+1) \bmod 3$, then the equation $x \circ \alpha = \beta$ has a unique solution $x = \gamma$. Note that $\gamma \neq \alpha$ since $\alpha \neq \beta$ and $\circ$ is idempotent. If $\gamma < \alpha$, then the pair (α, j), (β, k) occurs in the block $B_{\gamma,\alpha,j}$ and in no other block. If $\gamma > \alpha$, then, since $\circ$ is symmetric, the pair (α, j), (β, k) occurs in the block $B_{\alpha,\gamma,j}$ and in no other block.
3. If $j = (k+1) \bmod 3$, then the equation $x \circ \beta = \alpha$ has a unique solution $x = \gamma$. Note that $\gamma \neq \beta$ since $\alpha \neq \beta$ and $\circ$ is idempotent. If $\gamma < \beta$, then the pair (α, j), (β, k) occurs in the block $B_{\gamma,\beta,k}$ and in no other block. If $\gamma > \beta$, then, since $\circ$ is symmetric, the pair (α, j), (β, k) occurs in the block $B_{\beta,\gamma,k}$ and in no other block.

The discussion above, together with Theorem 6.1, establishes the following existence result.

Theorem 6.12. *There exists an* $\mathsf{STS}(v)$ *for all* $v \equiv 3 \pmod 6$, $v \geq 9$.

We illustrate the construction with an example.

Example 6.13. We construct an $\mathsf{STS}(15)$. Suppose we use the symmetric idempotent quasigroup of order 5 constructed in Example 6.9. This quasigroup is defined on the set $\{0,1,2,3,4\}$. The point set of the design we are going to construct is $Y = \{0,1,2,3,4\} \times \{0,1,2\}$. For convenience, we will write the elements of Y as $00, 01, 02, 10, 11, 12, \ldots, 40, 41, 42$.

There are 35 blocks in the $\mathsf{STS}(15)$. We present the five blocks A_x $(0 \leq x \leq 4)$ followed by the 30 blocks $B_{x,y,i}$ $(0 \leq x < y \leq 4, 0 \leq i \leq 2)$ in Figure 6.2. ∎

6.2.2 The Skolem Construction

The Skolem construction is a modification of the Bose construction. Recall that a symmetric idempotent quasigroup of even order does not exist. The Skolem construction instead uses symmetric quasigroups that are half-idempotent. Suppose that $X = \{0, \ldots, n-1\}$, where n is even. A quasigroup $(X, \circ)$ is called a *half-idempotent quasigroup* provided that

$$x \circ x = \begin{cases} x & \text{if } 0 \leq x < \frac{n}{2} \\ x - \frac{n}{2} & \text{if } \frac{n}{2} \leq x < n. \end{cases}$$

In other words, when we look down the diagonal of the operation table, we see the entries

{00,01,02}	{10,11,12}	{20,21,22}
{30,31,32}	{40,41,42}	
{00,10,31}	{01,11,32}	{02,12,30}
{00,20,11}	{01,21,12}	{02,22,10}
{00,30,41}	{01,31,42}	{02,32,40}
{00,40,21}	{01,41,22}	{02,42,20}
{10,20,41}	{11,21,42}	{12,22,40}
{10,30,21}	{11,31,22}	{12,32,20}
{10,40,01}	{11,41,02}	{12,42,00}
{20,30,01}	{21,31,02}	{22,32,00}
{20,40,31}	{21,41,32}	{22,42,30}
{30,40,11}	{31,41,12}	{32,42,10}

Fig. 6.2. The 35 Blocks of an STS(15)

$$0, 1, \ldots, \frac{n}{2} - 1, 0, 1, \ldots, \frac{n}{2} - 1$$

in that order.

We will construct a symmetric half-idempotent quasigroup for every even order n. Consider the group $(\mathbb{Z}_n, +)$. As was the case for n odd, $(\mathbb{Z}_n, +)$ is a symmetric quasigroup. We will be able to construct a symmetric half-idempotent quasigroup by a simple modification of $(\mathbb{Z}_n, +)$.

It is not hard to see that the list of values

$$(x + x \bmod n : x \in \mathbb{Z}_n)$$

contains every even residue in $\mathbb{Z}_n$ exactly twice when n is even. In fact, the main diagonal of the operation table of $(\mathbb{Z}_n, +)$ is (in order)

$$0, 2, \ldots, n-2, 0, 2, \ldots, n-2.$$

Hence, it is sufficient to relabel the elements of $\mathbb{Z}_n$ in such a way that the main diagonal of the operation table becomes (in order)

$$0, 1, \ldots, \frac{n}{2} - 1, 0, 1, \ldots, \frac{n}{2} - 1.$$

A permutation π that accomplishes this is as follows:

$$\pi(x) = \begin{cases} \frac{x}{2} & \text{if } x \text{ is even} \\ \frac{x+n-1}{2} & \text{if } x \text{ is odd.} \end{cases}$$

Therefore, the quasigroup operation can be defined to be

$$x \circ y = \pi((x + y) \bmod n).$$

Example 6.14. Suppose $n = 6$. The permutation π is defined as $\pi(0) = 0$, $\pi(1) = 3$, $\pi(2) = 1$, $\pi(3) = 4$, $\pi(4) = 2$, and $\pi(5) = 5$. The resulting symmetric half-idempotent quasigroup has the following operation table:

0	3	1	4	2	5
3	1	4	2	5	0
1	4	2	5	0	3
4	2	5	0	3	1
2	5	0	3	1	4
5	0	3	1	4	2

∎

The discussion above establishes the following theorem.

Theorem 6.15. *There exists a symmetric half-idempotent quasigroup of order n if and only if n is even.*

Now we proceed to the Skolem construction. Let $v = 6t + 1$, $t \geq 1$. Suppose $(\{0,\ldots,2t-1\}, \circ)$ is a symmetric half-idempotent quasigroup of (even) order $2t$. Define $Y = (\{0,\ldots,2t-1\} \times \mathbb{Z}_3) \cup \{\infty\}$. ($Y$ will be the set of points in the STS(v) that we construct.) For $0 \leq x \leq t-1$, define a block

$$A_x = \{(x,0),(x,1),(x,2)\}.$$

Then for every $x, y \in \{0,\ldots,2t-1\}$, $x < y$, and for every $i \in \mathbb{Z}_3$, define a block

$$B_{x,y,i} = \{(x,i),(y,i),(x \circ y,(i+1) \bmod 3)\}.$$

Finally, for $0 \leq x \leq t-1$ and for every $i \in \mathbb{Z}_3$, define a block

$$C_{x,i} = \{\infty,(x+t,i),(x,(i+1) \bmod 3)\}.$$

Then define the set of blocks to be

$$\begin{aligned}\mathcal{B} = &\{A_x : 0 \leq x \leq t-1\}\\ &\bigcup \{B_{x,y,i} : x,y \in \mathbb{Z}_{2t}, x < y, i \in \mathbb{Z}_3\}\\ &\bigcup \{C_{x,i} : 0 \leq x \leq t-1, i \in \mathbb{Z}_3\}.\end{aligned}$$

We will show that $(Y, \mathcal{B})$ is an STS(v). Clearly there are v points in Y, and every block in $\mathcal{B}$ contains three points. Hence, it suffices to show that every pair of points occurs in exactly one block.

First, consider a pair of points (α, j), ∞. If $\alpha \leq t-1$, then this pair occurs in the block $C_{\alpha,(j-1) \bmod 3}$ and in no other block. If $\alpha \geq t$, then this pair occurs in the block $C_{\alpha-t,j}$ and in no other block.

Next, consider the pair of points (α, j), (β, k). If $\alpha = \beta \leq t-1$, this pair occurs in the block A_α and in no other block. Suppose $\alpha = \beta \geq t$. Then $j \neq k$, so without loss of generality we have $k = (j+1) \bmod 3$. The equation $\alpha \circ x = \alpha$ has a unique solution $x = \gamma$. If $\gamma > \alpha$, then this pair occurs in the block $B_{\alpha,\gamma,j}$ and in no other block. If $\gamma < \alpha$, then, since $\circ$ is symmetric, the pair (α, j), (β, k) occurs in the block $B_{\gamma,\alpha,j}$ and in no other block.

Hence we can proceed to the case where $\alpha \neq \beta$. Without loss of generality, suppose that $\alpha < \beta$.

We consider three cases:

1. If $k = j$, then this pair occurs in the block $B_{\alpha,\beta,j}$ and in no other block.
2. If $k = (j+1) \bmod 3$, then the equation $x \circ \alpha = \beta$ has a unique solution $x = \gamma$. Note that $\gamma \neq \alpha$ since $\alpha < \beta$ and $\alpha \circ \alpha \leq \alpha$ for any α. If $\gamma < \alpha$, then the pair $(\alpha, j), (\beta, k)$ occurs in the block $B_{\gamma,\alpha,j}$ and in no other block. If $\gamma > \alpha$, then, since $\circ$ is symmetric, the pair $(\alpha, j), (\beta, k)$ occurs in the block $B_{\alpha,\gamma,j}$ and in no other block.
3. If $j = (k+1) \bmod 3$, then the equation $x \circ \beta = \alpha$ has a unique solution $x = \gamma$. We have $\gamma = \beta$ if and only if $\beta = \alpha + t$. If this happens, then the pair $(\alpha, j), (\beta, k)$ occurs in the block $C_{\alpha,k}$ and in no other block. If $\gamma < \beta$, then the pair $(\alpha, j), (\beta, k)$ occurs in the block $B_{\gamma,\beta,k}$ and in no other block. If $\gamma > \beta$, then, since $\circ$ is symmetric, the pair $(\alpha, j), (\beta, k)$ occurs in the block $B_{\beta,\gamma,k}$ and in no other block.

Thus we have proved the following theorem.

Theorem 6.16. *There exists an* $\mathsf{STS}(v)$ *for all* $v \equiv 1 \bmod 6$, $v \geq 7$.

Finally, combining Lemma 6.11 with Theorems 6.12 and 6.16, we obtain our main result.

Theorem 6.17. *There exists an* $\mathsf{STS}(v)$ *if and only if* $v \equiv 1, 3 \bmod 6$, $v \geq 7$.

We illustrate the Skolem construction with an example.

Example 6.18. We construct an $\mathsf{STS}(19)$. Suppose we use the symmetric half-idempotent quasigroup of order 6 constructed in Example 6.14. This quasigroup is defined on the set $\{0,1,2,3,4,5\}$. The point set of the design is $Y = (\{0,1,2,3,4,5\} \times \{0,1,2\}) \cup \{\infty\}$. We will write the elements of Y as $00, 01, 02, 10, 11, 12, \ldots, 50, 51, 52, \infty$.

There are 57 blocks in the $\mathsf{STS}(19)$. We present the three blocks A_x $(0 \leq x \leq 2)$ followed by the 45 blocks $B_{x,y,i}$ $(0 \leq x < y \leq 5, 0 \leq i \leq 2)$ and the nine blocks $C_{x,i}$ $(0 \leq x \leq 2, 0 \leq i \leq 2)$ in Figure 6.3. ∎

6.3 Orthogonal Latin Squares

Definition 6.19. *Suppose that L_1 is a Latin square of order n with entries from X and L_2 is a Latin square of order n with entries from Y. We say that L_1 and L_2 are* orthogonal Latin squares *provided that, for every $x \in X$ and for every $y \in Y$, there is a unique cell (i, j) such that $L_1(i, j) = x$ and $L_2(i, j) = y$.*

An equivalent way to define orthogonality of Latin squares is to consider the *superposition* of L_1 and L_2 in which every cell (i, j) is filled in with the ordered pair $(L_1(i, j), L_2(i, j))$. Then L_1 and L_2 are orthogonal if and only if their superposition contains every ordered pair in $X \times Y$.

In general, it is not easy to construct orthogonal Latin squares. To begin, we exhibit a few examples for small orders.

{00, 01, 02}	{10, 11, 12}	{20, 21, 22}
{00, 10, 31}	{01, 11, 32}	{02, 12, 30}
{00, 20, 11}	{01, 21, 12}	{02, 22, 10}
{00, 30, 41}	{01, 31, 42}	{02, 32, 40}
{00, 40, 21}	{01, 41, 22}	{02, 42, 20}
{00, 50, 51}	{01, 51, 52}	{02, 52, 50}
{10, 20, 41}	{11, 21, 42}	{12, 22, 40}
{10, 30, 21}	{11, 31, 22}	{12, 32, 20}
{10, 40, 51}	{11, 41, 52}	{12, 42, 50}
{10, 50, 01}	{11, 51, 02}	{12, 52, 00}
{20, 30, 51}	{21, 31, 52}	{22, 32, 50}
{20, 40, 01}	{21, 41, 02}	{22, 42, 00}
{20, 50, 31}	{21, 51, 32}	{22, 52, 30}
{30, 40, 31}	{31, 41, 32}	{32, 42, 30}
{30, 50, 11}	{31, 51, 12}	{32, 52, 10}
{40, 50, 41}	{41, 51, 42}	{42, 52, 40}
{∞, 30, 01}	{∞, 31, 02}	{∞, 32, 00}
{∞, 40, 11}	{∞, 41, 12}	{∞, 42, 10}
{∞, 50, 21}	{∞, 51, 22}	{∞, 52, 20}

Fig. 6.3. The 57 Blocks of an STS(19)

Example 6.20. Orthogonal Latin squares of order 3.

$$L_1 = \begin{array}{|c|c|c|}\hline 1&2&3\\\hline 2&3&1\\\hline 3&1&2\\\hline\end{array}, \quad L_2 = \begin{array}{|c|c|c|}\hline 1&2&3\\\hline 3&1&2\\\hline 2&3&1\\\hline\end{array}.$$

The superposition of L_1 and L_2 is as follows:

$$\begin{array}{|c|c|c|}\hline (1,1)&(2,2)&(3,3)\\\hline (2,3)&(3,1)&(1,2)\\\hline (3,2)&(1,3)&(2,1)\\\hline\end{array}.$$

It is easy to verify that all nine ordered pairs $(i, j) \in \{1,2,3\} \times \{1,2,3\}$ occur in the superposition of L_1 and L_2. ∎

It is not hard to verify that there is no Latin square that is orthogonal to the square given in Example 6.2. However, orthogonal Latin squares of order 4 do exist, as shown in the next example.

Example 6.21. Orthogonal Latin squares of order 4.

$$L_1 = \begin{array}{|c|c|c|c|}\hline 1&3&4&2\\\hline 4&2&1&3\\\hline 2&4&3&1\\\hline 3&1&2&4\\\hline\end{array}, \quad L_2 = \begin{array}{|c|c|c|c|}\hline 1&4&2&3\\\hline 3&2&4&1\\\hline 4&1&3&2\\\hline 2&3&1&4\\\hline\end{array}.$$

∎

Here is one more example of orthogonal Latin squares.

Example 6.22. Orthogonal Latin squares of order 8.

$$L_1 = \begin{array}{|c|c|c|c|c|c|c|c|}\hline 1&2&3&4&5&6&7&8\\\hline 2&1&4&3&6&5&8&7\\\hline 3&4&1&2&7&8&5&6\\\hline 4&3&2&1&8&7&6&5\\\hline 5&6&7&8&1&2&3&4\\\hline 6&5&8&7&2&1&4&3\\\hline 7&8&5&6&3&4&1&2\\\hline 8&7&6&5&4&3&2&1\\\hline \end{array}\,, \quad L_2 = \begin{array}{|c|c|c|c|c|c|c|c|}\hline 1&2&3&4&5&6&7&8\\\hline 3&4&1&2&7&8&5&6\\\hline 5&6&7&8&1&2&3&4\\\hline 7&8&5&6&3&4&1&2\\\hline 6&5&8&7&2&1&4&3\\\hline 8&7&6&5&4&3&2&1\\\hline 2&1&4&3&6&5&8&7\\\hline 4&3&2&1&8&7&6&5\\\hline \end{array}\,.$$

∎

Orthogonal Latin squares of order 1 exist, but they are not very interesting. It is not difficult to see that there do not exist orthogonal Latin squares of order 2. Over 200 years ago, the mathematician Euler conjectured that there do not exist orthogonal Latin squares of order n if $n \equiv 2 \pmod 4$. Euler's conjecture was proved true for order 6 by Tarry in 1900, essentially by means of an exhaustive search. (It was not until the mid-1980s, however, that a short theoretical proof of this result was found.) On the other hand, for all $n > 2$, $n \neq 6$, there exist orthogonal Latin squares of order n. This disproof of Euler's conjecture was published in the late 1950s by Bose, Shrikhande, and Parker, and it was reported on the front page of the *New York Times*. We will give a simplified proof of this result in Section 6.8.

We will look at several construction methods for orthogonal Latin squares. First, we give a construction that works for all odd $n > 1$.

Theorem 6.23. *If $n > 1$ is odd, then there exist orthogonal Latin squares of order n.*

Proof. We define two Latin squares of order n with entries from $\mathbb{Z}_n$:

$$L_1(i,j) = (i+j) \bmod n$$
$$L_2(i,j) = (i-j) \bmod n.$$

L_1 and L_2 are easily seen to be Latin squares for any positive integer n. Let's prove that they are orthogonal when n is odd. Suppose that $(x,y) \in \mathbb{Z}_n \times \mathbb{Z}_n$. We want to find a unique cell (i,j) such that $L_1(i,j) = x$ and $L_2(i,j) = y$. In other words, we want to solve the system

$$i + j \equiv x \pmod n$$
$$i - j \equiv y \pmod n$$

for i and j. Since n is odd, 2 has a multiplicative inverse modulo n, and the system has the unique solution

$$i = (x + y)2^{-1} \bmod n$$
$$j = (x - y)2^{-1} \bmod n.$$

Hence, L_1 and L_2 are orthogonal. □

Example 6.24. We construct orthogonal Latin squares of order 5 using Theorem 6.23:

$L_1 =$

0	1	2	3	4
1	2	3	4	0
2	3	4	0	1
3	4	0	1	2
4	0	1	2	3

, $L_2 =$

0	4	3	2	1
1	0	4	3	2
2	1	0	4	3
3	2	1	0	4
4	3	2	1	0

.

■

Suppose that L and M are Latin squares of order m and n (respectively) defined on symbol sets X and Y (respectively). We define the *direct product* of L and M, denoted $L \times M$, to be the $mn \times mn$ array defined as follows:

$$(L \times M)((i_1, i_2), (j_1, j_2)) = (L(i_1, j_1), M(i_2, j_2)).$$

Note that $L \times M$ is one Latin square; it is not the superposition of two Latin squares.

Lemma 6.25. *If L and M are Latin squares of order m and n (respectively) defined on symbol sets X and Y (respectively), then $L \times M$ is a Latin square of order mn defined on symbol set $X \times Y$.*

Proof. Consider a row of $L \times M$, say row (i_1, i_2). Let $x \in X$ and let $y \in Y$. We will show how to find the symbol (x, y) in row (i_1, i_2) of $L \times M$. Since L is a Latin square, there is a unique column j_1 such that $L(i_1, j_1) = x$. Since M is a Latin square, there is a unique column j_2 such that $M(i_2, j_2) = y$. Then $(L \times M)((i_1, i_2), (j_1, j_2)) = (x, y)$.

Similarly, every column of $L \times M$ contains every symbol in $X \times Y$, so $L \times M$ is a Latin square. □

Example 6.26. An example to illustrate the direct product. Suppose L and M are as follows:

$L =$

3	1	2
2	3	1
1	2	3

, $M =$

1	2
2	1

.

Then $L \times M$ is as follows:

(3,1)	(1,1)	(2,1)	(3,2)	(1,2)	(2,2)
(2,1)	(3,1)	(1,1)	(2,2)	(3,2)	(1,2)
(1,1)	(2,1)	(3,1)	(1,2)	(2,2)	(3,2)
(3,2)	(1,2)	(2,2)	(3,1)	(1,1)	(2,1)
(2,2)	(3,2)	(1,2)	(2,1)	(3,1)	(1,1)
(1,2)	(2,2)	(3,2)	(1,1)	(2,1)	(3,1)

The direct product $L \times M$ contains many copies of L and M within it. The Latin square $L \times M$ can be partitioned into m^2 disjoint $n \times n$ subarrays, each of which is a copy of M on the symbol set $\{x\} \times Y$, where $x \in X$. $L \times M$ can also be partitioned into n^2 disjoint $m \times m$ subarrays, each of which is a copy of L on the symbol set $X \times \{y\}$, where $y \in Y$.

We next prove that the direct product construction preserves orthogonality.

Theorem 6.27 (Direct Product). *If there exist orthogonal Latin squares of orders n_1 and n_2, then there exist orthogonal Latin squares of order $n_1 n_2$.*

Proof. Suppose that L_1 and L_2 are orthogonal Latin squares of order n_1 on symbol set X, and M_1 and M_2 are orthogonal Latin squares of order n_2 on symbol set Y. We will show that $L_1 \times M_1$ and $L_2 \times M_2$ are orthogonal Latin squares of order $n_1 n_2$. $L_1 \times M_1$ and $L_2 \times M_2$ are both Latin squares by Lemma 6.25, so we just have to prove that they are orthogonal.

Consider an ordered pair of symbols, $((x_1, y_1), (x_2, y_2))$. We want to find a unique cell $((i_1, i_2), (j_1, j_2))$ such that

$$(L_1 \times M_1)((i_1, i_2), (j_1, j_2)) = (x_1, y_1), \quad \text{and}$$
$$(L_2 \times M_2)((i_1, i_2), (j_1, j_2)) = (x_2, y_2).$$

This is equivalent to

$$\begin{aligned} L_1(i_1, j_1) &= x_1, \\ M_1(i_2, j_2) &= y_1, \\ L_2(i_1, j_1) &= x_2, \quad \text{and} \\ M_2(i_2, j_2) &= y_2. \end{aligned}$$

The first and third equations determine (i_1, j_1) uniquely because L_1 and L_2 are orthogonal; and the second and fourth equations determine (i_2, j_2) uniquely because M_1 and M_2 are orthogonal. The desired cell, $((i_1, i_2), (j_1, j_2))$, is therefore determined uniquely. □

Examples 6.21 and 6.22, together with Theorems 6.23 and 6.27, are sufficient to prove the following result.

Theorem 6.28. *There exist orthogonal Latin squares of order n if $n \not\equiv 2 \pmod 4$.*

Proof. If n is odd, then apply Theorem 6.23. Next suppose $n \geq 4$ is a power of two, say $n = 2^i$, where $i \geq 2$. The cases $i = 2$ and $i = 3$ were done in Examples 6.21 and 6.22. For $i \geq 4$, we can construct orthogonal Latin squares of order 2^i, by induction on i, applying Theorem 6.27 with $n_1 = 4$ and $n_2 = 2^{i-2}$.

Finally, suppose that n is even, $n \not\equiv 2 \pmod 4$, and n is not a power of two. Then we can write $n = 2^i n'$, where $i \geq 2$ and $n' > 1$ is odd. Apply Theorem 6.27 with $n_1 = 2^i$ and $n_2 = n'$. Since we have already constructed orthogonal Latin squares of orders 2^i and n', the result follows. □

6.4 Mutually Orthogonal Latin Squares

A set of s Latin squares of order n, say $L_1, \ldots, L_s$, are said to be *mutually orthogonal Latin squares* if L_i and L_j are orthogonal for all $1 \leq i < j \leq s$. We will abbreviate the term "mutually orthogonal Latin squares" to "*MOLS*". A set of s MOLS of order n will be denoted s $\mathsf{MOLS}(n)$.

One fundamental problem is to determine the maximum number of MOLS of order n. This quantity is denoted $N(n)$. Since any two Latin squares of order 1 are orthogonal, we say that $N(1) = \infty$. For all $n > 1$, however, it is possible to prove a finite upper bound on $N(n)$.

Theorem 6.29. *There do not exist n $\mathsf{MOLS}(n)$ if $n > 1$ (i.e., $N(n) \leq n - 1$ for $n > 1$).*

Proof. Suppose that $L_1, \ldots, L_s$ are mutually orthogonal Latin squares of order $n > 1$. Without loss of generality, we can assume that $L_1, \ldots, L_s$ are all defined on symbol set $\{1, \ldots, n\}$. Furthermore, we can assume that the first row of each of these squares is

1	2	$\cdots$	n

(This is justified by observing that within any L_i we can relabel the symbols so the first row is as specified. The relabeling does not affect the orthogonality of the squares.)

Now consider the s values $L_1(2,1), \ldots, L_s(2,1)$ (this is where we require the assumption $n \geq 2$). We first note that these s values are all distinct, as follows: Suppose that $L_i(2,1) = L_j(2,1) = x$, say. Then we have the ordered pair (x, x) occurring in the superposition of L_i and L_j in cell $(1, x)$ and again in cell $(2, 1)$. This contradicts the orthogonality of L_i and L_j.

Next we observe that $L_i(2,1) \neq 1$ for $1 \leq i \leq s$. This follows from the fact that $L_i(1,1) = 1$ and no symbol can occur in two cells in any column of a Latin square.

Combining our two observations, we see that $L_1(2,1), \ldots, L_s(2,1)$ are in fact s distinct elements from the set $\{2, \ldots, n\}$. Hence, $s \leq n - 1$. □

6.4.1 MOLS and Affine Planes

The cases where $N(n) = n - 1$ are particularly interesting because they correspond to affine planes. First, we show how to construct $n - 1$ $\mathsf{MOLS}(n)$ from an affine plane of order n. Suppose that $(X, \mathcal{A})$ is an affine plane of order n (i.e., an $(n^2, n, 1)$-BIBD). Recall from Theorem 5.9 that an affine plane is resolvable. Each of the $n + 1$ parallel classes contains n disjoint blocks, and Theorem 5.21 says that any two blocks from different parallel classes intersect in exactly one common point. Suppose for $1 \leq i \leq n + 1$ that the blocks in Π_i (the ith parallel class) are named $A_{i,j}, 1 \leq j \leq n$. We are going to construct $n - 1$ mutually orthogonal Latin squares of order n, which we name $L_1, \ldots, L_{n-1}$. These Latin squares are constructed using the following formula:

$$L_x(i, j) = k \text{ if and only if } A_{n,i} \cap A_{n+1,j} \in A_{x,k}$$

for $1 \le x \le n-1, 1 \le i \le n, 1 \le j \le n$.

Let us begin by showing that each L_x is a Latin square. First, given a symbol k and a row i, we want to find a unique column j such that $L_x(i, j) = k$. There is a unique point $y \in A_{n,i} \cap A_{x,k}$ because any two blocks in Π_n and Π_x intersect in a unique point. Then, there is a unique j such that $y \in A_{n+1,j}$ because Π_{n+1} is a parallel class. Hence $L_x(i, j) = k$.

Next, given a symbol k and a column j, we want to find a unique row i such that $L_x(i, j) = k$. There is a unique point $y \in A_{n+1,j} \cap A_{x,k}$ because any two blocks in Π_{n+1} and Π_x intersect in a unique point. Then, there is a unique i such that $y \in A_{n,i}$ because Π_n is a parallel class. Hence $L_x(i, j) = k$.

Now we show that L_x and L_y are orthogonal if $x \neq y$. Let k and ℓ be two symbols. We want to find a unique cell (i, j) such that

$$L_x(i, j) = k \quad \text{and}$$
$$L_y(i, j) = \ell.$$

This is equivalent to saying that

$$A_{n,i} \cap A_{n+1,j} \in A_{x,k} \quad \text{and}$$
$$A_{n,i} \cap A_{n+1,j} \in A_{y,\ell}.$$

There is a unique point $z \in A_{x,k} \cap A_{y,\ell}$ because any two blocks in Π_x and Π_y intersect in a unique point. Now, there is a unique i such that $z \in A_{n,i}$ because Π_n is a parallel class. Similarly, there is a unique j such that $z \in A_{n+1,j}$ because Π_{n+1} is a parallel class. Thus we have found the desired cell (i, j), and we have proved that L_x and L_y are orthogonal if $x \neq y$.

Example 6.30. We begin with the affine plane of order 3 constructed in Example 1.4:

$$X = \{1, 2, 3, 4, 5, 6, 7, 8, 9\}, \quad \text{and}$$
$$\mathcal{A} = \{123, 456, 789, 147, 258, 369, 159, 267, 348, 168, 249, 357\}.$$

Suppose we name the blocks as follows:

$$\begin{array}{llll} A_{1,1} = \{1,2,3\} & A_{2,1} = \{1,4,7\} & A_{3,1} = \{1,5,9\} & A_{4,1} = \{1,6,8\} \\ A_{1,2} = \{4,5,6\} & A_{2,2} = \{2,5,8\} & A_{3,2} = \{2,6,7\} & A_{4,2} = \{2,4,9\} \\ A_{1,3} = \{7,8,9\} & A_{2,3} = \{3,6,9\} & A_{3,3} = \{3,4,8\} & A_{4,3} = \{3,5,7\}. \end{array}$$

Then the Latin squares L_1 and L_2 are

$$L_1 = \begin{array}{|c|c|c|} \hline 1 & 3 & 2 \\ \hline 2 & 1 & 3 \\ \hline 3 & 2 & 1 \\ \hline \end{array}, \quad L_2 = \begin{array}{|c|c|c|} \hline 1 & 3 & 2 \\ \hline 3 & 2 & 1 \\ \hline 2 & 1 & 3 \\ \hline \end{array}.$$

■

The construction can in fact be reversed. Suppose we begin with $n-1$ MOLS(n), defined on symbol set $\{1,\ldots,n\}$, say $L_1,\ldots,L_{n-1}$. We will construct an affine plane having point set $X=\{1,\ldots,n\}\times\{1,\ldots,n\}$. The blocks are constructed as follows. For $1\leq x\leq n-1, 1\leq k\leq n$, define

$$A_{x,k}=\{(i,j):L_x(i,j)=k\}.$$

For $1\leq k\leq n$, define

$$A_{n,k}=\{(k,j):1\leq j\leq n\},$$

and for $1\leq k\leq n$, define

$$A_{n+1,k}=\{(i,k):1\leq i\leq n\}.$$

Finally, let

$$\mathcal{A}=\{A_{x,k}:1\leq x\leq n+1, 1\leq k\leq n\}.$$

We will show that $(X,\mathcal{A})$ is an affine plane of order n. Clearly $|X|=n^2$, and it is also not hard to see that every block contains n points. It remains to show that every pair of points occurs in a unique block. Consider a pair $(i_1,j_1),(i_2,j_2)$. If $i_1=i_2$, then this pair occurs in the block A_{n,i_1} and in no other block. If $j_1=j_2$, then this pair occurs in the block A_{n+1,j_1} and in no other block. Hence, we can assume that $i_1\neq i_2$ and $j_1\neq j_2$. We will show that any such pair occurs in at most one block in the design. Since the number of blocks is n^2+n, it then follows that each such pair occurs in exactly one block.

Suppose that $\{(i_1,j_1),(i_2,j_2)\}\subseteq A_{x_1,k_1}$ and $\{(i_1,j_1),(i_2,j_2)\}\subseteq A_{x_2,k_2}$, where $(x_1,k_1)\neq(x_2,k_2)$. Then we have

$$\begin{aligned}L_{x_1}(i_1,j_1)&=k_1,\\L_{x_1}(i_2,j_2)&=k_1,\\L_{x_2}(i_1,j_1)&=k_2,\quad\text{and}\\L_{x_2}(i_2,j_2)&=k_2.\end{aligned}$$

If $x_1=x_2$, then $k_1=k_2$, so we conclude that $x_1\neq x_2$. But then the two squares L_{x_1} and L_{x_2} are not orthogonal because the superposition contains the ordered pair (k_1,k_2) in cell (i_1,j_1) and again in cell (i_2,j_2). This contradiction completes the proof that $(X,\mathcal{A})$ is an affine plane of order n.

Example 6.31. Suppose we begin with the orthogonal Latin squares of order 3 from Example 6.30:

$$L_1=\begin{array}{|c|c|c|}\hline 1&3&2\\\hline 2&1&3\\\hline 3&2&1\\\hline\end{array},\ L_2=\begin{array}{|c|c|c|}\hline 1&3&2\\\hline 3&2&1\\\hline 2&1&3\\\hline\end{array}.$$

The blocks of the affine plane constructed from these orthogonal Latin squares are as follows:

$$\begin{array}{ll} A_{1,1} = \{(1,1),(2,2),(3,3)\} & A_{2,1} = \{(1,1),(2,3),(3,2)\} \\ A_{1,2} = \{(1,3),(2,1),(3,2)\} & A_{2,2} = \{(1,3),(2,2),(3,1)\} \\ A_{1,3} = \{(1,2),(2,3),(3,1)\} & A_{2,3} = \{(1,2),(2,1),(3,3)\} \\ & \\ A_{3,1} = \{(1,1),(1,2),(1,3)\} & A_{4,1} = \{(1,1),(2,1),(3,1)\} \\ A_{3,2} = \{(2,1),(2,2),(2,3)\} & A_{4,2} = \{(1,2),(2,2),(3,2)\} \\ A_{3,3} = \{(3,1),(3,2),(3,3)\} & A_{4,3} = \{(1,3),(2,3),(3,3)\}. \end{array}$$

■

The preceding discussion establishes that an affine plane of order $n \geq 2$ is equivalent to $n-1$ MOLS(n). We know from Theorem 5.10 that an affine plane of order n exists if and only if a projective plane of order n exists. Therefore we have the following result.

Theorem 6.32. *Let $n \geq 2$. Then the existence of any one of the following designs implies the existence of the other two designs:*

1. *$n-1$ MOLS(n);*
2. *an affine plane of order n;*
3. *a projective plane of order n.*

6.4.2 MacNeish's Theorem

The direct product construction (Theorem 6.27) can be generalized to sets of s MOLS in an obvious way. Further, it is possible to form the direct product of more than two Latin squares, again in an obvious manner. Orthogonality is preserved, and the following theorem results.

Theorem 6.33. *If there exist s MOLS(n_i), $1 \leq i \leq \ell$, then there exist s MOLS(n), where $n = n_1 \times n_2 \times \cdots \times n_\ell$.*

It is possible to construct many interesting examples of sets of MOLS by using Theorem 6.32 in conjunction with the direct product. The following theorem, known as MacNeish's Theorem, makes use of the fact that an affine plane of order q exists for every prime power q.

Theorem 6.34 (MacNeish's Theorem). *Suppose that n has prime power factorization $n = {p_1}^{e_1} \cdots {p_\ell}^{e_\ell}$, where the p_i's are distinct primes and $e_i \geq 1$ for $1 \leq i \leq \ell$. Let*

$$s = \min\{{p_i}^{e_i} - 1 : 1 \leq i \leq \ell\}.$$

Then there exist s MOLS(n).

Proof. For $1 \leq i \leq \ell$, there exists an affine plane of order ${p_i}^{e_i}$. Hence, there exist ${p_i}^{e_i} - 1$ MOLS$({p_i}^{e_i})$, for $1 \leq i \leq \ell$, by Theorem 6.32. Therefore there exist s MOLS$({p_i}^{e_i})$ for $1 \leq i \leq \ell$. Apply Theorem 6.33 to obtain the desired result. □

There are many corollaries of Theorem 6.34 that can be proven. Here is a specific result that we will use later.

Corollary 6.35. *If* $n \equiv 1, 5, 7,$ *or* $11 \pmod{12}$, *then there exist four* $\mathsf{MOLS}(n)$. *If* $n \equiv 4$ *or* $8 \pmod{12}$, *then there exist three* $\mathsf{MOLS}(n)$.

Proof. Suppose that n has prime power factorization $n = {p_1}^{e_1} \cdots {p_\ell}^{e_\ell}$. By Theorem 6.34, three $\mathsf{MOLS}(n)$ will exist if ${p_i}^{e_i} \geq 4$ for $1 \leq i \leq \ell$. The only situations in which ${p_i}^{e_i} < 4$ are when $(p_i, e_i) = (2,1)$ or $(3,1)$. In other words, if the prime power factorization of n does not contain the specific terms 2^1 or 3^1, then three $\mathsf{MOLS}(n)$ exist. By a similar argument, if the prime power factorization of n does not contain the specific terms 2^1, 2^2, or 3^1, then four $\mathsf{MOLS}(n)$ exist.

Now, if $n \equiv 1, 5, 7,$ or $11 \pmod{12}$, then $\gcd(n, 6) = 1$, so there are no terms involving 2 or 3 in the factorization of n. It follows that four MOLS of these orders exist.

If $n \equiv 4$ or $8 \pmod{12}$, then $n \equiv 0 \pmod 4$ and $n \not\equiv 0 \pmod 3$. Therefore there is no term involving 3 in the factorization of n, and the term involving 2 has an exponent that is at least 2. Therefore three MOLS of these orders exist. □

6.5 Orthogonal Arrays

In this section, we discuss an equivalent formulation of MOLS called an orthogonal array.

Definition 6.36. *Let* $k \geq 2$ *and* $n \geq 1$ *be integers. An* orthogonal array $\mathsf{OA}(k, n)$ *is an* $n^2 \times k$ *array,* A*, with entries from a set* X *of cardinality* n *such that, within any two columns of* A*, every ordered pair of symbols from* X *occurs in exactly one row of* A*.*

Note that an $\mathsf{OA}(2, n)$ exists trivially for all integers $n \geq 1$.

6.5.1 Orthogonal Arrays and MOLS

It is not difficult to construct an $\mathsf{OA}(s+2, n)$ from s $\mathsf{MOLS}(n)$. This is done as follows. Suppose without loss of generality that these s Latin squares are named $L_1, \ldots, L_s$, are defined on symbol set $\{1, \ldots, n\}$, and have rows and columns labeled $\{1, \ldots, n\}$. For every $i, j \in \{1, \ldots, n\}$, construct an $(s+2)$-tuple

$$(i, j, L_1(i,j), \ldots, L_s(i,j)).$$

Then form an array A whose rows consist of these n^2 $(s+2)$-tuples. We will show that A is an $\mathsf{OA}(s+2, n)$.

We need to show that every ordered pair of symbols occurs in any two columns a and b, where $1 \leq a < b \leq s+2$. We consider several cases:

1. If $a = 1$ and $b = 2$, then clearly we get every ordered pair.
2. If $a = 1$ and $b \geq 3$, then we get every ordered pair because every row of L_b is a permutation of $\{1,\ldots,n\}$.
3. If $a = 2$ and $b \geq 3$, then we get every ordered pair because every column of L_b is a permutation of $\{1,\ldots,n\}$.
4. If $a \geq 3$, then we get every ordered pair because L_a and L_b are orthogonal.

Example 6.37. An OA(4,3) constructed from the orthogonal Latin squares of order 3 presented in Example 6.20.

1	1	1	1
1	2	2	2
1	3	3	3
2	1	2	3
2	2	3	1
2	3	1	2
3	1	3	2
3	2	1	3
3	3	2	1

■

The construction can easily be reversed; if A is an OA(k,n) with $k \geq 3$, then we can construct $k-2$ MOLS(n) from it. Suppose without loss of generality that A is defined on symbol set $\{1,\ldots,n\}$. Label the columns of A by the integers $1,\ldots,k$, and label the rows of A by the integers $1,\ldots,n^2$. We construct $k-2$ MOLS(n), which we name $L_1,\ldots,L_{k-2}$, as follows: For $1 \leq h \leq k-2$ and $1 \leq r \leq n^2$, define

$$L_h(A(r,1),A(r,2)) = A(r,h+2).$$

We will show that $L_1,\ldots,L_{k-2}$ are orthogonal Latin squares of order n.

We begin by showing that each L_h is a Latin square. First, every cell of L_h contains one and only one entry because every ordered pair occurs exactly once in columns 1 and 2 of A. Next, let us show that each row i of each L_h is a permutation of $\{1,\ldots,n\}$. The entries in row i of L_h are in fact the symbols in the set

$$\{A(r,h+2) : A(r,1) = i\}.$$

These symbols are all distinct because every ordered pair occurs exactly in columns 1 and $h+2$ of A. A similar argument proves that each column i of each L_h is a permutation of $\{1,\ldots,n\}$. Hence the L_h's are all Latin squares.

It remains to prove orthogonality. But L_h and L_g are orthogonal because every ordered pair occurs exactly once in columns $h+2$ and $g+2$ of A.

As an example, if we begin with the OA(4,3) from Example 6.37, and apply this construction, then we recover the orthogonal Latin squares of order 3 from Example 6.20.

The discussion above proves the following theorem.

Theorem 6.38. *Suppose that $n \geq 1$ and $k \geq 3$ are integers. Then $k-2$ $\mathsf{MOLS}(n)$ exist if and only if an $\mathsf{OA}(k,n)$ exists.*

6.5.2 Some Constructions for Orthogonal Arrays

Because orthogonal arrays are equivalent to MOLS, any construction for MOLS can be expressed as a construction of orthogonal arrays, and vice versa. Presenting constructions for orthogonal arrays is sometimes more convenient, however. We consider some constructions in this section.

Suppose that n is a prime power. Then there is an affine plane of order n (Theorem 5.4), and hence there are $n-1$ $\mathsf{MOLS}(n)$ (Theorem 6.32). Finally, Theorem 6.38 tells us that there is an $\mathsf{OA}(n+1,n)$. This is a bit of a circuitous route, so we now give a direct construction for orthogonal arrays having a prime power number of symbols.

Theorem 6.39. *Suppose q is a prime power and $2 \leq k \leq q$. Then there exists an $\mathsf{OA}(k,q)$.*

Proof. Let $a_1,\ldots,a_k$ be k distinct elements in $\mathbb{F}_q$. Define two vectors in $(\mathbb{F}_q)^k$ as follows:

$$\mathbf{v}_1 = (1,\ldots,1) \quad \text{and}$$
$$\mathbf{v}_2 = (a_1,\ldots,a_k).$$

Now, define an array A, having rows indexed by $\mathbb{F}_q \times \mathbb{F}_q$, where row (i,j) is the k-tuple $i\mathbf{v}_1 + j\mathbf{v}_2$.

We prove that A is an $\mathsf{OA}(k,q)$ (the proof is very similar to the proof of Theorem 5.4). Let $1 \leq c < d \leq k$, and let $x,y \in \mathbb{F}_q$. We want to find the unique row (i,j) of A such that $A((i,j),c) = x$ and $A((i,j),d) = y$. This gives us the following system of two equations in $\mathbb{F}_q$ in the two unknowns i and j:

$$i + ja_c = x,$$
$$i + ja_d = y.$$

Subtracting the second equation from the first, we obtain

$$j(a_c - a_d) = x - y.$$

Since $a_c - a_d \neq 0$, there exists a multiplicative inverse $(a_c - a_d)^{-1} \in \mathbb{F}_q$. Then we have the following:

$$j = (a_c - a_d)^{-1}(x-y).$$

Back-substituting, we can solve for i:

$$i = x - ja_c = x - a_c(a_c - a_d)^{-1}(x-y).$$

Hence, A is an $\mathsf{OA}(k,q)$. □

We can "extend" an $\mathsf{OA}(q,q)$ constructed using the theorem above by adjoining an additional column in such a way that an $\mathsf{OA}(q+1,q)$ is obtained.

Theorem 6.40. *Suppose q is a prime power. Then there exists an* $\mathsf{OA}(q+1,q)$.

Proof. Construct an $\mathsf{OA}(q,q)$ as described in Theorem 6.39. Then adjoin one more column, column $q+1$, in which $A((i,j),q+1) = j$ for all i,j. The resulting array is an $\mathsf{OA}(q+1,q)$. □

We next give a construction for an $\mathsf{OA}(4,n)$ for all $n \equiv 10 \pmod{12}$. Such an integer n can be written in the form $n = 3m+1$, where $m \equiv 3 \pmod 4$. Define $X = \mathbb{Z}_{2m+1} \cup \Omega$, where $\Omega = \{\infty_i : 1 \le i \le m\}$. Begin with the following $4m+1$ four-tuples:

$$\begin{array}{ll}(0,0,0,0), & \\ (0,2i,i,\infty_i), & 1 \le i \le m, \\ (0,2i-1,\infty_i,m+i), & 1 \le i \le m, \\ (0,\infty_i,2m+1-i,i), & 1 \le i \le m, \quad \text{and} \\ (\infty_i,0,i,2m+1-i), & 1 \le i \le m.\end{array}$$

Next, develop each of these $4m+1$ four-tuples through the group $\mathbb{Z}_{2m+1}$ using the convention that $\infty_i + j = \infty_i$ for all $j \in \mathbb{Z}_{2m+1}$ and all i, $1 \le i \le m$. Call the resulting set of $(4m+1)(2m+1)$ four-tuples A_1.

Now let A_2 be an $\mathsf{OA}(4,m)$ on the symbol set Ω. (Note that m is odd, so there exist orthogonal Latin squares of order m from Theorem 6.23. Therefore an $\mathsf{OA}(4,m)$ exists from Theorem 6.38.) A_2 contains m^2 four-tuples.

The $(4m+1)(2m+1)+m^2 = (3m+1)^2$ four-tuples in $A_1 \cup A_2$ form an $\mathsf{OA}(4,3m+1)$. This orthogonal array has the following permutation α as an automorphism:

$$\alpha = (0\,1\,2\,\cdots\,2m)(\infty_1)\cdots(\infty_m).$$

The $(3m+1)^2$ four-tuples in this $\mathsf{OA}(4,3m+1)$ are comprised of $4m+1$ orbits each consisting of $2m+1$ rows and m^2 orbits each consisting of one row. In order to verify that we have constructed an $\mathsf{OA}(4,3m+1)$, we need to show for each choice of two columns that every orbit of ordered pairs is contained in exactly one of the orbits of four-tuples, within the specified columns. It is not hard to show that there are exactly m^2+4m+1 orbits of ordered pairs with respect to the group $G = \{\alpha^i : 0 \le i \le 2m\}$. The orbits of ordered pairs consist of $4m+1$ orbits of size $2m+1$ and m^2 orbits of size 1. Orbit representatives are as follows:

$$\begin{array}{ll}(0,i), & 0 \le i \le 2m, \\ (0,\infty_i), & 1 \le i \le m, \\ (\infty_i,0), & 1 \le i \le m, \quad \text{and} \\ (\infty_i,\infty_j), & 1 \le i,j \le m.\end{array}$$

With this information, it is straightforward to verify that we have an $\mathsf{OA}(4,3m+1)$. Therefore we have the following result.

$L_1 =$

0	∞_1	1	∞_2	2	∞_3	3	6	5	4
4	1	∞_1	2	∞_2	3	∞_3	0	6	5
∞_3	5	2	∞_1	3	∞_2	4	1	0	6
5	∞_3	6	3	∞_1	4	∞_2	2	1	0
∞_2	6	∞_3	0	4	∞_1	5	3	2	1
6	∞_2	0	∞_3	1	5	∞_1	4	3	2
∞_1	0	∞_2	1	∞_3	2	6	5	4	3
1	2	3	4	5	6	0	∞_1	∞_2	∞_3
2	3	4	5	6	0	1	∞_2	∞_3	∞_1
3	4	5	6	0	1	2	∞_3	∞_1	∞_2

$L_2 =$

0	4	∞_1	5	∞_2	6	∞_3	1	2	3
∞_3	1	5	∞_1	6	∞_2	0	2	3	4
1	∞_3	2	6	∞_1	0	∞_2	3	4	5
∞_2	2	∞_3	3	0	∞_1	1	4	5	6
2	∞_2	3	∞_3	4	1	∞_1	5	6	0
∞_1	3	∞_2	4	∞_3	5	2	6	0	1
3	∞_1	4	∞_2	5	∞_3	6	0	1	2
6	0	1	2	3	4	5	∞_1	∞_2	∞_3
5	6	0	1	2	3	4	∞_3	∞_1	∞_2
4	5	6	0	1	2	3	∞_2	∞_3	∞_1

Fig. 6.4. Orthogonal Latin Squares of Order 10

Theorem 6.41. *For all positive integers* $n \equiv 10 \pmod{12}$, *there exists an* $\mathsf{OA}(4, n)$, *and hence there exist orthogonal Latin squares of order* n *for all such* n.

We illustrate this construction by exhibiting orthogonal Latin squares of order 10 in Figure 6.4. These are obtained from an $\mathsf{OA}(4, 10)$ constructed using the technique described above.

6.6 Transversal Designs

Another type of new design equivalent to sets of MOLS is called a transversal design. We define these objects now.

Definition 6.42. *Let* $k \geq 2$ *and* $n \geq 1$. *A* transversal design $\mathsf{TD}(k, n)$ *is a triple* $(X, \mathcal{G}, \mathcal{B})$ *such that the following properties are satisfied:*

1. X *is a set of* kn *elements called* points,
2. $\mathcal{G}$ *is a partition of* X *into* k *subsets of size* n *called* groups,
3. $\mathcal{B}$ *is a set of* k*-subsets of* X *called* blocks,
4. *any group and any block contain exactly one common point, and*
5. *every pair of points from distinct groups is contained in exactly one block.*

Note that the "groups" in a transversal design are just subsets of points; they are not algebraic groups. Also, a $\mathsf{TD}(2,n)$ exists trivially for all integers $n \geq 1$.

We first show how to construct a $\mathsf{TD}(k,n)$ from an $\mathsf{OA}(k,n)$. Let A be an $\mathsf{OA}(k,n)$ on symbol set $\{1,\ldots,n\}$. Label the columns of A as $1,\ldots,k$, and label the rows of A as $1,\ldots,n^2$. Define

$$X = \{1,\ldots,n\} \times \{1,\ldots,k\}.$$

For $1 \leq i \leq k$, define

$$G_i = \{1,\ldots,n\} \times \{i\},$$

and then define

$$\mathcal{G} = \{G_i : 1 \leq i \leq k\}.$$

For $1 \leq r \leq n^2$, define

$$B_r = \{(A(r,i),i) : 1 \leq i \leq k\},$$

and define

$$\mathcal{B} = \{B_r : 1 \leq r \leq n^2\}.$$

Then it is essentially trivial to prove that $(X,\mathcal{G},\mathcal{B})$ is a $\mathsf{TD}(k,n)$.

Example 6.43. Given the $\mathsf{OA}(4,3)$ constructed in Example 6.37, we obtain a $\mathsf{TD}(4,3)$. The blocks of this transversal design are shown in Figure 6.5. ∎

$$\begin{aligned}
B_1 &= \{(1,1),(1,2),(1,3),(1,4)\}\\
B_2 &= \{(1,1),(2,2),(2,3),(2,4)\}\\
B_3 &= \{(1,1),(3,2),(3,3),(3,4)\}\\
B_4 &= \{(2,1),(1,2),(2,3),(3,4)\}\\
B_5 &= \{(2,1),(2,2),(3,3),(1,4)\}\\
B_6 &= \{(2,1),(3,2),(1,3),(2,4)\}\\
B_7 &= \{(3,1),(1,2),(3,3),(2,4)\}\\
B_8 &= \{(3,1),(2,2),(1,3),(3,4)\}\\
B_9 &= \{(3,1),(3,2),(2,3),(1,4)\}.
\end{aligned}$$

Fig. 6.5. The Blocks of a $\mathsf{TD}(4,3)$

The construction can be reversed: given a $\mathsf{TD}(k,n)$, we can use it to construct an $\mathsf{OA}(k,n)$. Suppose $(X,\mathcal{G},\mathcal{B})$ is a $\mathsf{TD}(k,n)$. By relabeling the points if necessary, we can assume that $X = \{1,\ldots,n\} \times \{1,\ldots,k\}$ and $\mathcal{G} = \{G_i : 1 \leq i \leq k\}$, where $G_i = \{1,\ldots,n\} \times \{i\}$ for $1 \leq i \leq k$. For each block $B \in \mathcal{B}$ and for $1 \leq i \leq k$, let $(b_i,i) \in B \cap G_i$ (recall that each block intersects each group in a unique point). Then, for each $B \in \mathcal{B}$, form the k-tuple

$$(b_1, \ldots, b_k).$$

Construct an array A whose rows consist of all these k-tuples; it is easy to show that A is an $\mathsf{OA}(k, n)$.

As an example, if we begin with the $\mathsf{TD}(4,3)$ from Example 6.43 and apply this construction, then we recover the $\mathsf{OA}(4,3)$ that we started with.

Gathering together the results of this section and Theorem 6.38, we have the following.

Theorem 6.44. *Suppose that $n \geq 2$ and $k \geq 3$. Then the existence of any one of the following designs implies the existence of the other two designs:*

1. $k - 2$ $\mathsf{MOLS}(n)$,
2. *an* $\mathsf{OA}(k, n)$,
3. *a* $\mathsf{TD}(k, n)$.

6.7 Wilson's Construction

In this section, we describe a powerful recursive construction for MOLS due to Wilson. It is in fact a generalization of the direct product construction for MOLS that we presented in Section 6.1. Wilson's construction is most easily presented in terms of transversal designs. We will get to it shortly, but first we recast the direct product construction in the language of transversal designs.

Let $(X, \mathcal{G}, \mathcal{A})$ be a $\mathsf{TD}(k, t)$, where $G_1, \ldots, G_k$ are the groups. Define

$$Y = X \times \{1, \ldots, m\},$$

and, for $1 \leq i \leq k$, define

$$H_i = G_i \times \{1, \ldots, m\}.$$

Let $\mathcal{H} = \{H_i : 1 \leq i \leq k\}$. Y and $\mathcal{H}$ will be the points and groups (respectively) of the $\mathsf{TD}(k, mt)$ that we are constructing.

We now define the blocks of this transversal design. For every block $A \in \mathcal{A}$, construct a set of m^2 blocks as follows. For $1 \leq i \leq k$, let $\{a_i\} = A \cap G_i$. Then let $\mathcal{B}_A$ be the set of m^2 blocks of a $\mathsf{TD}(k, m)$ in which the groups are

$$\{a_i\} \times \{1, \ldots, m\},$$

$1 \leq i \leq k$. Then define

$$\mathcal{B} = \bigcup_{A \in \mathcal{A}} \mathcal{B}_A.$$

We claim that $(Y, \mathcal{H}, \mathcal{B})$ is a transversal design. The main task is to show that any two points x and y from different groups occur in a unique block. Suppose that $x = (g, a)$ and $y = (h, b)$, where $g \in G_i$, $h \in G_j$, $i \neq j$, and $a, b \in \{1, \ldots, m\}$. There is a unique block $A \in \mathcal{A}$ such that $g, h \in A$ because g

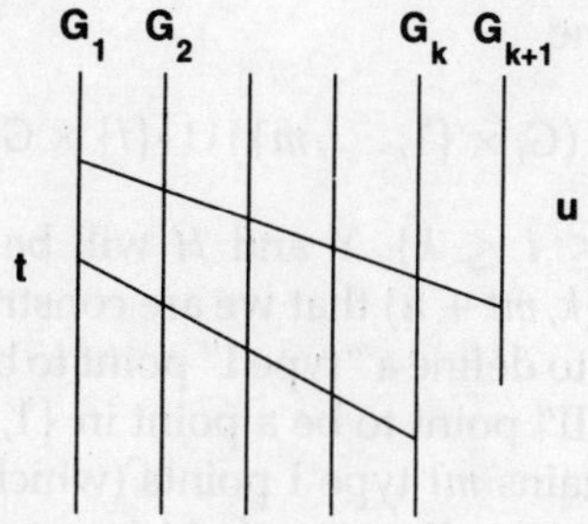

Fig. 6.6. A Truncated Transversal Design

and h occur in different groups in $\mathcal{G}$. Then it is easily seen that x and y occur in a unique block in $\mathcal{B}_A$ and in no other block in $\mathcal{B}$.

As we mentioned, the construction above is exactly the same as the direct product construction for MOLS. We will proceed next to a description of Wilson's construction. Wilson's construction uses a type of design called a truncated transversal design, which is formed from a transversal design by deleting some points from one of the groups. More specifically, let $(X, \mathcal{G}, \mathcal{B})$ be a $\mathsf{TD}(k+1, t)$, where $k \geq 2$. Pick a group $G \in \mathcal{G}$, and suppose that $1 \leq u \leq t$. Let $G' \subseteq G$, $|G'| = u$. Then define

$$
\begin{aligned}
Y &= (X \backslash G) \cup G' \\
\mathcal{H} &= (\mathcal{G} \backslash \{G\}) \cup \{G'\} \\
\mathcal{C} &= \{B \in \mathcal{B} : B \cap G' \neq \emptyset\} \cup \{B \backslash \{x\} : B \in \mathcal{B}, B \cap G = \{x\}, x \in G \backslash G'\}.
\end{aligned}
$$

The set system $(Y, \mathcal{H}, \mathcal{C})$ is a *truncated transversal design*. If $u < t$, then this design has $kt + u$ points, k groups of size t and one group of size u, $t(t-u)$ blocks of size k, and tu blocks of size $k+1$. (If $t = u$, then the design is just a $\mathsf{TD}(k+1, t)$ because we have deleted no points.)

We now present the statement and proof of Wilson's construction for MOLS.

Theorem 6.45 (Wilson's Construction for MOLS). *Let $k \geq 2$ and suppose that the following transversal designs exist: a $\mathsf{TD}(k, m)$, a $\mathsf{TD}(k, m+1)$, a $\mathsf{TD}(k+1, t)$, and a $\mathsf{TD}(k, u)$, where $1 \leq u \leq t$. Then there exists a $\mathsf{TD}(k, mt + u)$.*

Proof. First construct a truncated transversal design from a $\mathsf{TD}(k+1, t)$ by deleting $t - u$ points from some group, as described above. Let $(X, \mathcal{G}, \mathcal{A})$ be the resulting truncated transversal design, where $G_1, \ldots, G_k$ are k groups of size t and G_{k+1} is a group of size u.

In Figure 6.6, the groups of this truncated transversal design are drawn vertically, and two representative blocks are indicated.

Define

$$
Y = ((X \backslash G_{k+1}) \times \{1, \ldots, m\}) \cup (\{1, \ldots, k\} \times G_{k+1}).
$$

Then, for $1 \leq i \leq k$, define

$$H_i = (G_i \times \{1, \ldots, m\}) \cup (\{i\} \times G_{k+1}),$$

and let $\mathcal{H} = \{H_i : 1 \leq i \leq k\}$. Y and $\mathcal{H}$ will be the points and groups (respectively) of the $\mathsf{TD}(k, mt + u)$ that we are constructing.

It will be convenient to define a "type I" point to be a point in $(X \backslash G_{k+1}) \times \{1, \ldots, m\}$, and a "type II" point to be a point in $\{1, \ldots, k\} \times G_{k+1}$. Observe that each group H_i contains mt type I points (which consist of m copies of each point in G_i) and u type II points (which consist of one copy of each point in G_{k+1}).

We now define the blocks of this transversal design. For every block $A \in \mathcal{A}$, construct a set of blocks $\mathcal{B}_A$ according to the following recipe:

1. Suppose $|A| = k$. For $1 \leq i \leq k$, let $\{a_i\} = A \cap G_i$. Then let $\mathcal{B}_A$ be the set of m^2 blocks of a $\mathsf{TD}(k, m)$ in which the groups are

   $$\{a_i\} \times \{1, \ldots, m\}$$

   for $1 \leq i \leq k$.
   Observe that the blocks in $\mathcal{B}_A$ contain only type I points.
2. Suppose $|A| = k + 1$. For $1 \leq i \leq k + 1$, let $\{a_i\} = A \cap G_i$. There exists a $\mathsf{TD}(k, m + 1)$ in which the groups are

   $$(\{a_i\} \times \{1, \ldots, m\}) \cup \{(i, a_{k+1})\},$$

   for $1 \leq i \leq k$, and in which

   $$\{(1, a_{k+1}), \ldots, (k, a_{k+1})\}$$

   is a block. Delete this block, and let $\mathcal{B}_A$ be the set of $(m + 1)^2 - 1$ blocks that remain.
   In Figure 6.7, we show how two representative blocks in the truncated transversal design are "expanded into" transversal designs.
   Observe that each group of the $\mathsf{TD}(k, m + 1)$ consists of m type I points and one type II point. However, no block in $\mathcal{B}_A$ contains more than one type II point; this is because we deleted the block $\{(1, a_{k+1}), \ldots, (k, a_{k+1})\}$, which was the only block in the transversal design that contained more than one type II point.

Finally, there exists a $\mathsf{TD}(k, u)$ in which the groups are

$$\{i\} \times G_{k+1}$$

for $1 \leq i \leq k$. Let $\mathcal{B}^*$ denote the blocks of this transversal design. (Observe that the blocks in $\mathcal{B}^*$ contain only type II points.)

The block set of the $\mathsf{TD}(k, mt + u)$ is defined to be

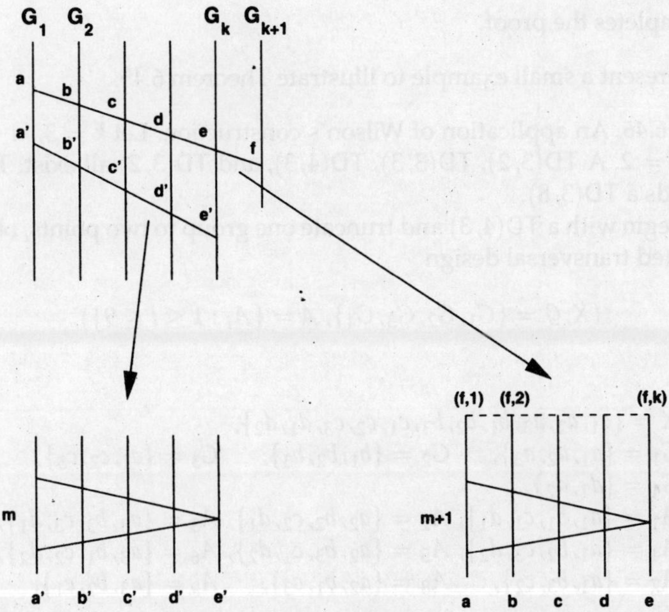

Fig. 6.7. Wilson's Construction (Detail)

$$\mathcal{B} = \left(\bigcup_{A \in \mathcal{A}} \mathcal{B}_A \right) \cup \mathcal{B}^*.$$

Let us sketch a proof that $(Y, \mathcal{H}, \mathcal{B})$ is a transversal design. The main task is to show that any two points x and y from different groups of $\mathcal{H}$ occur in a unique block. There are three different cases to consider according to the types of the two points x and y.

1. Suppose x and y are both of type I. Let $x = (g, a)$ and $y = (h, b)$, where $g \in G_i$, $h \in G_j$, $i \neq j$, and $a, b \in \{1, \dots, m\}$. There is a unique block $A \in \mathcal{A}$ such that $g, h \in A$. Then x and y occur in a unique block in $\mathcal{B}_A$ and in no other block in $\mathcal{B}$.
2. Suppose x is of type I and y is of type II. Let $x = (g, a)$ and $y = (j, h)$, where $g \in G_i$, $h \in G_{k+1}$, $a \in \{1, \dots, m\}$, and $j \in \{1, \dots, k\} \backslash \{i\}$. There is a unique block $A \in \mathcal{A}$ such that $g, h \in A$, and it must be the case that $|A| = k + 1$. x and y occur in a unique block in $\mathcal{B}_A$ and in no other block in $\mathcal{B}$.
3. Suppose x and y are both of type II. Let $x = (i, g)$ and $y = (j, h)$, where $g, h \in G_{k+1}$, $i, j \in \{1, \dots, k\}$, and $i \neq j$. Then x and y occur in a unique block in $\mathcal{B}^*$ and in no other block in $\mathcal{B}$ (note that we observed earlier that the blocks in $\mathcal{B}^*$ are the only ones that contain more than one type II point).

This completes the proof. □

We present a small example to illustrate Theorem 6.45.

Example 6.46. An application of Wilson's construction. Let $k = 3$, $m = 2$, $t = 3$, and $u = 2$. A $\mathsf{TD}(3,2)$, $\mathsf{TD}(3,3)$, $\mathsf{TD}(4,3)$, and $\mathsf{TD}(3,2)$ all exist. Theorem 6.45 yields a $\mathsf{TD}(3,8)$.

We begin with a $\mathsf{TD}(4,3)$ and truncate one group to two points, obtaining a truncated transversal design

$$(X, \mathcal{G} = \{G_1, G_2, G_3, G_4\}, \mathcal{A} = \{A_i : 1 \leq i \leq 9\}),$$

where

$$\begin{array}{lll}
X = \{a_1, a_2, a_3, b_1, b_2, b_3, c_1, c_2, c_3, d_1, d_2\}, & & \\
G_1 = \{a_1, a_2, a_3\}, & G_2 = \{b_1, b_2, b_3\}, & G_3 = \{c_1, c_2, c_3\}, \\
G_4 = \{d_1, d_2\}, & & \\
A_1 = \{a_1, b_1, c_1, d_1\}, & A_2 = \{a_2, b_2, c_2, d_1\}, & A_3 = \{a_3, b_3, c_3, d_1\}, \\
A_4 = \{a_1, b_2, c_3, d_2\}, & A_5 = \{a_2, b_3, c_1, d_2\}, & A_6 = \{a_3, b_1, c_2, d_2\}, \\
A_7 = \{a_1, b_3, c_2\}, & A_8 = \{a_2, b_1, c_3\}, & A_9 = \{a_3, b_2, c_1\}.
\end{array}$$

The groups of the $\mathsf{TD}(3,8)$ are

$$\begin{aligned}
H_1 &= \{(a_1,1),(a_1,2),(a_2,1),(a_2,2),(a_3,1),(a_3,2),(1,d_1),(1,d_2)\} \\
H_2 &= \{(b_1,1),(b_1,2),(b_2,1),(b_2,2),(b_3,1),(b_3,2),(2,d_1),(2,d_2)\} \\
H_3 &= \{(c_1,1),(c_1,2),(c_2,1),(c_2,2),(c_3,1),(c_3,2),(3,d_1),(3,d_2)\}.
\end{aligned}$$

Each block in $\mathcal{A}$ gives rise to a certain set of blocks in the $\mathsf{TD}(3,8)$. For example, consider $A_1 = \{a_1, b_1, c_1, d_1\}$. Since $|A_1| = 4 = k+1$, the blocks $\mathcal{B}_{A_1}$ are obtained from the blocks of a $\mathsf{TD}(3,3)$ having groups

$$\{(a_1,1),(a_1,2),(1,d_1)\}, \{(b_1,1),(b_1,2),(2,d_1)\}, \text{ and } \{(c_1,1),(c_1,2),(3,d_1)\}.$$

We construct such a transversal design, making sure that

$$\{(1,d_1),(2,d_1),(3,d_1)\}$$

is one of the blocks. Then this block is deleted and the remaining eight blocks comprise $\mathcal{B}_{A_1}$. For example, we could take the following eight blocks:

$$\begin{array}{lll}
\{(a_1,1),(b_1,1),(c_1,1)\} & \{(a_1,2),(b_1,2),(c_1,2)\} & \\
\{(a_1,1),(b_1,2),(3,d_1)\} & \{(a_1,2),(2,d_1),(c_1,1)\} & \{(1,d_1),(b_1,1),(c_1,2)\} \\
\{(a_1,1),(2,d_1),(c_1,2)\} & \{(a_1,2),(b_1,1),(3,d_1)\} & \{(1,d_1),(b_1,2),(c_1,1)\}.
\end{array}$$

Another block in $\mathcal{A}$ is $A_7 = \{a_1, b_3, c_2\}$. Since $|A_7| = 3 = k$, the blocks in $\mathcal{B}_{A_7}$ are the blocks of a $\mathsf{TD}(3,2)$ having groups $\{(a_1,1),(a_1,2)\}$, $\{(b_3,1),(b_3,2)\}$, and $\{(c_2,1),(c_2,2)\}$. For example, we could take the following four blocks:

$$\{(a_1,1),(b_3,1),(c_2,1)\}\ \{(a_1,2),(b_3,2),(c_2,1)\}$$
$$\{(a_1,2),(b_3,1),(c_2,2)\}\ \{(a_1,1),(b_3,2),(c_2,2)\}.$$

We apply this process to each of the nine blocks $A_1,\ldots,A_9$.

Finally, we adjoin the four blocks of a $\mathsf{TD}(3,2)$ on groups $\{(1,d_1),(1,d_2)\}$, $\{(2,d_1),(2,d_2)\}$, and $\{(3,d_1),(3,d_2)\}$. For example, we could take the following four blocks:

$$\{(1,d_1),(2,d_1),(3,d_1)\}\ \{(1,d_1),(2,d_2),(3,d_2)\}$$
$$\{(1,d_2),(2,d_1),(3,d_2)\}\ \{(1,d_2),(2,d_2),(3,d_1)\}.$$

The resulting set of 64 blocks yields the desired $\mathsf{TD}(3,8)$. ∎

In view of Theorem 6.44, the following corollary is obtained by rephrasing Theorem 6.45 in the language of MOLS.

Theorem 6.47. *Suppose* $s \geq 1$ *and there exist* s $\mathsf{MOLS}(m)$, s $\mathsf{MOLS}(m+1)$, s $\mathsf{MOLS}(u)$, *and* $s+1$ $\mathsf{MOLS}(t)$, *where* $1 \leq u \leq t$. *Then there exist* s $\mathsf{MOLS}(mt+u)$.

6.8 Disproof of the Euler Conjecture

As an application of Wilson's construction, we will complete the proof that there exist orthogonal Latin squares of order n for all positive integers $n \neq 2$ or 6. (This is not the "original" disproof of the Euler conjecture from 1958; Wilson's construction permits the proof to be simplified considerably.)

The following corollary will be useful.

Corollary 6.48. *Suppose* $t \equiv 1,5 \pmod 6$, u *is odd, and* $0 \leq u \leq t$. *Then there exist orthogonal Latin squares of order* $3t+u$.

Proof. We apply Theorem 6.47 with $s=2$ and $m=3$, noting that orthogonal Latin squares of orders 3, 4, and u exist (Theorem 6.28), as do three $\mathsf{MOLS}(t)$ (Corollary 6.35). □

We need one more example as a special case before proceeding to the general result.

Example 6.49. Orthogonal Latin squares of order 14.

We present a set of 17 four-tuples of elements in $\mathbb{Z}_{11} \cup \{\infty_1,\infty_2,\infty_3\}$. These rows are to be developed modulo 11 using the convention that $\infty_i + j = \infty_i$ for $i=1,2,3$ and $j \in \mathbb{Z}_{11}$. (In other words, the permutation

$$\alpha = (0\ 1\ 2\ \cdots\ 10)(\infty_1)(\infty_2)(\infty_3)$$

is an automorphism of this orthogonal array.) Then adjoin nine more rows that form an $\mathsf{OA}(4,3)$ on the symbols $\infty_1,\infty_2,\infty_3$. The result is an $\mathsf{OA}(4,14)$, which is equivalent to the desired orthogonal Latin squares.

Here are the 17 starting rows:

$$\begin{array}{cccc}
0 & 0 & 0 & 0 \\
0 & 4 & 1 & 6 \\
4 & 0 & 6 & 1 \\
6 & 1 & 0 & 4 \\
1 & 6 & 4 & 0 \\
\infty_1 & 4 & 0 & 1 \\
\infty_2 & 6 & 0 & 2 \\
\infty_3 & 9 & 0 & 8 \\
4 & \infty_1 & 1 & 0 \\
6 & \infty_2 & 2 & 0 \\
9 & \infty_3 & 8 & 0 \\
1 & 0 & \infty_1 & 4 \\
2 & 0 & \infty_2 & 6 \\
8 & 0 & \infty_3 & 9 \\
0 & 1 & 4 & \infty_1 \\
0 & 2 & 6 & \infty_2 \\
0 & 8 & 9 & \infty_3
\end{array}$$

∎

Theorem 6.50. *Suppose $n \equiv 2 \pmod 4$, $n \neq 2, 6$. Then there exist orthogonal Latin squares of order n.*

Proof. We already did the cases where $n \equiv 10 \pmod{12}$ in Theorem 6.41, so we can assume that $n \equiv 2, 6, 14, 18, 26,$ or $30 \pmod{36}$. For each of these six residue classes modulo 36, we present a construction that is an application of Corollary 6.48 by writing n in the form $n = 3t + u$ in an appropriate manner:

$$\begin{aligned}
36s + 2 &= 3(12s - 1) + 5, & s \geq 1 \\
36s + 6 &= 3(12s + 1) + 3, & s \geq 1 \\
36s + 14 &= 3(12s + 1) + 11, & s \geq 1 \\
36s + 18 &= 3(12s + 5) + 3, & s \geq 0 \\
36s + 26 &= 3(12s + 7) + 5, & s \geq 0 \\
36s + 30 &= 3(12s + 7) + 9, & s \geq 1.
\end{aligned}$$

The only values of n not covered by the constructions above are $n = 2, 6, 14,$ and 30. The first two values of n are exceptions, the case $n = 14$ is done in Example 6.49, and $n = 30$ can be handled by the direct product construction because $30 = 3 \times 10$ and orthogonal Latin squares of orders 3 and 10 exist.

□

Our main existence result is an immediate consequence of Theorems 6.28 and 6.50.

Theorem 6.51. *Suppose n is a positive integer and $n \neq 2$ or 6. Then there exist orthogonal Latin squares of order n.*

6.9 Notes and References

Bose's construction for Steiner triple systems was given in [13], and Skolem's modification is from [98]. Our description of these constructions is based on [77]. The book "Triple Systems" by Colbourn and Rosa [32] is an enormous work devoted to BIBDs with block size 3. It is essential reading for anyone interested in that topic.

The construction of a pair of orthogonal Latin squares of all orders $n \neq 2, 6$ was accomplished by Bose and Shrikhande [15] and Bose, Shrikhande, and Parker [16]. Theorem 6.41 is from [16]. A short proof of the nonexistence of a pair of orthogonal Latin squares of order 6 can be found in Stinson [101].

Wilson's construction for MOLS (a generalization of Theorem 6.45) is presented in [121]. An extensive table of MOLS of orders up to 10,000 can be found in [1]. Colbourn [24] provides a good summary of construction methods for MOLS, and Colbourn and Dinitz [28] describe how the tables in [1] were constructed. Some updated results can be found in Colbourn and Dinitz [29].

6.10 Exercises

6.1 (a) Suppose that $(X, \mathcal{A})$ is a $(v,3,1)$-BIBD and $(X, \circ)$ is any quasigroup of order v. Define $Y = X \times \{1,2,3\}$. For $1 \leq i \leq 3$ and for any $A \in \mathcal{A}$, define $A_i = \{(x,i) : x \in A\}$. Define

$$\mathcal{B}_1 = \{A_i : 1 \leq i \leq 3\}$$

and define

$$\mathcal{B}_2 = \{(x,1),(y,2),(x \circ y,3) : x,y \in X\}.$$

Prove that $(Y, \mathcal{B}_1 \cup \mathcal{B}_2)$ is a $(3v,3,1)$-BIBD.

(b) Describe how to construct a $(3v-2,3,1)$-BIBD from any $(v,3,1)$-BIBD and any quasigroup of order $v-1$.

(c) Describe how to construct a $(3v-6,3,1)$-BIBD from any $(v,3,1)$-BIBD and any quasigroup of order $v-3$.

6.2 (a) Describe how to construct an idempotent quasigroup of every even order $t > 2$.

(b) Explicitly construct idempotent quasigroups of orders 4 and 6.

(c) Describe a construction for a $(3t,3,2)$-BIBD from any idempotent quasigroup of order t. Illustrate your construction in the case $t = 4$.

6.3 Suppose $(X, \circ)$ is a quasigroup. We say that $(X, \circ)$ is a *Steiner quasigroup* if $(X, \circ)$ is symmetric and idempotent and $(x \circ y) \circ y = x$ for all $x, y \in X$.

(a) Suppose that $(X, \mathcal{A})$ is a Steiner triple system of order n. Define a binary operation $\circ$ on X as follows:

$$x \circ y = \begin{cases} x & \text{if } x = y \\ z & \text{if } x \neq y \text{ and } \{x, y, z\} \in \mathcal{A}. \end{cases}$$

Prove that $(X, \circ)$ is a Steiner quasigroup of order n.

(b) Suppose that $(X, \circ)$ is a Steiner quasigroup of order n. Define

$$\mathcal{A} = \{\{x, y, x \circ y\} : x, y \in X, x \neq y\}.$$

Prove that $(X, \mathcal{A})$ is a Steiner triple system of order n.

6.4 Suppose that there are s MOLS(n). Prove that there are $s - 1$ MOLS(n), all of which are idempotent.
Hint: This can be done by permuting rows, columns, and symbols in $s - 1$ of the s MOLS in a certain way. You should begin by choosing one of the s orthogonal Latin squares, picking a symbol x, and considering the set of n cells in this Latin square that contain x, say C. Then, in each of the remaining $s - 1$ MOLS, the cells in C must contain one occurrence of each symbol.

6.5 (a) Suppose that there is a $(v, k, 1)$-BIBD, and suppose there are $s - 1$ MOLS(k), all of which are idempotent. Prove that there are $s - 1$ MOLS(v), all of which are idempotent.
(b) Using a suitable BIBD, prove that there exist three MOLS(21), all of which are idempotent.

6.6 (a) Let $1 < m < n$ be integers. A Latin square L of order n has a *subsquare* of order m if there is an $m \times m$ subarray of L, say M, which is itself a Latin square on a subset of m symbols. Prove that $m \leq 2n$ if a Latin square of order n has a subsquare of order m.
(b) Let $1 < m < n$ be integers. Suppose that $L_1, \ldots, L_s$ are MOLS of order n. Suppose that $L_1, \ldots, L_s$ each have a subsquare of order m situated in the same positions (say, without loss of generality, in the upper left corners). Prove that these subsquares are necessarily s MOLS(m) and $m \leq (s+1)n$.

6.7 Prove that the following sets of MOLS exist by citing appropriate theorems or constructions.
(a) 8 MOLS(99).
(b) 7 MOLS(96).
(c) 5 MOLS(57).

6.8 Prove that there exist three MOLS(n) if $n \equiv 0 \bmod 36$.
Hint: Consider the factorization of n into prime powers.

6.9 A *magic square of order n* is an n by n array formed from the integers $1, \ldots, n^2$ such that the sum of the entries in any row or column is a fixed integer, say S.
(a) Prove that $S = (n^3 + n)/2$.
(b) Suppose that L and M are orthogonal Latin squares on symbol set $\{0, \ldots, n-1\}$. Define an n by n array $A = (a_{i,j})$ by the formula

$$a_{i,j} = n\,L(i,j) + M(i,j) + 1.$$

Prove that A is a magic square of order n.

(c) Construct orthogonal Latin squares of order 4 and then use them to construct a magic square of order 4.

6.10 A Latin square is *self-orthogonal* if it is orthogonal to its transpose. Let $a, b \in \mathbb{Z}_n$. Suppose that we define an n by n array $M = (m_{i,j})$, with symbols in $\mathbb{Z}_n$, by the rule

$$m_{i,j} = ai + bj \bmod n.$$

(a) Give a complete proof that M is a self-orthogonal Latin square of order n provided that $\gcd(a, n) = 1$, $\gcd(b, n) = 1$, and $\gcd(a^2 - b^2, n) = 1$.

(b) Construct a self-orthogonal Latin square of order 7.

7

Pairwise Balanced Designs I: Designs with Specified Block Sizes

7.1 Definitions and Basic Results

Pairwise balanced designs were defined in Section 1.3. These are among the most important and most studied types of designs. We will spend quite a bit of time in this chapter looking at pairwise balanced designs with specified block sizes. Interestingly, these have applications to the construction of infinite families of BIBDs with fixed block sizes.

We begin with a definition of pairwise balanced designs with specified block sizes.

Definition 7.1. *Suppose $v \geq 2$, $\lambda \geq 1$, and $K \subseteq \{n \in \mathbb{Z} : n \geq 2\}$. A (v, K, λ)-pairwise balanced design (which we abbreviate to (v, K, λ)-PBD) is a set system $(X, \mathcal{A})$ such that the following properties are satisfied:*

1. $|X| = v$,
2. $|A| \in K$ *for all* $A \in \mathcal{A}$, *and*
3. *every pair of distinct points is contained in exactly λ blocks.*

A $(v, K, 1)$-PBD is often denoted simply as a (v, K)-PBD.

Recall that a pairwise balanced design on v points is allowed to have blocks of size v. It is clear that a (v, k, λ)-BIBD is a $(v, \{k\}, \lambda)$-PBD. Conversely, if $k < v$, then a $(v, \{k\}, \lambda)$-PBD is a (v, k, λ)-BIBD.

We begin by presenting some simple constructions for pairwise balanced designs from other types of designs. Transversal designs and truncated transversal designs provide a convenient way of constructing certain pairwise balanced designs with $\lambda = 1$.

Lemma 7.2. *Suppose that $k \geq 2$ and there is a TD$(k+1, t)$. Then the following pairwise balanced designs exist:*

1. *a $(kt + u, \{k, k+1, t, u\})$-PBD for all u such that $2 \leq u \leq t - 1$,*
2. *a $(kt + 1, \{k, k+1, t\})$-PBD, and*

3. *a* $((k+1)t, \{k+1, t\})$-PBD.

Proof. To prove 1, delete $t-u$ points from one group of a $\mathsf{TD}(k+1,t)$. Then take all the groups and blocks of the truncated transversal design to be blocks of a PBD.

To prove 2, delete $t-1$ points from one group of a $\mathsf{TD}(k+1,t)$. Then take all the groups and blocks of the truncated transversal design (except for the group of size one) to be blocks of a PBD.

To prove 3, take all the groups and blocks of the transversal design to be blocks of a PBD. □

Resolvable $(v,k,1)$-BIBDs also can be used to produce pairwise balanced designs with $\lambda = 1$.

Lemma 7.3. *Suppose there is a resolvable* $(v,k,1)$-BIBD. *Then there exists a* $(v+r, \{k+1, r\})$-PBD, *where* $r = (v-1)/(k-1)$.

Proof. We use the same technique as in the proof of Theorem 5.10. Let $\Pi_1, \dots, \Pi_r$ denote the parallel classes in the BIBD. Let $\infty_1, \dots, \infty_r$ be r new points, and adjoin ∞_i to each block in the parallel class Π_i. Finally, let $\{\infty_1, \dots, \infty_r\}$ be a new block. □

Note that, if we start with an affine plane, then the resulting pairwise balanced design has only one block size (because $k+1=r$) and therefore it is in fact a BIBD (namely, a projective plane; see Theorem 5.10).

As a corollary of Lemma 7.3, we can obtain the following result.

Corollary 7.4. *For all even integers* $v \geq 4$, *there exists a* $(2v-1, \{3, v-1\})$-PBD.

Proof. Apply Lemma 7.3 with $k=2$, noting that a resolvable $(v,2,1)$-BIBD exists for all even $v \geq 4$ by Theorem 5.2. □

Example 7.5. An $(11, \{3,5\})$-PBD. We begin with the resolvable $(6,2,1)$-BIBD presented in Example 5.3 having parallel classes as follows:

$$\begin{aligned}
\Pi_0 &= \{\{\infty,0\},\{1,4\},\{2,3\}\},\\
\Pi_1 &= \{\{\infty,1\},\{2,0\},\{3,4\}\},\\
\Pi_2 &= \{\{\infty,2\},\{3,1\},\{4,0\}\},\\
\Pi_3 &= \{\{\infty,3\},\{4,2\},\{0,1\}\},\\
\Pi_4 &= \{\{\infty,4\},\{0,3\},\{1,2\}\}.
\end{aligned}$$

The blocks of the resulting $(11, \{3,5\})$-PBD are

$$\begin{array}{l}
\{\infty,0,\infty_1\},\ \{1,4,\infty_1\},\ \{2,3,\infty_1\},\\
\{\infty,1,\infty_2\},\ \{2,0,\infty_2\},\ \{3,4,\infty_2\},\\
\{\infty,2,\infty_3\},\ \{3,1,\infty_3\},\ \{4,0,\infty_3\},\\
\{\infty,3,\infty_4\},\ \{4,2,\infty_4\},\ \{0,1,\infty_4\},\\
\{\infty,4,\infty_5\},\ \{0,3,\infty_5\},\ \{1,2,\infty_5\},\\
\{\infty_1,\infty_2,\infty_3,\infty_4,\infty_5\}.
\end{array}$$

■

7.2 Necessary Conditions and PBD-Closure

In this section, we first discuss necessary numerical conditions for existence of (v, K)-PBDs. Then we present some definitions and results pertaining to the important idea of PBD-closure.

Definition 7.6. *Suppose $K \subseteq \{n \in \mathbb{Z} : n \geq 2\}$, and define*

$$\mathbb{B}(K) = \{v : \text{there exists a } (v, K)\text{-PBD}\}.$$

Furthermore, define

$$\alpha(K) = \gcd\{k - 1 : k \in K\}$$

and

$$\beta(K) = \gcd\{k(k-1) : k \in K\}.$$

Note that $K \subseteq \mathbb{B}(K)$ because a (trivial) $(k, \{k\})$-PBD exists for any integer $k \geq 2$.

Our next lemma provides some necessary numerical conditions for v to be an element of $\mathbb{B}(K)$. This lemma can be thought of as a generalization of Theorems 1.8 and 1.9.

Lemma 7.7. *Suppose $K \subseteq \{n \in \mathbb{Z} : n \geq 2\}$ and suppose that $v \geq 3$ is an integer. Then $v \in \mathbb{B}(K)$ only if*

$$v - 1 \equiv 0 \pmod{\alpha(K)}$$

and

$$v(v-1) \equiv 0 \pmod{\beta(K)}.$$

Proof. Suppose $v \in \mathbb{B}(K)$. Then there exists a (v, K)-PBD. Let X be the set of points in this design, suppose $x \in X$, and let r_x denote the number of blocks containing x. Let $A_1, \dots, A_{r_x}$ denote the blocks that contain x. Then

$$\sum_{i=1}^{r_x} (|A_i| - 1) = v - 1.$$

Clearly $|A_i| - 1 \equiv 0 \pmod{\alpha(K)}$ for all i, $1 \leq i \leq r_x$, and hence $v - 1 \equiv 0 \pmod{\alpha(K)}$. This proves the first condition.

To prove the second condition, let $A_1, \dots, A_b$ be all the blocks in the pairwise balanced design. Then

$$\sum_{i=1}^{b} |A_i|(|A_i| - 1) = v(v-1).$$

Clearly $|A_i|(|A_i| - 1) \equiv 0 \pmod{\beta(K)}$ for all i, $1 \leq i \leq b$. Hence, $v(v-1) \equiv 0 \pmod{\beta(K)}$. This completes the proof. □

If $|K| = 1$ (say $K = \{k\}$) and $v > k$, then a (v, K)-PBD is a $(v, k, 1)$-BIBD, and the conditions in Lemma 7.7 become

$$v - 1 \equiv 0 \pmod{k - 1}$$

and

$$v(v - 1) \equiv 0 \pmod{k(k - 1)}.$$

These are precisely the conditions that the BIBD parameters r and b (respectively) be integers.

Here is a small example to illustrate the application of Lemma 7.7 when $|K| > 1$.

Example 7.8. Suppose $K = \{3, 4, 6\}$. Then it is easy to compute

$$\alpha(K) = \gcd\{2, 3, 5\} = 1$$

and

$$\beta(K) = \gcd\{6, 12, 30\} = 6.$$

The necessary conditions in Lemma 7.7 simplify to $v \equiv 0 \text{ or } 1 \pmod 3$. It follows that

$$\mathbb{B}(\{3, 4, 6\}) \subseteq \{n \in \mathbb{Z} : n \equiv 0 \text{ or } 1 \pmod 3, n \geq 3\}.$$

∎

We now introduce the very useful notion of a PBD-closed set.

Definition 7.9. *Suppose that $K \subseteq \{n \in \mathbb{Z} : n \geq 2\}$. We say that K is a* PBD-closed set *if $\mathbb{B}(K) = K$, i.e., if $v \in K$ whenever there exists a (v, K)-PBD.*

The following lemma is simple but important. It is commonly called "breaking up blocks".

Lemma 7.10 (Breaking up Blocks). *Suppose $K \subseteq \{n \in \mathbb{Z} : n \geq 2\}$. Then $\mathbb{B}(K)$ is PBD-closed.*

Proof. Suppose $K \subseteq \{n \in \mathbb{Z} : n \geq 2\}$, and let $(X, \mathcal{A})$ be any $(v, \mathbb{B}(K))$-PBD. We want to prove that there is a (v, K)-PBD. For all $|A| \in \mathcal{A}$ there is a $(|A|, K)$-PBD, say $(A, \mathcal{B}_A)$. Define

$$\mathcal{B} = \bigcup_{A \in \mathcal{A}} \mathcal{B}_A.$$

Then it is easy to see that $(X, \mathcal{B})$ is a (v, K)-PBD. □

Lemma 7.10 implies some easy corollaries. The fact that $\mathbb{B}(K)$ is PBD-closed implies the following result.

Corollary 7.11. *Suppose $K \subseteq \{n \in \mathbb{Z} : n \geq 2\}$. Then $\mathbb{B}(\mathbb{B}(K)) = \mathbb{B}(K)$.*

For an integer $k \geq 2$, define

$$\mathsf{V}_k = \{k\} \cup \{v : \text{ there exists a } (v,k,1)\text{-BIBD}\}.$$

The next result is obtained by taking $K = \{k\}$ in Corollary 7.11 and noting that $\mathsf{V}_k = \mathbb{B}(\{k\})$.

Corollary 7.12. *For any integer $k \geq 2$, V_k is PBD-closed.*

We do a small example to illustrate Corollary 7.12.

Example 7.13. We construct a $(21,3,1)$-BIBD by applying Corollary 7.12 with $k = 3$. Observe that a TD$(3,7)$ yields a $(21,\{3,7\})$-PBD containing 49 blocks of size three and three blocks of size seven. Replace each block A of size seven by the seven blocks of a $(7,3,1)$-BIBD on point set A. The result is a $(21,3,1)$-BIBD. ∎

Corollary 7.12 says that the set of all v-values of $(v,k,1)$-BIBDs is PBD-closed, for any fixed integer $k \geq 2$. We now prove the interesting and important result that the set of r-values of $(v,k,1)$-BIBDs is also PBD-closed. First, however, we must introduce the concept of a group-divisible design.

Definition 7.14. *Let $v > 2$ be a positive integer. A* group-divisible design *(which we abbreviate to* GDD*) is a triple $(X, \mathcal{G}, \mathcal{A})$ such that the following properties are satisfied:*

1. *X is a finite set of elements called* points,
2. *$\mathcal{G}$ is a partition of X into at least two nonempty subsets called* groups *(note that groups of size one are allowed),*
3. *$\mathcal{A}$ is a set of subsets of X called* blocks *such that $|A| \geq 2$ for all $A \in \mathcal{A}$,*
4. *a group and a block contain at most one common point, and*
5. *every pair of points from distinct groups is contained in exactly one block.*

Transversal designs and truncated transversal designs are examples of group-divisible designs. The following lemmas record some simple ways of obtaining pairwise balanced designs from group-divisible designs and vice versa.

Lemma 7.15. *If $(X, \mathcal{G}, \mathcal{A})$ is a group-divisible design, then $(X, \mathcal{B})$ is a pairwise balanced design with $\lambda = 1$, where*

$$\mathcal{B} = \mathcal{A} \cup \{G \in \mathcal{G} \colon |G| \geq 2\}.$$

Lemma 7.16. *Suppose that $(X, \mathcal{G}, \mathcal{A})$ is a group-divisible design. Suppose that $\infty \notin X$, define $Y = X \cup \{\infty\}$, and define*

$$\mathcal{B} = \mathcal{A} \cup \{G \cup \{\infty\} : G \in \mathcal{G}\}.$$

Then $(Y, \mathcal{B})$ is a pairwise balanced design with $\lambda = 1$.

Lemma 7.17. *If $(X, \mathcal{A})$ is a pairwise balanced design with $\lambda = 1$, then $(X, \mathcal{G}, \mathcal{A})$ is a group-divisible design, where*

$$\mathcal{G} = \{\{x\} : x \in X\}.$$

Next, we state and prove a useful equivalence between $(v,k,1)$-BIBDs and certain group-divisible designs.

Lemma 7.18. *Suppose that $v > k > 1$. Then there exists a $(v,k,1)$-BIBD if and only if there exists a group-divisible design having $v-1$ points, r groups of size $k-1$, and blocks of size k (where, as usual, $r = (v-1)/(k-1)$).*

Proof. Given a $(v,k,1)$-BIBD, choose any point x. Form the groups of the desired group-divisible design by taking the blocks that contain x and deleting x from them. The blocks of the group-divisible design are all the remaining blocks in the BIBD.

The converse follows from Lemma 7.16. Note that the resulting pairwise balanced design has only blocks of size k and therefore it is a BIBD. □

Example 7.19. A $(9,3,1)$-BIBD was presented in Example 1.4:

$$\begin{aligned} X &= \{1,2,3,4,5,6,7,8,9\}, \quad \text{and} \\ \mathcal{A} &= \{123,456,789,147,258,369,159,267,348,168,249,357\}. \end{aligned}$$

If we delete the point 1, say, then we obtain the following GDD $(Y, \mathcal{G}, \mathcal{B})$:

$$\begin{aligned} Y &= \{2,3,4,5,6,7,8,9\}, \\ \mathcal{G} &= \{23,47,59,68\}, \quad \text{and} \\ \mathcal{B} &= \{456,789,258,369,267,348,249,357\}. \end{aligned}$$

This GDD contains four groups of size two and eight blocks of size three. ∎

Corollary 7.12 asserts that the set V_k is PBD-closed. For any integer $k \geq 2$, define

$$\mathsf{R}_k = \{r : \text{ there exists an } (r(k-1)+1,k,1)\text{-BIBD}\}.$$

We show that R_k is PBD-closed in the next theorem.

Theorem 7.20. R_k *is PBD-closed for any integer $k \geq 2$.*

Proof. Let $(X, \mathcal{A})$ be any (v, R_k)-PBD. We want to prove that $v \in \mathsf{R}_k$. In other words, we want to show that there exists a $(v(k-1)+1,k,1)$-BIBD.

For every block $A \in \mathcal{A}$, there exists an $(|A|(k-1)+1,k,1)$-BIBD. By Lemma 7.18, this BIBD is equivalent to a group-divisible design having $|A|(k-1)$ points, $|A|$ groups of size $k-1$, and blocks of size k. Let I be some set of size $k-1$. We can construct this group-divisible design on point set $A \times I$ such that the groups are $\{x\} \times I$, $x \in A$. Let $\mathcal{B}_A$ denote the blocks of this group-divisible design.

Define

$$Y = X \times I,$$
$$\mathcal{H} = \{\{x\} \times I : x \in X\}, \quad \text{and}$$
$$\mathcal{B} = \bigcup_{A \in \mathcal{A}} \mathcal{B}_A.$$

We will prove that $(Y, \mathcal{H}, \mathcal{B})$ is a group-divisible design having $v(k-1)$ points, v groups of size $k-1$, and blocks of size k. Then, by Lemma 7.18, there exists a $(v(k-1)+1, k, 1)$-BIBD, as required.

It is clear that $(Y, \mathcal{H}, \mathcal{B})$ has $v(k-1)$ points, v groups of size $k-1$, and blocks of size k. Thus we need only to verify that any two points from different groups occur in a unique block in $\mathcal{B}$. Consider two points, say (x,i) and (y,j), where $x, y \in X$, $x \neq y$, and $i, j \in I$. There is a unique block $A \in \mathcal{A}$ such that $x, y \in A$. Then, there is a unique block $B_0 \in \mathcal{B}_A$ such that $(x,i), (y,j) \in B_0$. B_0 is the unique block in $\mathcal{B}$ that contains (x,i) and (y,j), and the proof is complete. □

We present a small example to illustrate.

Example 7.21. Let $k = 3$. Since there exists a $(7,3,1)$-BIBD, it follows that $3 \in \mathsf{R}_3$. The fact that R_3 is PBD-closed, together with the existence of the same $(7,3,1)$-BIBD, establishes that $7 \in \mathsf{R}_3$. In other words, we can construct a $(15,3,1)$-BIBD by means of the construction given in the proof of Theorem 7.20.

Suppose we begin with the following $(7,3,1)$-BIBD:

$$X = \{1,2,3,4,5,6,7\}, \text{and}$$
$$\mathcal{A} = \{123, 145, 167, 246, 257, 347, 356\}.$$

For every block $A = \{x,y,z\} \in \mathcal{A}$, we replace A by the four blocks of a group-divisible design having points $A \times \{0,1\}$ and groups $\{x\} \times \{0,1\}$, $\{y\} \times \{0,1\}$, and $\{z\} \times \{0,1\}$. We can use the following set of four blocks:

$$\mathcal{B}_{\{xyz\}} = \{\{x_0, y_0, z_0\}, \{x_0, y_1, z_1\}, \{x_1, y_0, z_1\}, \{x_1, y_1, z_0\}\},$$

where we write points (x,i) in the form x_i in order to save space. We carry out this process for each of the seven blocks in $\mathcal{A}$, obtaining a set of 28 blocks, which are the blocks of a group-divisible design having seven groups of size two. We then add a new point to each group, to obtain the seven blocks in $\mathcal{B}_\infty$. The resulting set of 35 blocks, shown in Figure 7.1, form a $(15,3,1)$-BIBD. ∎

$$\begin{aligned}
\mathcal{B}_{\{123\}} &= \{\{1_0,2_0,3_0\},\{1_0,2_1,3_1\},\{1_1,2_0,3_1\},\{1_1,2_1,3_0\}\}\\
\mathcal{B}_{\{145\}} &= \{\{1_0,4_0,5_0\},\{1_0,4_1,5_1\},\{1_1,4_0,5_1\},\{1_1,4_1,5_0\}\}\\
\mathcal{B}_{\{167\}} &= \{\{1_0,6_0,7_0\},\{1_0,6_1,7_1\},\{1_1,6_0,7_1\},\{1_1,6_1,7_0\}\}\\
\mathcal{B}_{\{246\}} &= \{\{2_0,4_0,6_0\},\{2_0,4_1,6_1\},\{2_1,4_0,6_1\},\{2_1,4_1,6_0\}\}\\
\mathcal{B}_{\{257\}} &= \{\{2_0,5_0,7_0\},\{2_0,5_1,7_1\},\{2_1,5_0,7_1\},\{2_1,5_1,7_0\}\}\\
\mathcal{B}_{\{347\}} &= \{\{3_0,4_0,7_0\},\{3_0,4_1,7_1\},\{3_1,4_0,7_1\},\{3_1,4_1,7_0\}\}\\
\mathcal{B}_{\{356\}} &= \{\{3_0,5_0,6_0\},\{3_0,5_1,6_1\},\{3_1,5_0,6_1\},\{3_1,5_1,6_0\}\}\\
\mathcal{B}_{\infty} &= \{\{\infty,0_0,0_1\},\{\infty,1_0,1_1\},\{\infty,2_0,2_1\},\{\infty,3_0,3_1\}\\
&\quad \{\infty,4_0,4_1\},\{\infty,5_0,5_1\},\{\infty,6_0,6_1\}\}.
\end{aligned}$$

Fig. 7.1. A $(15,3,1)$-BIBD

7.3 Steiner Triple Systems

Recall that we constructed Steiner triple systems (i.e., $(v,3,1)$-BIBDs) of all permissible orders in Section 6.2. In this section, we give a different proof of the same result using PBD-closure techniques.

We know that the set of v-values of $(v,3,1)$-BIBDs is PBD-closed, as is the set of r-values of $(v,3,1)$-BIBDs. If we construct some small $(v,3,1)$-BIBDs (e.g., $(7,3,1)$- and $(9,3,1)$-BIBDs), then we can construct larger $(v,3,1)$-BIBDs by first constructing pairwise balanced designs with block sizes $3,7,9$, etc. (This approach was illustrated in Example 7.13.) However, it turns out to be easier to use the PBD-closure of R_3 to construct Steiner triple systems. This is because the set R_3 contains several small numbers (e.g., $3,4,6$, and 7, as we will show) and it is easier to construct pairwise balanced designs when there are more allowable block sizes. In particular, the fact that R_3 contains the consecutive integers 3 and 4 makes it an easy task to construct (v,R_3)-PBDs using truncated transversal designs.

Recall from Lemma 6.11 that an $\mathsf{STS}(v)$ exists only if $v \equiv 1$ or $3 \pmod 6$, $v \geq 7$. Defining $r = (v-1)/2$, these conditions can be restated as $r \equiv 0$ or $1 \pmod 3$, $r \geq 3$. Therefore we have that

$$\mathsf{R}_3 \subseteq \{n \geq 3 : n \equiv 0,1 \pmod 3\}.$$

We will in fact prove that $\mathsf{R}_3 = \{n \geq 3 : n \equiv 0,1 \pmod 3\}$. The strategy is as follows:

1. Find some small elements of the set R_3. We will show by direct constructions that $\{3,4,6\} \subseteq \mathsf{R}_3$.
2. Try to construct $(v,\{3,4,6\})$-PBDs for as many values of v as possible. We will (mainly) use truncated transversal designs to show that

$$\mathbb{B}(\{3,4,6\}) = \{n \geq 3 : n \equiv 0,1 \pmod 3\}.$$

3. Since R_3 is PBD-closed, we conclude that

$$R_3 = \{n \geq 3 : n \equiv 0, 1 \pmod 3\}.$$

We proceed to carry out these three steps now.

Step 1

- A projective plane of order 2 exists, which is a $(7,3,1)$-BIBD. Hence, $3 \in R_3$.
- An affine plane of order 3 exists, which is a $(9,3,1)$-BIBD. Hence, $4 \in R_3$.
- Example 3.45 displayed a $(13,3,1)$-difference family; this implies that $6 \in R_3$.

Hence we have shown that $\{3,4,6\} \subseteq R_3$.

Step 2

We first show that three particular integers are in $\mathbb{B}(\{3,4,6\})$ by means of direct constructions.

Lemma 7.22. $\{7, 18, 19\} \subseteq \mathbb{B}(\{3,4,6\})$.

Proof.

- $7 \in \mathbb{B}(\{3\})$ because a $(7,3,1)$-BIBD exists.
- $18 \in \mathbb{B}(\{3,6\})$ by Lemma 7.2 because a $\mathsf{TD}(3,6)$ exists.
- Adjoining a point to the groups of a $\mathsf{TD}(3,6)$, we obtain a $(19,\{3,7\})$-PBD (see Lemma 7.16). Then, because we have shown above that $7 \in \mathbb{B}(\{3\})$, it follows that $19 \in \mathbb{B}(\{3\})$.

□

The following simple lemma will now allow us to complete the determination of $\mathbb{B}(\{3,4,6\})$.

Lemma 7.23. *Suppose that $t \equiv 0$ or $1 \pmod 3$, $t \geq 3$, and $t \neq 6$. Then the following PBDs exist in which all block sizes are congruent to 0 or 1 modulo 3.*

1. *If $u \equiv 0$ or $1 \pmod 3$ and $3 \leq u \leq t$, then there is a $(3t+u, \{3,4,t,u\})$-PBD.*
2. *If $u \in \{0,1\}$, then there is a $(3t+u, \{3,4,t\})$-PBD.*

Proof. Apply Lemma 7.2 using the fact that a $\mathsf{TD}(4,t)$ exists if and only if $t \neq 2, 6$. □

Now we prove the main result of Step 2.

Theorem 7.24. $\mathbb{B}(\{3,4,6\}) = \{n \geq 3 : n \equiv 0, 1 \pmod 3\}$.

Proof. First, we compute $\alpha(\{3,4,6\}) = 1$ and $\beta(\{3,4,6\}) = 6$. Therefore, by Lemma 7.7, it follows that

$$\mathbb{B}(\{3,4,6\}) \subseteq \{n \geq 3 : n \equiv 0,1 \pmod 3\}.$$

By constructing appropriate $(v, \{3,4,6\})$-PBDs, we will show that

$$\mathbb{B}(\{3,4,6\}) = \{n \geq 3 : n \equiv 0,1 \pmod 3\}.$$

Our proof is by induction. Suppose that $v_0 \equiv 0$ or $1 \pmod 3$, $v_0 \geq 3$, and, as an induction hypothesis, suppose that $v \in \mathbb{B}(\{3,4,6\})$ for $v \equiv 0$ or $1 \pmod 3$, $3 \leq v \leq v_0$. Clearly this is true for $v_0 = 3, 4$, and 6, which we can take as base cases for the induction. Now, assuming that $v_0 \geq 7$, we will prove that a $(v_0, \{3,4,6\})$-PBD exists.

If $v_0 \in \{7, 18, 19\}$, then apply Lemma 7.22. Otherwise, write v_0 in the form $v_0 = 9s + j$, where $j \in \{0,1,3,4,6,7\}$, and apply Lemma 7.23 with values t and u as indicated in the following table:

v_0	$= 3t + u$	
$9s$	$= 3(3s)$,	$s \geq 1, s \neq 2$
$9s+1$	$= 3(3s)+1$,	$s \geq 1, s \neq 2$
$9s+3$	$= 3(3s+1)$,	$s \geq 1$
$9s+4$	$= 3(3s+1)+1$,	$s \geq 1$
$9s+6$	$= 3(3s+1)+3$,	$s \geq 1$
$9s+7$	$= 3(3s+1)+4$,	$s \geq 1$.

We therefore construct a $(v_0, \{3,4,t\})$-PBD. By induction, we have that $t \in \mathbb{B}(\{3,4,6\})$, and hence Lemma 7.10 shows that $v_0 \in \mathbb{B}(\{3,4,6\})$. This completes the proof. □

Step 3

At this point, we have shown the following:

$$\begin{aligned} &\{3,4,6\} \subseteq \mathsf{R}_3, && \text{in Step 1} \\ &\mathbb{B}(\{3,4,6\}) = \{n \geq 3 : n \equiv 0,1 \pmod 3\}, && \text{in Step 2, and} \\ &\mathsf{R}_3 \subseteq \{n \geq 3 : n \equiv 0,1 \pmod 3\}. \end{aligned}$$

Clearly, $\{3,4,6\} \subseteq \mathsf{R}_3$ implies that $\mathbb{B}(\{3,4,6\}) \subseteq \mathbb{B}(\mathsf{R}_3)$. Also, $\mathbb{B}(\mathsf{R}_3) = \mathsf{R}_3$ because R_3 is PBD-closed. It therefore follows that

$$\begin{aligned} \{n \geq 3 : n \equiv 0,1 \pmod 3\} &= \mathbb{B}(\{3,4,6\}) \\ &\subseteq \mathbb{B}(\mathsf{R}_3) \\ &= \mathsf{R}_3 \\ &\subseteq \{n \geq 3 : n \equiv 0,1 \pmod 3\}. \end{aligned}$$

Hence,

$$\mathsf{R}_3 = \{n \geq 3 : n \equiv 0,1 \pmod 3\},$$

and we have proven the following theorem.

Theorem 7.25. *There exists an* STS(v) *(i.e., a* $(v,3,1)$-BIBD*) if and only if* $v \equiv 1$ *or* 3 (mod 6), $v \geq 7$.

7.4 $(v,4,1)$-BIBDs

In this section, we use similar techniques to study $(v,4,1)$-BIBDs. The necessary numerical conditions for the existence of a $(v,4,1)$-BIBD are that $v \equiv 1$ or 4 (mod 12), $v \geq 13$. Defining $r = (v-1)/3$, these conditions can be restated as $r \equiv 0$ or 1 (mod 4), $r \geq 4$. Therefore we have that

$$\mathsf{R}_4 \subseteq \{n \geq 4 : n \equiv 0,1 \pmod 4\}.$$

We will prove that $\mathsf{R}_4 = \{n \geq 4 : n \equiv 0,1 \pmod 4\}$.

We will carry out the proof in three steps in a fashion similar to the proof in Section 7.3.

Step 1

The first step is to find some small elements of the set R_4. We have the following:

- A projective plane of order 3 exists. This design is a $(13,4,1)$-BIBD, so $4 \in \mathsf{R}_4$.
- An affine plane of order 4 exists. This design is a $(16,4,1)$-BIBD, so $5 \in \mathsf{R}_4$.
- Example 7.26 displays a $(25,4,1)$-difference family. This yields a $(25,4,1)$-BIBD, and therefore $8 \in \mathsf{R}_4$.
- In Example 7.27, we present a group-divisible design with nine groups of size three and blocks of size four. Then, Lemma 7.16 establishes that there is a $(28,4,1)$-BIBD, and hence $9 \in \mathsf{R}_4$.
- Example 7.28 presents a $(37,4,1)$-difference family, which gives rise to a $(37,4,1)$-BIBD. This ensures that $12 \in \mathsf{R}_4$.

Example 7.26. A $(25,4,1)$-difference family in $(\mathbb{Z}_5 \times \mathbb{Z}_5, +)$. There are two base blocks, namely

$$\{(0,0),(0,1),(1,0),(2,2)\} \text{ and } \{(0,0),(0,2),(2,0),(4,4)\}.$$

∎

Example 7.27. A group-divisible design with nine groups of size three and blocks of size four. The set of points is $\mathbb{Z}_3 \times \mathbb{Z}_3 \times \mathbb{Z}_3$, and the nine groups of the GDD are $\{(x,y,z) : z \in \mathbb{Z}_3\}$, $x,y \in \mathbb{Z}_3$. The blocks are obtained by developing the following two base blocks through the additive group $\mathbb{Z}_3 \times \mathbb{Z}_3 \times \mathbb{Z}_3$:

$$\{(0,0,0),(0,2,0),(1,1,1),(2,1,1)\} \text{ and } \{(0,0,0),(1,0,2),(0,1,2),(1,1,0)\}.$$

∎

Example 7.28. A $(37,4,1)$-difference family in $(\mathbb{Z}_{37},+)$. There are three base blocks, namely

$$\{0,1,3,24\}, \{0,10,18,30\}, \text{ and } \{0,4,26,32\}.$$

∎

The preceding examples and discussion establish the following lemma.

Lemma 7.29. $\{4,5,8,9,12\} \subseteq R_4$.

Step 2

The second step is to construct $(v,\{4,5,8,9,12\})$-PBDs. We will mainly use truncated transversal designs to prove the following result.

Theorem 7.30.

$$\mathbb{B}(\{4,5,8,9,12\}) = \{n \geq 4, n \equiv 0,1 \pmod 4\}.$$

Proof. First, we compute $\alpha(\{4,5,8,9,12\}) = 1$ and $\beta(\{4,5,8,9,12\}) = 4$. Therefore, by Lemma 7.7, it follows that

$$\mathbb{B}(\{4,5,8,9,12\}) \subseteq \{n \geq 4 : n \equiv 0,1 \pmod 4\}.$$

By constructing appropriate $(v,\{4,5,8,9,12\})$-PBDs, we will show that

$$\mathbb{B}(\{4,5,8,9,12\}) = \{n \geq 4, n \equiv 0,1 \pmod 4\}.$$

Our proof is by induction. Suppose that $v_0 \equiv 0$ or $1 \pmod 4$, $v_0 \geq 4$, and, as an induction hypothesis, suppose that $v \in \mathbb{B}(\{4,5,8,9,12\})$ for $v \equiv 0$ or $1 \pmod 4$, $4 \leq v \leq v_0$. Clearly this is true for $v_0 = 4,5,8,9$, and 12, which we can take as base cases for the induction. Now, assuming that $v_0 \geq 13$, we will prove that a $(v_0,\{4,5,8,9,12\})$-PBD exists.

We will handle several small values of v_0 as special cases, namely

$$v_0 \in S = \{13,28,29,41,44,45,48,49\},$$

as follows:

- A $(13,4,1)$-BIBD (i.e., a projective plane of order 3) is a $(13,\{4\})$-PBD.
- A $(28,4,1)$-BIBD (see Example 7.27) is a $(28,\{4\})$-PBD.
- A TD$(4,7)$ with a new point added to each group yields a $(29,\{4,8\})$-PBD (see Lemma 7.16).
- A $(41,\{4,5,9\})$-PBD exists by truncating four points from a group of a TD$(5,9)$ (i.e., apply Lemma 7.2, noting that a TD$(5,9)$ exists from Theorem 6.34).
- A $(44,\{4,5,8,9\})$-PBD exists by truncating one point from a group of a TD$(5,9)$ and applying Lemma 7.2.

v_0	$= 4t + u$
$48s$	$= 4(12s-4)+16,\ s \geq 2$
$48s+1$	$= 4(12s-4)+17,\ s \geq 2$
$48s+4$	$= 4(12s+1),\ s \geq 1$
$48s+5$	$= 4(12s+1)+1,\ s \geq 1$
$48s+8$	$= 4(12s+1)+4,\ s \geq 1$
$48s+9$	$= 4(12s+1)+5,\ s \geq 1$
$48s+12$	$= 4(12s+1)+8,\ s \geq 1$
$48s+13$	$= 4(12s+1)+9,\ s \geq 1$
$48s+16$	$= 4(12s+4)$
$48s+17$	$= 4(12s+4)+1$
$48s+20$	$= 4(12s+5)$
$48s+21$	$= 4(12s+5)+1$
$48s+24$	$= 4(12s+5)+4$
$48s+25$	$= 4(12s+5)+5$
$48s+28$	$= 4(12s+5)+8,\ s \geq 1$
$48s+29$	$= 4(12s+5)+9,\ s \geq 1$
$48s+32$	$= 4(12s+8)$
$48s+33$	$= 4(12s+8)+1$
$48s+36$	$= 4(12s+8)+4$
$48s+37$	$= 4(12s+8)+5$
$48s+40$	$= 4(12s+8)+8$
$48s+41$	$= 4(12s+8)+9,\ s \geq 1$
$48s+44$	$= 4(12s+8)+12,\ s \geq 1$
$48s+45$	$= 4(12s+8)+13,\ s \geq 1$

Table 7.1. Constructions for Truncated Transversal Designs

- A TD(5,9) yields a $(45, \{5,9\})$-PBD (apply Lemma 7.15).
- A TD(4,12) yields a $(48, \{4,12\})$-PBD (apply Lemma 7.15).
- A TD(4,12) with a new point added to each group yields a $(49, \{4,13\})$-PBD (see Lemma 7.16). Since there is a $(13, \{4\})$-PBD (i.e., a $(13,4,1)$-BIBD), a $(49, \{4\})$-PBD exists by Lemma 7.10.

If $v_0 \notin S$, then write v_0 in the form $v_0 = 48s + j$, where $j \equiv 0$ or $1 \pmod 4$ and $0 \leq j \leq 45$. Then we construct a truncated transversal design by deleting $t - u$ points from a group of a TD$(5,t)$, where the values t and u are as indicated in Table 7.1. In each case, we have $v_0 = 4t + u$, where $t \equiv 1, 4, 5,$ or $8 \pmod{12}$, $0 \leq u \leq t$, and $u \equiv 0$ or $1 \pmod 4$.

For these values of t and u, we can apply Lemma 7.2, noting that a TD$(5,t)$ exists from Corollary 6.35.

The pairwise balanced design that results is a $(v_0, \{4,5,8,9,12,t,u\})$-PBD where $t \equiv 0$ or $1 \pmod 4$ and $4 \leq u \leq 17$ or a $(v_0, \{4,5,8,9,12,t\})$-PBD if $u = 0$ or 1. By induction, we have that $t \in \mathbb{B}(\{4,5,8,9,12\})$. If $u = 13, 16,$ or 17, then $u \in \mathbb{B}(\{4,5\})$. In every case, it follows from Lemma 7.10 that $v_0 \in \mathbb{B}(\{4,5,8,9,12\})$, and the proof is complete.

□

Step 3

Summarizing in a fashion similar to Step 3 in Section 7.3, the following existence result concerning $(v, 4, 1)$-BIBDs can be proven. (The reader can fill in the details.)

Theorem 7.31. *There exists a* $(v, 4, 1)$-BIBD *if and only if* $v \equiv 1$ *or* $4 \pmod{12}$ *and* $v \geq 13$.

7.5 Kirkman Triple Systems

We now turn our attention to resolvable $(v, 3, 1)$-BIBDs. A resolvable $(v, 3, 1)$-BIBD is known as a *Kirkman triple system* (of *order* v) in honor of the Rev. Thomas Kirkman, who posed the problem of constructing resolvable $(v, 3, 1)$-BIBDs in the mid-nineteenth century. The case $v = 15$ came to be known as the "15 schoolgirls problem", and several solutions were found. However, for general v, the problem remained unsolved for over 100 years.

A resolvable $(v, 3, 1)$-BIBD will be denoted a $\mathsf{KTS}(v)$. Using PBD techniques, we will give a complete proof that a $\mathsf{KTS}(v)$ exists for all integers $v \equiv 3 \pmod 6, v \geq 9$. First, we need to prove a variation of Theorem 7.20 that pertains to resolvable BIBDs. This theorem will make use of group-divisible designs that satisfy certain resolvability properties that we define now.

Definition 7.32. *Suppose that* $(X, \mathcal{G}, \mathcal{B})$ *is a group-divisible design. Let* $G \in \mathcal{G}$. *A* holey parallel class *with* hole G *is a subset of blocks* $\mathcal{B}_0 \subseteq \mathcal{B}$ *that partitions* $X \backslash G$.

Now suppose that $(X, \mathcal{G}, \mathcal{B})$ *is a group-divisible design with* r *groups of size* $k - 1$ *and blocks of size* k. *Denote the groups as* $G_1, \ldots, G_r$. *Suppose there exist* r *holey parallel classes, say* $\Pi_1, \ldots, \Pi_r$, *that satisfy the following properties.*

1. *For* $1 \leq i \leq r$, Π_i *is a holey parallel class with hole* G_i.
2. *Every block* $B \in \mathcal{B}$ *is in exactly one of the* Π_i*'s.*

Then $(X, \mathcal{G}, \mathcal{B})$ *is said to be a* k-frame *on* r *holes.*

Lemma 7.33. *There exists a resolvable* $(v, k, 1)$-BIBD *if and only if there exists a* k*-frame on* r *holes, where* $r = (v - 1)/(k - 1)$.

Proof. Let $\Pi_1, \ldots, \Pi_r$ be the parallel classes of a resolvable $(v, k, 1)$-BIBD, where $r = (v - 1)/(k - 1)$. Choose any point x, and form groups and blocks of a group-divisible design as in the proof of Lemma 7.18. We need to show that the set of blocks of this group-divisible design can be partitioned into holey parallel classes. For $1 \leq i \leq r$, there is a unique block $B_i \in \Pi_i$ such that $x \in B_i$. Define the ith group to be $G_i = B_i \backslash \{x\}$, and define the ith holey parallel class to be $\Pi'_i = \Pi_i \backslash \{B_i\}$. The result is easily seen to be a k-frame on r holes.

The converse is proven by adding a new point to each group, as in Theorem 7.18. A parallel class of the resulting BIBD is just a parallel class of the frame, together with its corresponding hole, augmented with the new point. □

Example 7.34. The following KTS(9) was presented in Example 1.4:

$$X = \{1,2,3,4,5,6,7,8,9\}, \quad \text{and}$$
$$\mathcal{A} = \{123,456,789,147,258,369,159,267,348,168,249,357\}.$$

The parallel classes are

$$\begin{aligned}\Pi_1 &= \{123,456,789\},\\ \Pi_2 &= \{147,258,369\},\\ \Pi_3 &= \{159,267,348\}, \quad \text{and}\\ \Pi_4 &= \{168,249,357\}.\end{aligned}$$

Take $x = 1$ in Lemma 7.33, and construct a 3-frame with four holes. The holes are 23, 47, 59, and 68, and the corresponding holey parallel classes are (respectively)

$$\begin{aligned}\Pi_1' &= \{456,789\},\\ \Pi_2' &= \{258,369\},\\ \Pi_3' &= \{267,348\}, \quad \text{and}\\ \Pi_4' &= \{249,357\}.\end{aligned}$$

We now prove an analogue of Theorem 7.20 for resolvable BIBDs.

Theorem 7.35. *Suppose $k \geq 2$, and define*

$$\mathsf{R}_k^* = \{r : \textit{there exists a resolvable } (r(k-1)+1,k,1)\text{-BIBD}\}.$$

Then R_k^* is PBD-closed.

Proof. Let $(X, \mathcal{A})$ be any (v, R_k^*)-PBD. We will show that there exists a resolvable $(v(k-1)+1,k,1)$-BIBD. The BIBD can be constructed exactly in the proof of Theorem 7.20, so our main task is to show that this BIBD is resolvable.

For every block $A \in \mathcal{A}$, there exists a resolvable $(|A|(k-1)+1,k,1)$-BIBD. By Lemma 7.33, this BIBD is equivalent to a k-frame on $|A|$ holes. We can construct this group-divisible design on point set $A \times I$ such that the groups are $\{x\} \times I$, $x \in A$, where $|I| = k-1$. For all $x \in A$, let $\Pi_{A,x}$ denote the holey parallel class with hole $\{x\} \times I$.

By Theorem 7.20, we obtain a group-divisible design on point set $Y = X \times I$ in which the groups are $\{\{x\} \times I : x \in X\}$ and where the blocks all have size k. For all $x \in X$, define

$$\Pi_x = \bigcup_{\{A \in \mathcal{A} : x \in A\}} \Pi_{A,x}.$$

It is not hard to see that Π_x is a holey parallel class with hole $\{x\} \times I$. It is also straightforward to see that each block of the group-divisible design occurs in exactly one of the Π_x's. Therefore we have constructed a k-frame on v holes. Applying Lemma 7.33, we have a resolvable $(v(k-1)+1, k, 1)$-BIBD, as desired. □

As mentioned earlier, the necessary numerical conditions for the existence of a KTS(v) are that $v \equiv 3 \pmod 6$. Defining $r = (v-1)/2$, this can be restated as $r \equiv 1 \pmod 3$. Therefore we have that

$$\mathsf{R}_3^* \subseteq \{n \geq 4 : n \equiv 1 \pmod 3\}.$$

We will give a proof in this section that $\mathsf{R}_3^* = \{n \geq 4 : n \equiv 1 \pmod 3\}$. We employ a three-step strategy similar to the one used in previous sections.

Step 1

We begin with a direct construction for an infinite class of Kirkman triple systems.

Lemma 7.36. *If $q \equiv 1 \pmod 6$ is a prime power, then there exists a* KTS$(2q+1)$.

Proof. Let $q = 6t+1$ and let $\alpha \in \mathbb{F}_q$ be a primitive element. Let $\theta = (\alpha^t + 1)2^{-1}$. Now define $X = (\mathbb{F}_q \times \{1,2\}) \cup \{\infty\}$. Start with the following set of blocks, which is in fact a parallel class:

$$\begin{aligned}
\Pi_0 = {} & \{\{\infty, (0,1), (0,2)\}\} \\
& \cup \{\{(\alpha^i,1), (\alpha^{i+t},1), (\theta\alpha^i,2)\} : 0 \leq i \leq t-1\} \\
& \cup \{\{(\alpha^i,1), (\alpha^{i+t},1), (\theta\alpha^i,2)\} : 2t \leq i \leq 3t-1\} \\
& \cup \{\{(\alpha^i,1), (\alpha^{i+t},1), (\theta\alpha^i,2)\} : 4t \leq i \leq 5t-1\} \\
& \cup \{\{(\theta\alpha^{i+t},2), (\theta\alpha^{i+3t},2), (\theta\alpha^{i+5t},2)\} : 0 \leq i \leq t-1\}.
\end{aligned}$$

The other parallel classes are obtained by developing this base class through $\mathbb{F}_q$. It can be shown that the resulting design is a KTS$(2q+1)$. □

Example 7.37. A KTS(15). $\alpha = 3$ is a primitive element in $\mathbb{Z}_7$, and then we compute $\theta = (3+1)2^{-1} = 2$. Then

$$\Pi_0 = \left\{ \begin{array}{l} \{\infty,(0,1),(0,2)\}, \{(1,1),(3,1),(2,2)\}, \{(2,1),(6,1),(4,2)\}, \\ \{(4,1),(5,1),(1,2)\}, \{(6,2),(5,2),(3,2)\} \end{array} \right\}.$$

■

Lemma 7.38. *There exist* KTS(v) *for* $v \in \{9, 15, 21, 39\}$; *hence* $\{4, 7, 10, 19\} \subseteq \mathsf{R}_3^*$.

Proof. An affine plane of order 3 is a KTS(9). KTS(15) and KTS(39) are special cases of Lemma 7.36. We construct a KTS(21) now. Let $X = \mathbb{Z}_7 \times \mathbb{Z}_3$. First, define

$$\Pi_0 = \left\{ \begin{array}{l} \{(0,0),(0,1),(0,2)\},\{(3,0),(6,0),(5,0)\},\{(3,1),(6,1),(5,1)\}, \\ \{(3,2),(6,2),(5,2)\},\{(2,0),(4,1),(1,2)\},\{(2,1),(4,2),(1,0)\}, \\ \{(2,2),(4,0),(1,1)\} \end{array} \right\},$$

and define $\Pi_i = (i, 0) + \Pi_0$ for $i \in \mathbb{Z}_7$. Next, define

$$\Psi_0 = \left\{ \begin{array}{l} \{(3,0),(6,1),(5,2)\},\{(4,0),(0,1),(6,2)\},\{(5,1),(1,2),(0,0)\}, \\ \{(6,2),(2,0),(1,1)\},\{(0,0),(3,1),(2,2)\},\{(1,0),(4,1),(3,2)\}, \\ \{(2,1),(5,2),(4,0)\} \end{array} \right\},$$

and define $\Psi_j = (0, j) + \Psi_0$ for $j \in \mathbb{Z}_3$. The ten sets Π_i $(i \in \mathbb{Z}_7)$ and Ψ_j $(j \in \mathbb{Z}_3)$ are the parallel classes of a resolvable (21, 3, 1)-BIBD. □

Step 2

The second step is to construct $(v, \{4, 7, 10, 19\})$-PBDs for all $v \equiv 1 \pmod 3$. In order to do this, we will make use of a powerful recursive construction for group-divisible designs known as "Wilson's construction for GDDs".

Theorem 7.39 (Wilson's Construction for GDDs). *Suppose that* $(X, \mathcal{G}, \mathcal{A})$ *is a group-divisible design. Let* w *be a positive integer and let* I *be a set of size* w. *Suppose that* $K \subseteq \{n \in \mathbb{Z} : n \geq 2\}$ *and, for every* $A \in \mathcal{A}$, *suppose that there is a group-divisible design having* $|A|$ *groups of size* w *and all block sizes in* K, *say*

$$(A \times I, \{\{x\} \times I : x \in A\}, \mathcal{B}_A).$$

Define

$$Y = X \times I,$$
$$\mathcal{H} = \{G \times I : G \in \mathcal{G}\}, \quad \textit{and}$$
$$\mathcal{B} = \bigcup_{A \in \mathcal{A}} \mathcal{B}_A.$$

Then $(Y, \mathcal{H}, \mathcal{B})$ *is a group-divisible design such that* $|B| \in K$ *for all* $B \in \mathcal{B}$.

Proof. Clearly $|B| \in K$ for all $B \in \mathcal{B}$, so we just need to verify that $(Y, \mathcal{H}, \mathcal{B})$ is a group-divisible design. Take two points from different groups, say (x, i) and (y, j), where $x \neq y$. There is a unique block $A \in \mathcal{A}$ such that $x, y \in A$. Then there is a unique block $B \in \mathcal{B}_A$ such that $(x, i), (y, j) \in B$. □

We have already used a form of Theorem 7.39 in the proof of Theorem 7.20, where the main step in the construction can be viewed as an application of Wilson's construction in which $w = k - 1$.

We will make essential use of the following corollary of Theorem 7.39, which provides constructions for group-divisible designs having blocks of size 4.

Corollary 7.40. *Suppose there is a* $\mathsf{TD}(5,t)$ *and* $0 \leq u \leq t$. *Then there exists a group-divisible design on* $3(4t+u)$ *points, having four groups of size* $3t$ *and one group of size* $3u$ *and blocks of size four.*

Proof. First, construct a truncated transversal design having four groups of size t and one group of size u and blocks of sizes four and five. Then, apply Theorem 7.39 with $w = 3$ and $K = \{4\}$. We require group-divisible designs with four and five groups of size three and blocks of size four. These are obtained from $(13,4,1)$- and $(16,4,1)$-BIBDs by applying Lemma 7.18. The result follows. □

We now construct the necessary $(v, \{4,7,10,19\})$-PBDs. We do this in several steps.

Lemma 7.41. *Suppose* $n \geq 0$ *is an integer such that*

$$n \notin T = \{0,1,2,3,6,7,\ldots,19,26,27,36,37,\ldots,43,66,67\}.$$

Then there exists a $(3n+1, \{3t+1, 3u+1, 4\})$-PBD *for some integers* $t, u \geq 0$.

Proof. Write $n = 24m + j$, where $4 \leq j \leq 27$ (this can be done uniquely). If $4 \leq j \leq 19$, then take $t = 6m+1$ and $u = j - 4$; if $20 \leq j \leq 27$, then take $t = 6m+5$ and $u = j - 20$. In each case, we have that $n = 4t + u$, and a $\mathsf{TD}(5,t)$ exists by Theorem 6.34 since $\gcd(t,2) = \gcd(t,3) = 1$. It is straightforward to verify that $t \geq u$ if and only if $n \notin T$. Therefore, it follows from Corollary 7.40 that, if $n \notin T$, then there is a group-divisible design on $3n$ points, which has four groups of size $3t$ and one group of size $3u$ and blocks of size four. Applying Lemma 7.16, we see that there is a $(3n+1, \{3t+1, 3u+1, 4\})$-PBD for these values of n. □

Lemma 7.42. *Suppose that*

$$n \in \{16,17,18,19,36,37,\ldots,43,66,67\}.$$

Then there exists a $(3n+1, \{3t+1, 3u+1, 4\})$-PBD *for some integers* $t, u \geq 0$.

Proof. For $16 \leq n \leq 19$, take $t = 4$; for $36 \leq n \leq 43$, take $t = 9$; and for $n = 66, 67$, take $t = 16$. Let $u = n - 4t$. In each case, we have that $n = 4t + u$, where $0 \leq u \leq t$, and a $\mathsf{TD}(5,t)$ exists by Theorem 6.34. Then, proceed as in the proof of Lemma 7.41. □

Lemma 7.43. *Suppose that*

$$n \in \{6,\ldots,15,26,27\}.$$

Then there exists a $(3n+1,\{4,7,10\})$*-PBD.*

Proof. For $n = 8,9,12,13$, there is a $(3n+1,4,1)$-BIBD by Theorem 7.30.

For $n = 7,10$, start with resolvable $(2n+1,3,1)$-BIBDs, which exist by Lemma 7.38. Then apply Lemma 7.3.

For $n = 15,27$, start with $(n+1,4,1)$-BIBDs, which exist by Theorem 7.30. Apply Lemma 7.18 to obtain a group-divisible design with $n/3$ groups of size three and blocks of size four. Then apply Theorem 7.39 with $w = 3$, using as ingredients group-divisible designs with four groups of size three and blocks of size four (these arise from $(13,4,1)$-BIBDs using Theorem 7.30). The result is a group-divisible design with $n/3$ groups of size nine and blocks of size four. Then apply Lemma 7.16 to obtain $(3n+1,\{4,10\})$-PBDs.

The cases $n = 14,26$ are done in a similar fashion. We begin with group-divisible designs having $n/2$ groups of size two and blocks of size four, which are presented in Examples 7.44 and 7.45, respectively. Then proceed exactly as in the cases $n = 15,27$, obtaining group-divisible designs with $n/2$ groups of size six and blocks of size four. Then apply Lemma 7.16 to obtain $(3n+1,\{4,7\})$-PBDs.

The final case is $n = 11$. A $(34,\{4,7\})$-PBD is constructed in Example 7.46. □

Example 7.44. A group-divisible design $(X,\mathcal{G},\mathcal{A})$ having seven groups of size two and blocks of size four. $X = \mathbb{Z}_{14}$,

$$\mathcal{G} = \{\{0,7\},\{1,8\},\ldots,\{6,13\}\},$$

and

$$\mathcal{A} = \{\{0,2,5,6\}+i : i \in \mathbb{Z}_{14}\}.$$

■

Example 7.45. A group-divisible design $(X,\mathcal{G},\mathcal{A})$ having 13 groups of size two and blocks of size four. $X = \mathbb{Z}_{26}$,

$$\mathcal{G} = \{\{0,13\},\{1,14\},\ldots,\{12,25\}\},$$

and

$$\mathcal{A} = \{\{0,6,8,9\}+i,\{0,4,11,16\}+i, : i \in \mathbb{Z}_{26}\}.$$

■

Example 7.46. A $(34,\{4,7\})$-PBD. Define $X = \mathbb{Z}_9 \times \mathbb{Z}_3$, and define the following four sets of blocks of sizes three and four:

$$\mathcal{A}_1 = \{\{(0,0),(2,1),(2,2),(3,2)\} + (i,j) : (i,j) \in \mathbb{Z}_9 \times \mathbb{Z}_3\}$$
$$\mathcal{A}_2 = \{\{(0,0),(3,1),(5,1)\} + (i,j) : (i,j) \in \mathbb{Z}_9 \times \mathbb{Z}_3\}$$
$$\mathcal{A}_3 = \{\{(0,0),(4,1),(8,1)\} + (i,j) : (i,j) \in \mathbb{Z}_9 \times \mathbb{Z}_3\}$$
$$\mathcal{A}_4 = \{\{(0,0),(3,0),(6,0)\} + (i,j) : i = 0,1,2, j \in \mathbb{Z}_3\}.$$

It is not hard to check that $(X, \mathcal{A}_1 \cup \mathcal{A}_2 \cup \mathcal{A}_3 \cup \mathcal{A}_4)$ is a $(27, \{3,4\})$-PBD. Now, $\mathcal{A}_4$ is a parallel class, and it is not difficult to show that each of $\mathcal{A}_2$ and $\mathcal{A}_3$ can be partitioned into three parallel classes. We obtain a total of seven parallel classes, which we name Π_i, $i = 1,\ldots,7$. Adjoin a new point ∞_i to each block in Π_i, for $1 \leq i \leq 7$, and denote the modified parallel classes as Π_i', $i = 1,\ldots,7$. Then create a new block of size seven, namely $\Omega = \{\infty_1 \ldots, \infty_7\}$. It is clear that $(Y, \mathcal{B})$ is a $(34, \{4,7\})$-PBD, where $Y = X \cup \Omega$ and

$$\mathcal{B} = \mathcal{A}_1 \bigcup \left(\bigcup_{i=1}^{7} \Pi_i' \right) \bigcup \{\Omega\}.$$

∎

Theorem 7.47. *There exists a* $(v, \{4,7,10,19\})$-PBD *for all* $v \equiv 1 \pmod 3$, $v \geq 4$.

Proof. The proof is by induction on v. Clearly there exists a $(v, \{4,7,10,19\})$-PBD if $v \in \{4,7,10,19\}$. Denote $n = (v-1)/3$. If $n \in T \setminus \{4,7,10,19\}$, then apply Lemma 7.42 or 7.43 to obtain the desired pairwise balanced design. If $n \notin T$, then apply Lemma 7.41 to obtain a $(3n+1, \{3t+1, 3u+1, 4\})$-PBD for some integers $t \geq 1$, $u \geq 0$. By induction, $3t+1, 3u+1 \in \mathbb{B}(\{4,7,10,19\})$ (or $3u+1 = 1$), so it follows from Lemma 7.10 that $v \in \mathbb{B}(\{4,7,10,19\})$ by ignoring blocks of size one if they are present. By induction, the proof is complete. □

Step 3

Summarizing, here is the main existence result concerning Kirkman triple systems.

Theorem 7.48. *There exists a* KTS(v) *if and only if* $v \equiv 3 \pmod 6$ *and* $v \geq 9$.

Proof. We have already discussed the necessary conditions. Sufficiency follows from Theorem 7.35, Lemma 7.38, and Theorem 7.47. □

7.6 Notes and References

Much recent information on pairwise balanced designs can be found in Section III of "The CRC Handbook of Combinatorial Designs" [27].

Haim Hanani was a pioneer in the use of pairwise balanced designs and their application to the construction of PBDs. Theorem 7.31 was proven by Hanani in [57].

Frames were first defined formally in Stinson [102], although the use of these objects is implicit in earlier work of Hanani. Furino, Miao, and Yin [46] is a monograph on frames and their application to the construction of resolvable designs.

Wilson wrote a series of three important papers [118, 119, 123] in which he proved that the necessary numerical conditions for existence of a (v,K)-PBD (Lemma 7.7) are asymptotically sufficient (i.e., the necessary conditions are sufficient for $v > c_K$, where c_K is a constant depending on K). Theorem 7.39 is from Wilson [122], and Theorem 7.48 is due to Ray-Chaudhuri and Wilson [84].

7.7 Exercises

7.1 Describe how to construct the following PBDs.
  (a) a $(31,\{4,10\})$-PBD.
  (b) a $(31,\{3,15\})$-PBD.
  (c) a $(31,\{3,5\})$-PBD.
  (d) a $(31,\{3,11\})$-PBD.
  (e) a $(36,\{5,8\})$-PBD.
  (f) a $(36,\{4,9\})$-PBD.
  (g) a $(36,\{3,4\})$-PBD.
  (h) a $(33,\{4,5,7\})$-PBD.
  (i) a $(49,\{4,5,9\})$-PBD.
  (j) a $(49,\{6,9\})$-PBD.
  (k) a $(49,\{3,6\})$-PBD.

7.2 Suppose that a $\mathsf{TD}(k,t)$ exists, and let $2 \le u \le k$. Prove that a $(k(t-1)+u, \{k,k-1,t,t-1,u\})$-PBD exists.
Hint: Delete points from a block of the given transversal design.

7.3 Using the facts that R_3 is PBD-closed, $3 \in \mathsf{R}_3$, and a $(9,3,1)$-BIBD exists, construct a $(19,3,1)$-BIBD.

7.4 Given any $(v,3,1)$-BIBD, say $(X,\mathcal{A})$, describe how to construct a $(2v+1,3,1)$-BIBD, say $(Y,\mathcal{B})$, where $X \subseteq Y$ and $\mathcal{A} \subseteq \mathcal{B}$.

7.5 Suppose there is a GDD, say $(X,\mathcal{G},\mathcal{A})$, such that all blocks have size k and all groups have size m. Denote $|X| = v$. Prove that the following hold:
  (a) $v \equiv 0 \pmod{m}$,
  (b) $v \ge mk$,
  (c) $v - m \equiv 0 \pmod{k-1}$, and
  (d) $v(v-m) \equiv 0 \pmod{k^2-k}$.

7.6 A GDD is *resolvable* if the set of blocks of the GDD can be partitioned into parallel classes. Prove that every group in a resolvable GDD has the same size.

7.7 Use any method you wish to construct a $(15,3,1)$-BIBD, and then construct a 3-frame on seven holes from this BIBD.

7.8 (a) Prove that $6 \notin \mathbb{B}(\{3,4\})$.
(b) Prove that $\mathbb{B}(\{3,4\}) \subseteq \{n \geq 3 : n \equiv 0,1 \pmod 3\}$.
(c) Give a complete proof that

$$\mathbb{B}(\{3,4\}) = \{n \geq 3 : n \equiv 0,1 \pmod 3, n \neq 6\}.$$

7.9 Let $K = \{3,4,5,6,8\}$. Assume that there exists a (v,K)-PBD for all $3 \leq v \leq 25$. Then use (truncated) transversal designs and induction to give a complete proof that there exists a (v,K)-PBD for all $v \geq 3$.
Hint: Use the fact that a $\mathsf{TD}(4,n)$ exists for all positive integers $n \neq 2,6$.

7.10 (a) A $\mathsf{GRS}(v,2,1)$, say R, is *standardized* if there exists a special symbol, say ∞, such that ∞ occurs in the cells on the main diagonal of R. Prove that any $\mathsf{GRS}(v,2,1)$ can be transformed into a standardized $\mathsf{GRS}(v,2,1)$ by means of an appropriate permutation of the columns of R.
(b) Define $\mathsf{S} = \{v-1 : \text{a standardized } \mathsf{GRS}(v,2,1) \text{ exists}\}$. Prove that S is PBD-closed.

8

Pairwise Balanced Designs II: Minimal Designs

In the previous chapter, we studied constructions of pairwise balanced designs whose block sizes are required to be elements of a specified set of integers. In this chapter, we consider the problem of determining the minimum number of blocks in pairwise balanced designs in which the maximum size of a block is specified or in which the size of a particular block is specified.

For a pairwise balanced design, $(X, \mathcal{A})$, we will generally denote $b = |\mathcal{A}|$ (i.e., b is the number of blocks in the PBD).

8.1 The Stanton-Kalbfleisch Bound

Theorem 8.1 (Stanton-Kalbfleisch Bound). *Let k and v be integers such that $2 \le k < v$. Suppose there is a $(v, \{2, \ldots, v-1\})$-PBD in which there exists a block containing exactly k points. Then*

$$b \geq \mathsf{SK}(k, v) = 1 + \frac{k^2(v-k)}{v-1}.$$

Proof. Suppose that $(X, \mathcal{A})$ is a $(v, \{2, \ldots, v-1\})$-PBD such that $A \in \mathcal{A}$ is a block containing exactly k points. Denote the blocks of $\mathcal{A}$ by $A_1, \ldots, A_b$, where $A_b = A$.

Now construct a set system $(Y, \mathcal{B})$ by deleting all the points in the block A_b as follows:

$$\begin{aligned} Y &= X \backslash A_b, \\ B_i &= A_i \backslash A_b, 1 \leq i \leq b-1, \quad \text{and} \\ \mathcal{B} &= \{B_i : 1 \leq i \leq b-1\}. \end{aligned}$$

$(Y, \mathcal{B})$ is a set system with $v - k$ points and $b - 1$ blocks in which every pair of points occurs in a unique block. (This set system may contain blocks of size one, so it need not be a PBD.)

For $1 \leq i \leq b-1$, denote $k_i = |B_i|$. Note that $k_i = |A_i|$ or $k_i = |A_i| - 1$ for $1 \leq i \leq b-1$. Furthermore, $k_i = |A_i| - 1$ if and only if A_i intersects A_b in a point.

Denote the points in Y by y_j, $1 \leq j \leq v-k$. For $1 \leq j \leq v-k$, define $r_j = |\{B_i \in \mathcal{B} : y_j \in B_i\}|$. Then a straightforward generalization of Theorem 1.8 shows that

$$\sum_{i=1}^{b-1} k_i = \sum_{j=1}^{v-k} r_j. \tag{8.1}$$

Now, in the pairwise balanced design $(X, \mathcal{A})$, every point y_j must occur in a unique block with each of the points in A_b. Hence $r_j \geq k$ for all j, $1 \leq j \leq v-k$. Substituting into (8.1), it follows that

$$\sum_{i=1}^{b-1} k_i \geq k(v-k). \tag{8.2}$$

Every pair of points in Y occurs in exactly one of the B_i's, so it follows that

$$\sum_{i=1}^{b-1} k_i(k_i - 1) = (v-k)(v-k-1). \tag{8.3}$$

Denote the mean of the integers $k_1, \ldots, k_{b-1}$ to be

$$\overline{k} = \frac{\sum_{i=1}^{b-1} k_i}{b-1}. \tag{8.4}$$

Now we study the quantity

$$S = \sum_{i=1}^{b-1} (k_i - \overline{k})^2.$$

We can use equations (8.3) and (8.4) to derive a formula for S:

$$\begin{aligned} S &= \sum_{i=1}^{b-1} k_i^2 - 2\overline{k} \sum_{i=1}^{b-1} k_i + (b-1)(\overline{k})^2 \\ &= \sum_{i=1}^{b-1} k_i(k_i - 1) - (2\overline{k} - 1) \sum_{i=1}^{b-1} k_i + (b-1)(\overline{k})^2 \\ &= (v-k)(v-k-1) - (2\overline{k} - 1)(b-1)(\overline{k}) + (b-1)(\overline{k})^2 \\ &= (v-k)(v-k-1) - (b-1)\overline{k}(\overline{k} - 1). \end{aligned}$$

Also, we observe that S is a sum of nonnegative terms, so clearly $S \geq 0$. Therefore we have that

$$0 \leq (v-k)(v-k-1) - (b-1)\overline{k}(\overline{k} - 1). \tag{8.5}$$

We have that $\overline{k} \geq 1$ because $k_i \geq 1$ for all i, and hence, from (8.2), we have that

$$\overline{k}(\overline{k}-1) \geq \left(\frac{k(v-k)}{b-1}\right)\left(\frac{k(v-k)}{b-1}-1\right).$$

Substituting into (8.5), we obtain

$$0 \leq (v-k)(v-k-1) - (b-1)\left(\frac{k(v-k)}{b-1}\right)\left(\frac{k(v-k)}{b-1}-1\right).$$

Dividing by a factor of $v-k$ and simplifying, we obtain

$$\begin{aligned} 0 &\leq v-k-1-\left(\frac{k}{b-1}\right)(k(v-k)-(b-1)) \\ &= v-1-\frac{k^2(v-k)}{b-1}. \end{aligned}$$

Hence,

$$b \geq 1+\frac{k^2(v-k)}{v-1}.$$

□

When $2 \leq k \leq v-2$, the case of equality in the bound above can be characterized in a very nice way. We prove the following theorem.

Theorem 8.2. *Suppose that k and v are integers such that $2 \leq k \leq v-2$. Then there is a $(v, \{2, \ldots, v-1\})$-PBD with $\mathsf{SK}(k,v)$ blocks and having a block containing exactly k points if and only if there is a resolvable $(v-k, (v-1)/k, 1)$-BIBD.*

Proof. Suppose there is a $(v, \{2, \ldots, v-1\})$-PBD with $\mathsf{SK}(k,v)$ blocks and having a block containing exactly k points. We use the same notation as in the proof of Theorem 8.1. Since all inequalities in the proof of Theorem 8.1 must be equalities, the following conditions hold:

- $b-1 = k^2(v-k)/(v-1)$,
- $k_i = \overline{k} = k(v-k)/(b-1)$ for $1 \leq i \leq b-1$, and
- $r_j = k$ for $1 \leq j \leq v-k$.

These conditions imply that $k(v-k)/(b-1) = (v-1)/k$. Because $v > k+1$, it follows that $v-k > (v-1)/k > 1$, and therefore the set system $(Y, \mathcal{B})$ is a $(v-k, (v-1)/k, 1)$-BIBD.

We now show that $(Y, \mathcal{B})$ is resolvable. For each point $x \in A_b$, let $\mathcal{A}(x)$ denote the blocks in $\mathcal{A}$ that contain the point x, and let $\mathcal{B}(x)$ denote the corresponding blocks in $\mathcal{B}$ (obtained by deleting x from each block in $\mathcal{A}(x)$). It is obvious that each set of blocks $\mathcal{B}(x)$ is a parallel class. Furthermore, every block in $\mathcal{B}$ is in exactly one of these $k = r$ parallel classes (if there were a block B_i in $\mathcal{B}$ that is not in one of these parallel classes, then each point in B_i would occur in more than r blocks, a contradiction). Therefore we have a resolution of $(Y, \mathcal{B})$.

Conversely, suppose $(Y, \mathcal{B})$ is a resolvable $(v-k, (v-1)/k, 1)$-BIBD for some integers v and k such that $2 \le k \le v-2$. This BIBD is a $(v', b', r', k', 1)$-BIBD, where

$$\begin{aligned} v' &= v - k, \\ k' &= \frac{v-1}{k}, \\ r' &= k, \quad \text{and} \\ b' &= \frac{k^2(v-k)}{v-1}. \end{aligned}$$

If we apply Lemma 7.3, then we obtain a $(v' + r', \{k'+1, r'\})$-PBD having $b' + 1$ blocks, where $r' = (v'-1)/(k'-1)$. This pairwise balanced design has v points, $\mathsf{SK}(k, v)$ blocks, and a block of size k, as desired. □

For $v \ge 4$, a *near-pencil* is a $(v, \{2, v-1\})$-PBD, say $(X, \mathcal{A})$, in which $\mathcal{A}$ contains one block of size $v-1$ and $v-1$ blocks of size two. A near-pencil on three points contains three blocks of size two. For all integers $v \ge 4$, a near-pencil on v points exists, and it has $v = \mathsf{SK}(v-1, v)$ blocks.

Given a near-pencil, the set system $(Y, \mathcal{B})$ (constructed in the proof of Theorem 8.2) would have one point and blocks of size one. It is not a BIBD, which is why we required that $v \ge k+2$ in Theorem 8.2.

Definition 8.3. *Let $g^k(v)$ denote the minimum number of blocks in any $(v, K, 1)$-*PBD *in which the largest block has size equal to k.*

Define the function

$$\mathsf{C}(k, v) = \frac{v^2 - v}{k^2 - k}.$$

Then we have the following upper bound on $g^k(v)$.

Theorem 8.4.

$$g^k(v) \ge \max\{\mathsf{C}(k, v), \mathsf{SK}(k, v)\}.$$

Proof. Suppose that $(X, \mathcal{A})$ is a pairwise balanced design on v points, having b blocks and such that the largest block has size k. First, Theorem 8.1 shows that $b \ge \mathsf{SK}(k, v)$. Second, it is a simple matter to see that $b \ge \mathsf{C}(k, v)$ because every block in $\mathcal{A}$ contains at most $\binom{k}{2}$ pairs and all the blocks together contain $\binom{v}{2}$ pairs. Finally, because $b \ge \mathsf{SK}(k, v)$ and $b \ge \mathsf{C}(k, v)$, it follows that b must be at least as big as the maximum of these two numbers. □

8.1.1 The Erdös-de Bruijn Theorem

We next state and prove a famous theorem, due to Erdös and de Bruijn, that characterizes the nontrivial pairwise balanced designs with $\lambda = 1$ having the minimum possible number of blocks.

Theorem 8.5 (Erdös-de Bruijn Theorem). *Let* $(X, \mathcal{A})$ *be a* $(v, \{2, \ldots, v-1\})$-PBD, *and suppose that the number of blocks in the PBD is denoted by* b. *Then* $b \geq v$. *Furthermore,* $b = v$ *if and only if* $(X, \mathcal{A})$ *is a projective plane or a near-pencil.*

Proof. Let k be the size of the largest block in $(X, \mathcal{A})$. If $k^2 - k + 1 < v$, then $k(k-1) < v - 1$, and hence

$$\mathsf{C}(k, v) = \frac{v(v-1)}{k(k-1)} > \frac{v(v-1)}{v-1} = v.$$

In this case, Theorem 8.4 implies that $b > v$. Therefore we can assume that $k^2 - k + 1 \geq v \geq k + 1$.

Now, let us consider the conditions under which $\mathsf{SK}(k, v) \leq v$:

$$\begin{aligned} & \mathsf{SK}(k,v) \leq v \\ \iff \quad & k^2(v-k) \leq (v-1)^2 \\ \iff \quad & v^2 - (k^2+2)v + k^3 + 1 \geq 0 \\ \iff \quad & (v-(k+1))(v-(k^2-k+1)) \geq 0. \end{aligned} \tag{8.6}$$

Given that $k^2 - k + 1 \geq v \geq k + 1$, the inequality (8.6) holds if and only if $v = k + 1$ or $v = k^2 - k + 1$. In other words, for $k^2 - k + 1 \geq v \geq k + 1$, $\mathsf{SK}(k, v) \geq v$, and $\mathsf{SK}(k, v) = v$ only if $v = k + 1$ or $v = k^2 - k + 1$. We further consider these two possible cases as follows:

1. If $v = k + 1$ and $b = v$, then $(X, \mathcal{A})$ is a near-pencil. Conversely, if $(X, \mathcal{A})$ is a near-pencil, then $b = v$.
2. Suppose $v = k^2 - k + 1$ and $b = v$. If $k = 2$, then $v = 3$ and we have a near-pencil. Therefore we can assume that $k \geq 3$, and we will show that $(X, \mathcal{A})$ is a projective plane of order $k - 1$ as follows. Theorem 8.2 implies that the design obtained by deleting a block of size k is a $(v - k, (v-1)/k, 1)$-BIBD. This design is a $((k-1)^2, k-1, 1)$-BIBD; i.e., an affine plane of order $k - 1$. Therefore $(X, \mathcal{A})$ is a projective plane of order $k - 1$. Conversely, if $(X, \mathcal{A})$ is a projective plane of order $k - 1$, then it has a longest block of size k, and $b = v$.

This completes the proof. □

The proof of Theorem 8.5 was algebraic. It is possible to give a proof of this result that is more analytic in nature. We illustrate this approach by graphing the function $\max\{\mathsf{C}(k, v), \mathsf{SK}(k, v)\}$ when $v = 21$ in Figure 8.1. This function is graphed for real values of k ranging from 2 to 20.

Several observations may be made from this graph that can then be generalized to arbitrary v. We list these observations now.

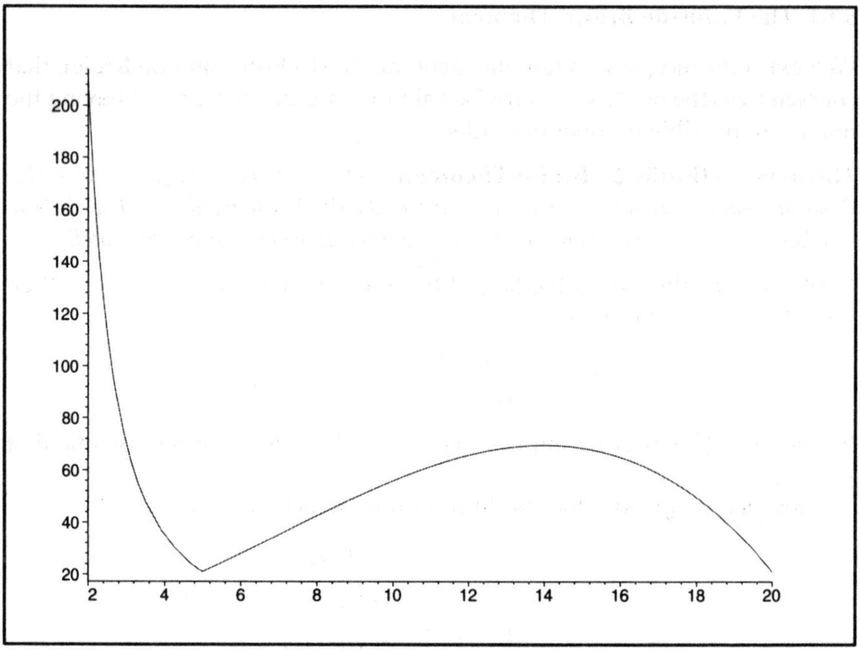

Fig. 8.1. Lower Bounds on $g^k(21)$ for $2 \leq k \leq 20$

Lemma 8.6. *Suppose that k and v are real numbers such that $2 \leq k \leq v-1$. Then the following hold.*

1. *If $k^2 - k + 1 < v$, then $\mathsf{C}(k,v) > \mathsf{SK}(k,v)$; and if $k^2 - k + 1 > v$, then $\mathsf{C}(k,v) < \mathsf{SK}(k,v)$.*
2. *When $k^2 - k + 1 = v$ (i.e., when $k = (-1 + \sqrt{4v-3})/2$, it holds that*

$$\frac{v(v-1)}{k(k-1)} = \mathsf{SK}(k,v) = v.$$

3. *For $k \geq 2$, $v(v-1)/(k(k-1))$ is a decreasing function of k.*
4. *For $2 \leq k \leq v-1$, $\mathsf{SK}(k,v)$ attains its maximum when $k = 2v/3$; $\mathsf{SK}(k,v)$ is an increasing function of k when $2 \leq k < 2v/3$; and $\mathsf{SK}(k,v)$ is a decreasing function of k when $2v/3 < k \leq v-1$.*
5. *$\mathsf{SK}(v-1,v) = v$.*

These properties are sufficient to give an alternate proof of Theorem 8.5; we leave the details of this proof (and the proofs of the above-mentioned properties) to the reader.

8.2 Improved Bounds

A strengthening of the Stanton-Kalbfleisch bound was given by Stinson. We prove this result now.

Theorem 8.7 (Stinson Bound). *Let k and v be integers such that $2 \le k < v$. Suppose there is a $(v, \{2, \ldots, v-1\})$-PBD in which there exists a block containing exactly k points. For any integer t, define*

$$f(t,k,v) = 1 + (v-k)\left(\frac{k(2t+1)-(v-1)}{t^2+t}\right).$$

Then $b \ge f(t,k,v)$.

Proof. We use the same notation as in the proof of Theorem 8.1. The proof is again based on (8.2) and (8.3). Let t be an integer and consider the quantity

$$S = \sum_{i=1}^{b-1} (k_i - t)(k_i - (t+1)).$$

We use equations (8.2) and (8.3) to derive an upper bound on S:

$$\begin{aligned} S &= \sum_{i=1}^{b-1} {k_i}^2 - (2t+1)\sum_{i=1}^{b-1} k_i + (b-1)(t^2+t) \\ &= \sum_{i=1}^{b-1} k_i(k_i-1) - 2t\sum_{i=1}^{b-1} k_i + (b-1)(t^2+t) \\ &\le (v-k)(v-k-1) - 2tk(v-k) + (b-1)(t^2+t). \end{aligned}$$

S is a sum of nonnegative terms, so clearly $S \ge 0$. Therefore we have that

$$0 \le (v-k)(v-k-1) - 2tk(v-k) + (b-1)(t^2+t). \tag{8.7}$$

Rearranging (8.7), the desired bound is obtained. □

Theorem 8.8. *Let t, k, and v be integers such that $2 \le k < v$. Then the function $f(t,k,v)$ is maximized when $t = \lfloor \frac{v-1}{k} \rfloor$.*

Proof. We compute

$$\begin{aligned} &f(t,k,v) - f(t-1,k,v) \\ &= (v-k)\left(\frac{k(2t+1)-(v-1)}{t(t+1)} - \frac{k(2t-1)-(v-1)}{t(t-1)}\right) \\ &= \left(\frac{v-k}{t}\right)\left(\frac{k(2t+1)-(v-1)}{t+1} - \frac{k(2t-1)-(v-1)}{t-1}\right). \end{aligned}$$

It therefore follows that

$$\begin{aligned}
& f(t,k,v) \geq f(t-1,k,v) \\
\iff\quad & \frac{k(2t+1)-(v-1)}{t+1} \geq \frac{k(2t-1)-(v-1)}{t-1} \\
\iff\quad & (k(2t+1)-(v-1))(t-1) \geq (k(2t-1)-(v-1))(t+1) \\
\iff\quad & (v-1)(2t+1-(2t-1)) \geq k((2t-1)(t+1)-(2t+1)(t-1)) \\
\iff\quad & 2(v-1) \geq 2tk \\
\iff\quad & t \leq \frac{v-1}{k}.
\end{aligned}$$

Because t is an integer, it follows that $f(t,k,v)$ is maximized when $t = \lfloor \frac{v-1}{k} \rfloor$. □

For future reference, define $\mathsf{St}(k,v) = f(\lfloor \frac{v-1}{k} \rfloor, k, v)$.

Theorem 8.9. *$\mathsf{St}(k,v) \geq \mathsf{SK}(k,v)$ for all integers k and v such that $2 \leq k < v$. Furthermore, $\mathsf{St}(k,v) = \mathsf{SK}(k,v)$ if and only if $v-1 \equiv 0 \pmod{k}$.*

Proof. We consider the conditions under which $f(t,k,v) \geq \mathsf{SK}(k,v)$. We have that

$$\begin{aligned}
& f(t,k,v) \geq \mathsf{SK}(k,v) \\
\iff\quad & (v-k)\left(\frac{k(2t+1)-(v-1)}{t^2+t}\right) \geq \frac{k^2(v-k)}{v-1} \\
\iff\quad & \frac{k(2t+1)-(v-1)}{t^2+t} \geq \frac{k^2}{v-1} \\
\iff\quad & k^2(t^2+t) \leq (v-1)(k(2t+1)-(v-1)) \\
\iff\quad & (k(t+1)-(v-1))(kt-(v-1)) \leq 0 \\
\iff\quad & \frac{v-1}{k} - 1 \leq t \leq \frac{v-1}{k}.
\end{aligned}$$

This last inequality is satisfied when $t = \lfloor \frac{v-1}{k} \rfloor$, and therefore it follows that $\mathsf{St}(k,v) \geq \mathsf{SK}(k,v)$. It is also easy to see that $\mathsf{St}(k,v) = \mathsf{SK}(k,v)$ if and only if $(v-1)/k$ is an integer. □

We can prove yet another bound based on the same inequalities. This bound is a strengthening of the inequality $b \geq \mathsf{C}(k,v)$, which applies when the longest block has size k (this inequality was derived in the proof of Theorem 8.4).

Theorem 8.10. *Let k and v be integers such that $2 \leq k < v$. Suppose there is a $(v, \{2, \ldots, v-1\})$-PBD in which the largest block contains exactly k points. Then*

$$b \geq \mathsf{C}^*(k,v) = \frac{v\left(2(k-1)\left\lceil \frac{v-1}{k-1} \right\rceil - (v-1)\right)}{k^2-k}.$$

Proof. Let $(X, \mathcal{A})$ be the hypothesized PBD. Denote the blocks $A_1, \ldots, A_b$, and define $k_i = |A_i|$, for $1 \leq i \leq b$. Let the points be denoted $x_1, \ldots, x_v$, and define $r_j = |\{i : x_j \in A_i\}|$ for $1 \leq j \leq v$.

Using the fact that all blocks have size at most k, it is easily seen that $r_j \geq (v-1)/(k-1)$, $1 \leq j \leq v$. Every r_j is an integer, so

$$r_j \geq \left\lceil \frac{v-1}{k-1} \right\rceil .$$

Therefore we have the inequality

$$\sum_{i=1}^{b} k_i \geq v \left\lceil \frac{v-1}{k-1} \right\rceil . \tag{8.8}$$

Every pair of points occurs in exactly one block, so it follows that

$$\sum_{i=1}^{b} k_i(k_i - 1) = v(v-1). \tag{8.9}$$

Now, consider the quantity

$$S = \sum_{i=1}^{b} (k_i - (k-1))(k_i - k).$$

Proceeding as in the proof of Theorem 8.7, we use equations (8.8) and (8.9), and the fact that $S \geq 0$, to derive an upper bound on S:

$$\begin{aligned} S &= \sum_{i=1}^{b} {k_i}^2 - (2k-1)\sum_{i=1}^{b} k_i + b(k^2 - k) \\ &= \sum_{i=1}^{b} k_i(k_i - 1) - (2k-2)\sum_{i=1}^{b} k_i + b(k^2 - k) \\ &\leq v(v-1) - 2(k-1)v \left\lceil \frac{v-1}{k-1} \right\rceil + b(k^2 - k). \end{aligned}$$

This yields the desired bound on b. □

Theorem 8.11. $\mathsf{C}^*(k,v) \geq \mathsf{C}(k,v)$ *for all integers* k *and* v *such that* $2 \leq k < v$. *Furthermore,* $\mathsf{C}^*(k,v) = \mathsf{C}(k,v)$ *if and only if* $v - 1 \equiv 0 \pmod{k-1}$.

Proof.

$$\begin{aligned} & & \mathsf{C}^*(k,v) &\geq \mathsf{C}(k,v) \\ &\iff & \frac{v\left(2(k-1)\left\lceil \frac{v-1}{k-1}\right\rceil - (v-1)\right)}{k^2-k} &\geq \frac{v^2 - v}{k^2 - k} \\ &\iff & 2(k-1)\left\lceil \frac{v-1}{k-1} \right\rceil - (v-1) &\geq v - 1 \\ &\iff & \left\lceil \frac{v-1}{k-1} \right\rceil &\geq \frac{v-1}{k-1}. \end{aligned}$$

This last inequality is true for all v and k, and equality holds if and only if $(v-1)/(k-1)$ is an integer. □

8.2.1 Some Examples

We illustrate the application of the bounds above in determining values $g^k(v)$ for small k and v. To be specific, Table 8.1 is a table of values of $g^k(v)$ for $3 \le v \le 9$. For each v, we look at all integers k such that $2 \le k \le v-1$. We tabulate the values of the four lower bounds $\mathsf{SK}(k,v)$, $\mathsf{St}(k,v)$, $\mathsf{C}(k,v)$, and $\mathsf{C}^*(k,v)$. (It is of course unnecessary to include the values of $\mathsf{SK}(k,v)$ and $\mathsf{C}(k,v)$ because we have proven that $\mathsf{St}(k,v) \ge \mathsf{SK}(k,v)$ and $\mathsf{C}^*(k,v) \ge \mathsf{C}(k,v)$. We include all four values mainly for the purposes of illustration so that the bounds can easily be compared.) We also include the exact value of $g^k(v)$.

k	v	SK	St	C	C*	$g^k(v)$
2	3	3	3	3	3	3
2	4	11/3	4	6	6	6
3	4	4	4	2	10/3	4
2	5	4	4	10	10	10
3	5	11/2	6	10/3	10/3	6
4	5	5	5	5/3	10/3	5
2	6	21/5	13/3	15	15	15
3	6	32/5	7	5	7	7
4	6	37/5	8	5/2	7/2	8
5	6	6	6	3/2	33/10	6
2	7	13/3	13/3	21	21	21
3	7	7	7	7	7	7
4	7	9	10	7/2	7/2	10
5	7	28/3	10	21/10	7/2	10
6	7	7	7	7/5	49/15	7
2	8	31/7	9/2	28	28	28
3	8	52/7	23/3	28/3	12	12
4	8	71/7	11	14/3	22/3	11
5	8	82/7	13	14/5	18/5	13
6	8	79/7	12	28/15	52/15	12
7	8	8	8	4/3	68/21	8
2	9	9/2	9/2	36	36	36
3	9	31/4	8	12	12	12
4	9	11	11	6	15/2	12
5	9	27/2	15	18/5	18/5	15
6	9	29/2	16	12/5	18/5	16
7	9	53/4	14	12/7	24/7	14
8	9	9	9	9/7	45/14	9

Table 8.1. Values of $g^k(v)$ for Small k and v

With one exception, it can be verified that every value of $g^k(v)$ in Table 8.1 is the ceiling of the maximum of the four lower bounds. In these cases, it suffices to give a construction of a PBD with the appropriate number of blocks.

The one exceptional parameter situation is when $k = 4$ and $v = 9$, where we claim that

$$g^4(9) = 12 = \lceil \max\{\mathsf{SK}(4,9), \mathsf{St}(4,9), \mathsf{C}(4,9), \mathsf{C}^*(4,9)\} \rceil + 1.$$

In order to prove that $g^4(9) = 12$, we need to construct a PBD with 12 blocks as well as prove that no PBD with 11 blocks exists. We can prove that $g^4(9) \neq 11$ by referring to Theorem 8.2. Note that $\mathsf{SK}(4,9) = 11$, so there exists a $(9, \{2, \ldots, 8\})$-PBD having a block of size four if and only if there is a resolvable $(5,2,1)$-BIBD. This is clearly impossible because $5 \not\equiv 0 \pmod 2$. Hence, we conclude that $g^4(9) > 11$.

For the values of k and v considered in Table 8.1, the construction of PBDs with $g^k(v)$ blocks is not too difficult. Several parameter situations can be handled by similar constructions. For example, the block sets of the PBDs with $k = 2$ consist of all 2-subsets of points; and the PBDs with $k = v - 1$ are near-pencils.

When $k = v - 2$, it is always possible to take a block B of size $v - 2$, a block of size three intersecting B, and take all remaining blocks to have size two. This yields a PBD with $2v - 4$ blocks. It is also easy to verify that $\mathsf{St}(v-2, v) = 2v - 4$ for all $v \geq 4$. Therefore $g^{v-2}(v) = 2v - 4$ for all $v \geq 4$.

The remaining cases have $3 \leq k \leq v - 3$. These PBDs can be constructed fairly easily by trial and error, and we list appropriate block sets in Table 8.2.

k	v	b	blocks
3	6	7	$\{123, 145, 246, 356, 16, 25, 34\}$
3	7	7	(7,3,1)-BIBD
4	7	10	$\{1234, 156, 257, 367, 17, 26, 35, 45, 46, 47\}$
3	8	12	$\{013 \bmod 8\} \cup \{04, 15, 26, 37\}$
4	8	11	$\{1234, 1567, 258, 368, 478, 26, 27, 35, 37, 45, 46\}$
5	8	13	$\{12345, 167, 268, 378, 18, 27, 36, 46, 47, 48, 56, 57, 58\}$
3	9	12	(9,3,1)-BIBD
4	9	12	$\{1234, 1567, 189, 258, 368, 478, 269, 379, 459, 27, 35, 46\}$
5	9	15	$\{12345, 167, 189, 268, 279, 369, 378\} \cup \{ij : i = 4,5; 6 \leq j \leq 9\}$
6	9	16	$\{123456, 178, 279, 389, 19, 28, 37\} \cup \{ij : 4 \leq i \leq 6; 7 \leq j \leq 9\}$

Table 8.2. Block Sets of some PBDs with $b = g^k(v)$ Blocks

8.3 Minimal PBDs and Projective Planes

In this section, we consider the problem of determining the minimum number of blocks in a $(v, \{2, \ldots, v-1\})$-PBD that is not a near-pencil. Equivalently, what is the minimum number of blocks in a $(v, \{2, \ldots, v-2\})$-PBD? Let us denote this quantity by $\mathsf{b}^*(v)$. Note that

$$\mathsf{b}^*(v) = \min\{g^k(v) : 2 \leq k \leq v-2\}.$$

Clearly $v \geq 4$ is necessary in order for $\mathsf{b}^*(v)$ to be defined. The following values of $\mathsf{b}^*(v)$ for $4 \leq v \leq 9$ are easily determined from Table 8.1. We record these values in the next lemma.

Lemma 8.12. $\mathsf{b}^*(4) = \mathsf{b}^*(5) = 6$, $\mathsf{b}^*(6) = \mathsf{b}^*(7) = 7$, $\mathsf{b}^*(8) = 11$, *and* $\mathsf{b}^*(9) = 12$.

Lemma 8.13. *For all integers $v \geq 4$, it holds that $\mathsf{b}^*(v+1) \geq \mathsf{b}^*(v)$.*

Proof. The stated result is true for $v = 4$ by Lemma 8.12, so we will assume that $v \geq 5$. Let $(X, \mathcal{A})$ be a $(v+1, \{2, \ldots, v-1\})$-PBD containing $\mathsf{b}^*(v+1)$ blocks. Let A denote a block in $\mathcal{A}$ having maximum cardinality. Let $x \in A$, and delete x from all blocks in $\mathcal{A}$. If any blocks of size one are created by this process, then delete them. This creates a PBD, say $(X \backslash \{x\}, \mathcal{B})$, on v points, having at most $\mathsf{b}^*(v+1)$ blocks.

If we can show that there are no blocks of size $v-1$ in $\mathcal{B}$, then we will be done. Suppose that $B \in \mathcal{B}$ has cardinality $v-1$. Then there are at least two blocks of cardinality $v-1$ in $(X, \mathcal{A})$, namely A and B. $|A \cap B| \leq 1$, so $|A \cup B| \geq 2v-3$. However, $|A \cup B| \leq v+1$, so $v \leq 4$. This contradicts the assumption that $v \geq 5$, and the proof is complete. □

Lemma 8.14. *Suppose that $v \geq 6$ and suppose that $k_0 = (1 + \sqrt{4v-3})/2$. Denote $k_1 = \lfloor k_0 \rfloor$ and $k_2 = \lceil k_0 \rceil$. Then*

$$\mathsf{b}^*(v) \geq \min\{\mathsf{C}(k_1, v), \mathsf{SK}(k_2, v), \mathsf{SK}(v-2, v)\}.$$

Proof. For $v \geq 6$, it holds that $k_0 < 2v/3 \leq v-2$. Therefore the result follows from Lemma 8.6. □

Theorem 8.15. *Suppose that $n \geq 2$ and v are integers such that $n^2 + 2 \leq v \leq n^2 + n + 1$. Then $\mathsf{b}^*(v) \geq n^2 + n + 1$. Furthermore, $\mathsf{b}^*(v) = n^2 + n + 1$ if there exists a projective plane of order n.*

Proof. First, we apply Lemma 8.14 with $v = n^2 + 2$. We have $k_1 = n$ and $k_2 = n + 1$. Then

$$\begin{aligned}
\mathsf{C}(k_1, v) &= \mathsf{C}(n, n^2 + 2) \\
&= n^2 + n + 4 + \frac{4n+2}{n^2 - n} \\
&> n^2 + n;
\end{aligned}$$

$$\begin{aligned}\mathsf{SK}(k_2,v) &= \mathsf{SK}(n+1,n^2+2)\\ &= n^2+n+\frac{2}{n^2+1}\\ &> n^2+n; \quad\text{and}\\ \mathsf{SK}(v-2,v) &= \mathsf{SK}(n^2,n^2+2)\\ &= 2n^2-1+\frac{2}{n^2+1}\\ &> n^2+n.\end{aligned}$$

Hence, $\mathsf{b}^*(n^2+2) > n^2+n$. Because b^* is an integer-valued function, it follows that $\mathsf{b}^*(n^2+2) \geq n^2+n+1$. Then Lemma 8.13 implies that $\mathsf{b}^*(v) \geq n^2+n+1$ for all $v \geq n^2+2$.

Now suppose there is a projective plane of order n and $n^2+2 \leq v \leq n^2+n+1$. We can delete any $n^2+n+1-v$ points from the projective plane and obtain a PBD on v points having n^2+n+1 blocks that is not a near-pencil. □

Lemma 8.16. *Suppose that* $v \geq 6$, *and let* $k_0 = (1+\sqrt{4v-3})/2$. *Denote* $k_1 = \lfloor k_0 \rfloor$ *and* $k_2 = \lceil k_0 \rceil$. *Then*

$$\mathsf{b}^*(v) \geq \min\{\mathsf{C}(k_1-1,v),\mathsf{C}^*(k_1,v),\mathsf{St}(k_2,v),\mathsf{SK}(k_2+1,v),\mathsf{SK}(v-2,v)\}.$$

Theorem 8.17. *Suppose that* $n \geq 2$ *and* v *are integers such that* $n^2-n+3 \leq v \leq n^2+1$. *Then* $\mathsf{b}^*(v) \geq n^2+n$. *Furthermore,* $\mathsf{b}^*(v) = n^2+n$ *if there exists a projective plane of order* n.

Proof. From Lemma 8.12, we have that $\mathsf{b}^*(5) = 6$ and $\mathsf{b}^*(9) = 12$. Projective planes of orders 2 and 3 exist, so the theorem is true when $n = 2, 3$. Therefore we can assume that $n \geq 4$.

We apply Lemma 8.16 with $v = n^2-n+3$. We have $k_1 = n$ and $k_2 = n+1$. Then

$$\begin{aligned}\mathsf{C}(k_1-1,v) &= \mathsf{C}(n-1,n^2-n+3)\\ &= n^2+n+7+\frac{14n-8}{(n-1)(n-2)}\\ &> n^2+n-1;\\ \mathsf{C}^*(k_1,v) &= \mathsf{C}^*(n,n^2-n+3)\\ &= n^2+n-1+\frac{6n-12}{n^2-n}\\ &> n^2+n-1;\\ \mathsf{St}(k_2,v) &= \mathsf{St}(n+1,n^2-n+3)\\ &= n^2+n-1+\frac{2n-6}{(n-1)(n-2)}\\ &> n^2+n-1;\end{aligned}$$

$$\begin{aligned}
\mathsf{SK}(k_2+1,v) &= \mathsf{SK}(n+2,n^2-n+2)\\
&= n^2+3n-1-\frac{12n-8}{n^2-n+3}\\
&> n^2+n-1; \quad \text{and}\\
\mathsf{SK}(v-2,v) &= \mathsf{SK}(n^2-n+1,n^2-n+3)\\
&= 2n^2-2n+1+\frac{2}{n^2-n+2}\\
&> n^2+n-1.
\end{aligned}$$

Hence, $\mathsf{b}^*(n^2-n+3) > n^2+n-1$. Because b^* is an integer-valued function, it follows that $\mathsf{b}^*(n^2-n+3) \geq n^2+n$. Then Lemma 8.13 implies that $\mathsf{b}^*(v) \geq n^2+n$ for all $v \geq n^2-n+3$.

Now suppose there is a projective plane of order n, and $n^2-n+3 \leq v \leq n^2+1$. We can delete n points from any block A of the projective plane, delete A, and delete any n^2+1-v additional points. We obtain a PBD on v points having n^2+n blocks that is not a near-pencil. □

Theorem 8.18. *Suppose that $n \geq 3$. Then $\mathsf{b}^*(n^2-n+2) \geq n^2+n-1$. Furthermore, $\mathsf{b}^*(n^2-n+2) = n^2+n-1$ if there exists a projective plane of order n.*

Proof. First, we apply Lemma 8.16 with $v = n^2-n+2$. We have $k_1 = n$ and $k_2 = n+1$. Then

$$\begin{aligned}
\mathsf{C}(k_1-1,v) &= \mathsf{C}(n-1,n^2-n+2)\\
&= n^2+n+5+\frac{10n-8}{(n-1)(n-2)}\\
&> n^2+n-2;\\
\mathsf{C}^*(k_1,v) &= \mathsf{C}^*(n,n^2-n+2)\\
&= n^2+n-1+\frac{4n-6}{n^2-n}\\
&> n^2+n-2;\\
\mathsf{St}(k_2,v) &= \mathsf{St}(n+1,n^2-n+2)\\
&= n^2+n-1;\\
\mathsf{SK}(k_2+1,v) &= \mathsf{SK}(n+2,n^2-n+2)\\
&= n^2+3n-1-\frac{13n-2}{n^2-n+1}\\
&> n^2+n-2; \quad \text{and}
\end{aligned}$$

$$\begin{aligned}\mathsf{SK}(v-2,v) &= \mathsf{SK}(n^2-n,n^2-n+2)\\ &= 2n^2-2n-5+\frac{2}{n^2-n+1}\\ &> n^2+n-2.\end{aligned}$$

Hence, $\mathsf{b}^*(n^2-n+2) \geq n^2+n-1$.

Now suppose there is a projective plane of order n, and let A_1 and A_2 be any two blocks in this design. A_1 and A_2 intersect in a point, say x. Pick a point $x_1 \in A_1 \backslash \{x\}$ and a point $x_2 \in A_2 \backslash \{x\}$. Delete all the points in A_1 and A_2 except for x_1 and x_2, and then delete A_1 and A_2. We obtain a PBD on n^2-n+2 points having n^2+n-1 blocks that is not a near-pencil. □

Summarizing the results in this section, we have the following theorem.

Theorem 8.19 (Erdös, Mullin, Sós, and Stinson). *Suppose that $v \geq 5$ is an integer. Then $\mathsf{b}^*(v) \geq B(v)$, where*

$$B(v) = \begin{cases} n^2+n+1 & \text{if } n^2+2 \leq v \leq n^2+n+1\\ n^2+n & \text{if } n^2-n+3 \leq v \leq n^2+1\\ n^2+n-1 & \text{if } v = n^2-n+2.\end{cases}$$

Furthermore, $\mathsf{b}^*(v) = B(v)$ if there exists a projective plane of order n, where

$$n^2-n+2 \leq v \leq n^2+n+1.$$

8.4 Minimal PBDs with $\lambda > 1$

We state and prove a theorem that generalizes Fisher's Inequality to nontrivial pairwise balanced designs. We already mentioned this result, in the special case of regular PBDs, in Theorem 1.34. Also, when $\lambda = 1$, the next theorem follows from Theorem 8.5.

Theorem 8.20. *In any nontrivial (v, K, λ)-PBD, $b \geq v$.*

Proof. We first prove the theorem for proper PBDs. We again use the proof technique introduced in Theorem 1.33. Let $(X, \mathcal{A})$ be a $(v, \{2,\dots,v-1\}, \lambda)$-PBD, where $X = \{x_1,\dots,x_v\}$ and $\mathcal{A} = \{A_1,\dots,A_b\}$. For $1 \leq j \leq b$, define $k_j = |A_j|$, and for $1 \leq i \leq v$, define

$$r_i = |\{j : x_i \in A_j\}|.$$

Let M be the incidence matrix of this PBD, and define $\mathbf{s}_j$ to be the jth row of M^T. Recall that $\mathbf{s}_1,\dots,\mathbf{s}_b$ are all v-dimensional vectors in the real vector space $\mathbb{R}^v$.

Define $S = \{\mathbf{s}_j : 1 \leq j \leq b\}$ and define $\mathbf{S} = \mathsf{span}(\mathbf{s}_j : 1 \leq j \leq b)$. As in Theorem 1.33, we will prove that $\mathbf{S} = \mathbb{R}^v$, which implies that $b \geq v$.

For $1 \leq i \leq v$, define $\mathbf{e}_i \in \mathbb{R}^v$ to be the vector with a "1" in the ith coordinate and "0"s in all other coordinates. We show that $\mathbf{e}_i \in \mathbf{S}$ for $1 \leq i \leq v$.

First, we observe that

$$\sum_{j=1}^{b} \mathbf{s}_j = (r_1, \ldots, r_v). \tag{8.10}$$

If we fix a value i, $1 \leq i \leq v$, then we have

$$\sum_{\{j:x_i \in A_j\}} \mathbf{s}_j = (r_i - \lambda)\mathbf{e}_i + (\lambda, \ldots, \lambda). \tag{8.11}$$

Next, sum equation (8.11) over all i, $1 \leq i \leq v$, to obtain

$$\sum_{i=1}^{v} \sum_{\{j:x_i \in A_j\}} \mathbf{s}_j = (r_1, \ldots, r_v) + (\lambda(v-1), \ldots, \lambda(v-1)). \tag{8.12}$$

Equations (8.10) and (8.12) imply that $(1, \ldots, 1) \in \mathbf{S}$:

$$(1, \ldots, 1) = \frac{1}{\lambda(v-1)} \left(\sum_{i=1}^{v} \sum_{\{j:x_i \in A_j\}} \mathbf{s}_j - \sum_{j=1}^{b} \mathbf{s}_j \right). \tag{8.13}$$

We can now substitute this back into equation (8.11). Fix a value of i, $1 \leq i \leq v$. Using the fact that $r_i > \lambda$ (which follows because a proper PBD does not contain a block of size v), we obtain the following:

$$\mathbf{e}_i = \frac{1}{r_i - \lambda} \left(\sum_{\{j:x_i \in A_j\}} \mathbf{s}_j - \frac{1}{v-1} \left(\sum_{i=1}^{v} \sum_{\{j:x_i \in A_j\}} \mathbf{s}_j - \sum_{j=1}^{b} \mathbf{s}_j \right) \right). \tag{8.14}$$

This implies that every basis vector $\mathbf{e}_i \in \mathbf{S}$, which completes the proof for proper PBDs.

Now assume that $(X, \mathcal{A})$ is a nontrivial PBD that contains exactly $\ell > 0$ blocks of size v. Deleting these ℓ blocks, we obtain a proper PBD, which therefore must contain at least v blocks. This means that $(X, \mathcal{A})$ has at least $v + \ell$ blocks, and the proof is complete. □

A pairwise balanced design with one block size is a BIBD. Of course, symmetric BIBDs are examples of (v, K, λ)-PBDs with $b = v$. Examples of pairwise balanced designs with $b = v$ and having more than one block size can be constructed from symmetric BIBDs as follows.

Theorem 8.21. *Suppose there is a symmetric* (v, k, λ)-BIBD. *Then there exists a* $(v, \{k, v+1-k\}, k-\lambda)$-PBD *with* $b = v$.

Proof. Suppose $(X, \mathcal{A})$ is a symmetric (v, k, λ)-BIBD. Let $x \in X$ be any point. Define $\mathcal{B} = \mathcal{B}_1 \cup \mathcal{B}_2$, where

$$\mathcal{B}_1 = \{A \in \mathcal{A} : x \notin A\}$$

and

$$\mathcal{B}_2 = \{(X \backslash A) \cup \{x\} : A \in \mathcal{A}, x \in A\}.$$

Clearly $\mathcal{B}$ contains exactly v blocks, every block in $\mathcal{B}_1$ has size k, and every block in $\mathcal{B}_2$ has size $v + 1 - k$. Therefore we need only to show that every pair of points occurs in exactly $k - \lambda$ blocks in $\mathcal{B}$.

For any $y \in X$, $y \neq x$, there are $k - \lambda$ blocks in $\mathcal{A}$ that contain x but do not contain y. These blocks give rise to the $k - \lambda$ blocks (all in $\mathcal{B}_2$) that contain x and y. Now consider two points $y, y' \in X \backslash \{x\}$. Suppose that there are μ blocks in $\mathcal{A}$ that contain y, y', and x. Then there are $\lambda - \mu$ blocks in $\mathcal{A}$ that contain y and y' but not x. These blocks yield $\lambda - \mu$ blocks in $\mathcal{B}_1$ that contain y and y'. Also, there are $k - 2\lambda + \mu$ blocks in $\mathcal{A}$ that contain x but neither y nor y'. These blocks yield $k - 2\lambda + \mu$ blocks in $\mathcal{B}_2$ that contain y and y'. In total, we have $k - \lambda$ blocks in $\mathcal{B}$ that contain y and y'. □

Example 8.22. $\{1,3,4,5,9\}$ is an $(11,5,2)$-difference set in $\mathbb{Z}_{11}$. This difference set generates a symmetric $(11,5,2)$-BIBD in which the points are the elements of $\mathbb{Z}_{11}$. Take $x = 0$; then the blocks of the symmetric $(11,5,2)$-BIBD are transformed as follows:

$$\begin{array}{lcl}
\{1,3,4,5,9\} & \rightarrow & \{1,3,4,5,9\} \\
\{2,4,5,6,10\} & \rightarrow & \{2,4,5,6,10\} \\
\{3,5,6,7,0\} & \rightarrow & \{0,1,2,4,8,9,10\} \\
\{4,6,7,8,1\} & \rightarrow & \{4,6,7,8,1\} \\
\{5,7,8,9,2\} & \rightarrow & \{5,7,8,9,2\} \\
\{68,9,10,3\} & \rightarrow & \{6,8,9,10,3\} \\
\{7,9,10,0,4\} & \rightarrow & \{0,1,2,3,5,6,8\} \\
\{8,10,0,1,5\} & \rightarrow & \{0,2,3,4,6,7,9\} \\
\{9,0,1,2,6\} & \rightarrow & \{0,3,4,5,7,8,10\} \\
\{10,1,2,3,7\} & \rightarrow & \{10,1,2,3,7\} \\
\{0,2,3,4,8\} & \rightarrow & \{0,1,5,6,7,9,10\}.
\end{array}$$

■

Example 8.23. For any $v \geq 4$, there exists a symmetric $(v, v-1, v-2)$-BIBD whose blocks are all the $(v-1)$-subsets of a v-set. If we apply the construction of Theorem 8.21 to this BIBD, the reader can check that we obtain a near-pencil, which has blocks of size two and $v - 1$. ■

The "λ-design Conjecture" is that every pairwise balanced design with $b = v$ either is a symmetric BIBD or can be constructed from a symmetric BIBD using Theorem 8.21. This conjecture is due to Ryser and Woodall and it remains open to this day, although many partial results are known.

Provided that $k \neq (v+1)/2$, the construction of Theorem 8.21 yields a pairwise balanced design with exactly two block sizes, and these block sizes sum to $v+1$. This property holds for any nontrivial pairwise balanced design with $b = v$ that has two block sizes, as we show in the following theorem.

Theorem 8.24 (Ryser-Woodall Theorem). *Suppose $(X, \mathcal{A})$ is a (v, K, λ)-PBD with $b = v$ that contains at least two block sizes. Then there are exactly two block sizes, say k_1 and k_2, and $k_1 + k_2 = v + 1$.*

Proof. We use notation as in the proof of Theorem 8.20. First, we note that $(X, \mathcal{A})$ cannot contain any blocks of size v (this follows from the proof of Theorem 8.20). Hence, $r_i > \lambda$ for all i, $1 \leq i \leq v$.

In this proof, we will use b in the context of blocks and v in the context of points. Of course $b = v$, as stated in the hypotheses.

Fix i, $1 \leq i \leq v$; then we can rewrite (8.11) as follows:

$$\frac{1}{r_i - \lambda} \sum_{\{j: x_i \in A_j\}} \mathbf{s}_j = \mathbf{e}_i + \frac{\lambda}{r_i - \lambda}(1, \ldots, 1). \tag{8.15}$$

For any j, $1 \leq j \leq b$, define

$$c_j = \sum_{\{i: x_i \in A_j\}} \frac{1}{r_i - \lambda}. \tag{8.16}$$

Now we apply (8.15) and (8.16) as follows:

$$\begin{aligned}
\sum_{j=1}^{b} c_j \mathbf{s}_j &= \sum_{j=1}^{b} \sum_{\{i: x_i \in A_j\}} \frac{1}{r_i - \lambda} \mathbf{s}_j \\
&= \sum_{i=1}^{v} \left(\frac{1}{r_i - \lambda} \sum_{\{j: x_i \in A_j\}} \mathbf{s}_j \right) \\
&= \sum_{i=1}^{v} \left(\mathbf{e}_i + \frac{\lambda}{r_i - \lambda}(1, \ldots, 1) \right) \\
&= \left(1 + \lambda \sum_{i=1}^{v} \frac{1}{r_i - \lambda} \right) (1, \ldots, 1).
\end{aligned}$$

Denoting

$$C = 1 + \lambda \sum_{i=1}^{v} \frac{1}{r_i - \lambda},$$

we have the following:

$$(1, \ldots, 1) = \sum_{j=1}^{b} \frac{c_j}{C} \mathbf{s}_j. \tag{8.17}$$

Now we derive another expression for the vector $(1,\ldots,1)$. Observe that

$$\sum_{i=1}^{v} \sum_{\{j:x_i\in A_j\}} \mathbf{s}_j = \sum_{j=1}^{b} \sum_{\{i:x_i\in A_j\}} \mathbf{s}_j = \sum_{j=1}^{b} |A_j|\mathbf{s}_j.$$

Therefore (8.13) implies the following:

$$(1,\ldots,1) = \frac{1}{\lambda(v-1)} \sum_{j=1}^{b} (|A_j|-1)\mathbf{s}_j. \tag{8.18}$$

Equations (8.18) and (8.17) give two expressions for the same vector as a linear combination of basis vectors (the vectors $\mathbf{s}_1,\ldots,\mathbf{s}_b$ form a basis because $b=v$). Therefore, corresponding coefficients in the two linear combinations must be identical, and it follows that

$$\frac{c_j}{C} = \frac{|A_j|-1}{\lambda(v-1)}$$

for $1 \le j \le b$. Denoting

$$\gamma = \frac{C}{\lambda(v-1)},$$

we have that

$$c_j = \gamma\left(|A_j|-1\right) \tag{8.19}$$

for all j, $1 \le j \le b$, where γ is a constant.

We now fix an integer h, $1 \le h \le b$, and sum (8.15) over all i such that $x_i \in A_h$:

$$\begin{aligned}\sum_{\{i:x_i\in A_h\}} \left(\frac{1}{r_i-\lambda} \sum_{\{j:x_i\in A_j\}} \mathbf{s}_j \right) &= \sum_{\{i:x_i\in A_h\}} \left(\mathbf{e}_i + \frac{\lambda}{r_i-\lambda}(1,\ldots,1) \right) \\ &= \mathbf{s}_h + \lambda c_h (1,\ldots,1) \\ &= \mathbf{s}_h + c_h \sum_{j=1}^{b} \frac{|A_j|-1}{v-1}\mathbf{s}_j,\end{aligned}$$

where we apply (8.18) in the last line.

On the other hand, we can evaluate the same double sum in a different way:

$$\sum_{\{i:x_i\in A_h\}} \left(\frac{1}{r_i-\lambda} \sum_{\{j:x_i\in A_j\}} \mathbf{s}_j \right) = \sum_{j=1}^{b} \sum_{\{i:x_i\in A_h\cap A_j\}} \frac{1}{r_i-\lambda}\mathbf{s}_j.$$

Thus we have the following equation:

$$\mathbf{s}_h + c_h \sum_{j=1}^{b} \frac{|A_j|-1}{v-1}\mathbf{s}_j = \sum_{j=1}^{b} \sum_{\{i:x_i\in A_h\cap A_j\}} \frac{1}{r_i-\lambda}\mathbf{s}_j. \tag{8.20}$$

The coefficients of $\mathbf{s}_h$ must be the same on both sides of this equation, so we have that

$$c_h = 1 + \frac{c_h(|A_h| - 1)}{v - 1}.$$

Substituting (8.19), we obtain the following:

$$\gamma(|A_h| - 1) = 1 + \frac{\gamma(|A_h| - 1)^2}{v - 1}.$$

Denote $x = |A_h| - 1$. Simplifying, we get the following quadratic equation in x:

$$x^2 - (v-1)x + \gamma^{-1}(v-1) = 0. \tag{8.21}$$

Since any block in the PBD has size $x + 1$, where x is a root of the quadratic equation (8.21), it follows that there are at most two block sizes. Since we hypothesized that there are at least two block sizes, we conclude that there are exactly two block sizes.

In general, the sum of the roots of a quadratic equation $x^2 + a_1 x + a_2 = 0$ is equal to $-a_1$. Therefore the sum of the roots of (8.21) is equal to $v - 1$. This implies that the sum of the two block sizes in the PBD is equal to $v + 1$, as desired. □

8.5 Notes and References

Theorem 8.5 was proven in 1948 by de Bruijn and Erdös [38]. Theorem 8.1 is due to Stanton and Kalbfleisch [99], and Theorem 8.7 is from Stinson [100]. Another important bound along these lines is the Rees Bound; see [85]. These various bounds are discussed and compared in Rees and Stinson [86].

Most of the results in Section 8.3 are adapted from Erdös, Mullin, Sós, and Stinson [44]. There is much literature on pairwise balanced designs with $\lambda = 1$ having "few lines". The monograph by Batten and Beutelspacher [4] is a good source of additional information on this topic.

Theorem 8.24 was proven independently by Ryser [90] and Woodall [125]. The λ-design Conjecture has been widely studied; see Singhi and Shrikhande [97] and Ionin and Shrikhande [63, 64] for more information.

8.6 Exercises

8.1 Suppose that $K \subseteq \{n \geq 2 : n \in \mathbb{Z}\}$ is a finite set. Denote the largest and smallest elements of K by k_1 and k_2, respectively.
   (a) Prove that there exists a (v, K)-PBD only if $v \geq k_1(k_2 - 1) + 1$.
   (b) Prove that there exists a $(k_1(k_2 - 1) + 1, K)$-PBD if and only if there exists a resolvable $(k_1(k_2 - 2) + 1, k_2 - 1, 1)$-BIBD.

8.2 Suppose that v is fixed and $2 \leq k \leq v - 1$ is a real number.

(a) Prove that $\mathsf{SK}(k,v)$ attains its maximum when $k = 2v/3$.
(b) Prove that $\mathsf{SK}(k,v) = \mathsf{C}(k,v)$ if and only if $k^2 - k + 1 = v$.
(c) Assume that $k^2 - k + 1 \geq v \geq k + 1$. Prove that $\mathsf{SK}(k,v) \geq v$.

8.3 Suppose there is a $(v, \{2, \ldots, v-1\})$-PBD with $\mathsf{St}(k,v)$ blocks that has a block containing exactly k points. Prove that there are at most three different block sizes in this PBD.

8.4 (a) Suppose that k is odd and $v = 2k + 1$. Prove that there exists a $(v, \{3, k\})$-PBD with $\mathsf{SK}(k,v)$ blocks that contains a block of size k.
(b) Suppose that k is even and $v = 2k + 1$. Prove that there does not exist any (v, K)-PBD with $\mathsf{SK}(k,v)$ blocks that contains a block of size k.

8.5 (a) Suppose that $k + 1 \leq v \leq 2k$. Prove that

$$\mathsf{St}(k,v) = 1 + \frac{(v-k)(3k-v+1)}{2}.$$

(b) Suppose that $k + 1 \leq v \leq 2k$ and $v - k$ is even. Use the existence of a resolvable $(v-k, 2, 1)$-BIBD to prove that there exists a $(v, \{2, 3, k\})$-PBD with $\mathsf{St}(k,v)$ blocks.
Hint: An essential step of the proof is to form $2k - v + 1$ parallel classes of singletons on $v - k$ points.
(c) Suppose that $k + 1 \leq v \leq 2k$ and $v - k$ is odd. Use the existence of a resolvable $(v - k + 1, 2, 1)$-BIBD to prove that there exists a $(v, \{2, 3, k\})$-PBD with $\mathsf{St}(k,v)$ blocks.
Hint: Delete a point from the BIBD, and then proceed in a manner similar to (b).

8.6 (a) Suppose that $(X, \mathcal{A})$ is a $(v, \{2, \ldots, v-1\})$-PBD in which the largest block contains exactly k points and in which there are exactly $\mathsf{C}^*(k,v)$ blocks. Prove that every block has size k or $k-1$.
(b) Denote $t = (v-1) \bmod (k-1)$ and suppose further that $t \neq 0$. Prove that every point x occurs in $k - t - 1$ blocks of size $k - 1$ and $r - k + t + 1$ blocks of size k, where $r = \lceil \frac{v-1}{k-1} \rceil$.
(c) Suppose there is a $(v+1, k, 1)$-BIBD. Prove that there exists a $(v, \{2, \ldots, v-1\})$-PBD in which the largest block contains exactly k points and in which there are exactly $\mathsf{C}^*(k,v)$ blocks.

8.7 Extend Table 8.1 to include all the cases when $v = 10$. For $2 \leq k \leq 9$, determine the values of the four relevant bounds and the exact values of $g^k(10)$.

8.8 Construct $(v, \{2, \ldots, v-2\})$-PBDs with $B(v)$ blocks for $10 \leq v \leq 15$.

8.9 Use Theorem 8.5 to prove that the λ-design Conjecture is valid when $\lambda = 1$.

8.10 (a) Prove that the only $(v, K, 2)$-PBD that can be constructed using Theorem 8.21 is a $(7, \{3, 5\}, 2)$-PBD with $b = 7$.
(b) Construct the PBD described in part (a).

9

t-Designs and t-wise Balanced Designs

9.1 Basic Definitions and Properties of t-Designs

Definition 9.1. *Let v, k, λ, and t be positive integers such that $v > k \geq t$. A t-(v, k, λ)-design is a design $(X, \mathcal{A})$ such that the following properties are satisfied:*

1. *$|X| = v$,*
2. *each block contains exactly k points, and*
3. *every set of t distinct points is contained in exactly λ blocks.*

The general term t-design is used to indicate any t-(v, k, λ)-design.

Note that we allow a t-(v, k, λ)-design to contain repeated blocks. (Of course, if $\lambda = 1$, then there cannot be any repeated blocks in a t-(v, k, λ)-design.) A t-(v, k, λ)-design without repeated blocks is called a *simple t-design*. When $\lambda > 1$, it is usually the case that constructing simple t-(v, k, λ)-designs is more difficult than constructing nonsimple ones.

If we take λ copies of every k-subset of a v-set, where $k < v$, we obtain a t-$\left(v, k, \lambda\binom{v-t}{k-t}\right)$-design. This t-design is not very exciting; we refer to it as a *trivial t-design*. In general, we are interested in constructing nontrivial designs.

Observe that a 2-(v, k, λ)-design is just a (v, k, λ)-BIBD. There are not nearly as many existence results known for t-designs with $t > 2$ as there are for BIBDs. We will be presenting some of the nicer construction methods for certain types of t-designs, but first we survey some basic properties of t-designs.

The proof of the following theorem follows immediately from the definition of a t-design.

Theorem 9.2. *Suppose that $(X, \mathcal{A})$ is a t-(v, k, λ)-design. Let $Z \subseteq X$, $|Z| = i < t$. Then $(X \backslash Z, \{A \backslash Z : Z \subseteq A \in \mathcal{A}\})$ is a $(t-i)$-$(v-i, k-i, \lambda)$-design.*

Example 9.3. It is known that there exists a 5-(12, 6, 1)-design (we will give a construction for this design in Example 9.29). Hence, from Theorem 9.2, there also exist 4-(11, 5, 1)-, 3-(10, 4, 1)-, and 2-(9, 3, 1)-designs. ▮

The following result is proven in the same manner as Theorems 1.8 and 1.9.

Theorem 9.4. *Suppose that $(X, \mathcal{A})$ is a t-(v, k, λ)-design. Suppose that $Y \subseteq X$, where $|Y| = s \leq t$. Then there are exactly*

$$\lambda_s = \frac{\lambda\binom{v-s}{t-s}}{\binom{k-s}{t-s}}$$

blocks in $\mathcal{A}$ that contain all the points in Y.

Proof. Let $\lambda_s(Y)$ denote the number of blocks containing all the points in Y. Define a set

$$I = \{(Z, A) : Z \subseteq X, |Z| = t - s, Y \cap Z = \emptyset, A \in \mathcal{A}, Y \cup Z \subseteq A\}.$$

We will compute $|I|$ in two different ways.

First, there are $\binom{v-s}{t-s}$ ways to choose Z. For each such Z, there are λ blocks A such that $Y \cup Z \subseteq A$. Hence,

$$|I| = \lambda\binom{v-s}{t-s}.$$

On the other hand, there are $\lambda_s(Y)$ ways to choose a block A such that $Y \subseteq A$. For each choice of A, there are $\binom{k-s}{t-s}$ ways to choose Z. Hence,

$$|I| = \lambda_s(Y)\binom{k-s}{t-s}.$$

Combining these two equations, we see that $\lambda_s(Y) = \lambda_s$, as desired. □

Observe that the number of blocks in a t-design is $\lambda_0 = \lambda\binom{v}{t}/\binom{k}{t}$ and each point occurs in $\lambda_1 = \lambda\binom{v-1}{t-1}/\binom{k-1}{t-1}$ blocks. In the case $t = 2$ (i.e., for a BIBD), λ_0 and λ_1 correspond to the parameters b and r, respectively. We will sometimes use the notations b and r for t-designs with other values of t as well.

Example 9.5. In a 5-(12, 6, 1)-design, we have that $\lambda_0 = 132$, $\lambda_1 = 66$, $\lambda_2 = 30$, $\lambda_3 = 12$, $\lambda_4 = 4$, and $\lambda_5 = 1$. ▮

The following is an immediate corollary of Theorem 9.4.

Corollary 9.6. *Suppose that $(X, \mathcal{A})$ is a t-(v, k, λ)-design, and $1 \leq s \leq t$. Then $(X, \mathcal{A})$ is an s-(v, k, λ_s)-design, where*

$$\lambda_s = \frac{\lambda\binom{v-s}{t-s}}{\binom{k-s}{t-s}}.$$

Theorem 9.4 can be generalized as follows.

Theorem 9.7. *Suppose that $(X, \mathcal{A})$ is a t-(v, k, λ)-design. Suppose that $Y, Z \subseteq X$, where $Y \cap Z = \emptyset$, $|Y| = i$, $|Z| = j$, and $i + j \leq t$. Then there are exactly*

$$\lambda_i^j = \frac{\lambda\binom{v-i-j}{k-i}}{\binom{v-t}{k-t}}$$

blocks in $\mathcal{A}$ that contain all the points in Y and none of the points in Z.

Proof. First we consider the case where $i = 0$. Let $\lambda_0^j(Z)$ denote the number of blocks that contain none of the points in Z. Using the Principle of Inclusion-Exclusion, we will obtain a formula for $\lambda_0^j(Z)$. For any $z \in Z$, define

$$\mathcal{A}_z = \{A \in \mathcal{A} : z \in A\}.$$

Then, for any $Z_0 \subseteq Z$, $|Z_0| = h$, it is clear that

$$\left| \bigcap_{z \in Z_0} \mathcal{A}_z \right| = \lambda_h.$$

The *Principle of Inclusion-Exclusion* asserts that

$$\left| \mathcal{A} \setminus \left(\bigcup_{z \in Z} \mathcal{A}_z \right) \right| = |\{A \in \mathcal{A} : A \cap Z = \emptyset\}| = \sum_{Z_0 \subseteq Z} (-1)^{|Z_0|} \left| \bigcap_{z \in Z_0} \mathcal{A}_z \right|.$$

From this, it follows immediately that

$$\lambda_0^j(Z) = \sum_{h=0}^{j} (-1)^h \binom{j}{h} \lambda_h.$$

Hence, $\lambda_0^j(Z)$ is a constant, say C (i.e., it is independent of the choice of Z). We have expressed C as a complicated-looking sum. C can be simplified using appropriate identities involving binomial coefficients; however, it is easier to proceed as follows.

Define a set

$$\mathcal{I} = \{(Z_0, A) : Z_0 \subseteq X, |Z_0| = j, A \in \mathcal{A}, Z_0 \cap A = \emptyset\}.$$

We will compute $|I|$ in two different ways.

First, there are $\binom{v}{j}$ ways to choose Z_0. For each such Z_0, there are C blocks A such that $Z_0 \cap A = \emptyset$. Hence,

$$|I| = C\binom{v}{j}.$$

On the other hand, there are λ_0 ways to choose a block A, and for each choice of A, there are $\binom{v-k}{j}$ ways to choose Z_0. Hence,

$$|I| = \lambda_0\binom{v-k}{j}.$$

Combining these two equations, we see that

$$\begin{aligned} C &= \frac{\lambda_0\binom{v-k}{j}}{\binom{v}{j}} \\ &= \frac{\lambda\binom{v}{t}\binom{v-k}{j}}{\binom{k}{t}\binom{v}{j}} \\ &= \frac{\lambda\ v!\ (v-k)!\ (k-t)!\ t!\ j!\ (v-j)!}{t!\ (v-t)!\ j!\ (v-k-j)!\ k!\, v!} \\ &= \frac{\lambda\ (v-k)!\ (k-t)!\ (v-j)!}{(v-t)!\ (v-k-j)!\ k!} \\ &= \frac{\lambda\binom{v-j}{k}}{\binom{v-t}{k-t}}, \end{aligned}$$

as desired.

Now we consider the case $i > 0$. This follows by applying the result proven above for $i = 0$ to the design $(X\backslash Y, \{A\backslash Y : Y \subseteq A \in \mathcal{A}\})$, which is a $(t-i)$-$(v-i, k-i, \lambda)$-design by Theorem 9.2. □

We have already mentioned that simple t-designs are, in general, more difficult to construct than nonsimple ones. We next present an easy nonconstructive proof that nontrivial t-designs exist for all permissible choices of $t < k$ and all $v > k + t$. (This proof does not yield simple designs, however.)

Theorem 9.8. *For all positive integers t, k, and v such that $t < k < v - t$, there exists a nontrivial t-(v, k, λ)-design for some positive integer λ.*

Proof. Let X be a v-set, and let $N = \binom{v}{t}$. Consider the N-dimensional vector space $\mathbb{Q}^N$ in which the coordinates are indexed by the t-subsets of X. For each k-subset $A \subseteq X$, define a vector $\mathbf{s}_A \in \mathbb{Q}^N$ in which the entry in the coordinate corresponding to a t-subset $Y \subseteq X$ is equal to 1 if $Y \subseteq A$ and 0 otherwise. We obtain a set of $\binom{v}{k}$ vectors in a $\binom{v}{t}$-dimensional vector space. Since $t < k < v - t$, it follows that

$$\binom{v}{t} < \binom{v}{k},$$

and hence there exists a linear dependence relation among this set of vectors. In other words, there exist rational numbers α_A ($A \subseteq X, |A| = k$) such that

$$\sum_{A \subseteq X, |A|=k} \alpha_A \mathbf{s}_A = (0, \ldots, 0).$$

Let D denote the least common multiple of the denominators of the numbers α_A, and define $\beta_A = D\alpha_A$ for all A. Then

$$\sum_{A \subseteq X, |A|=k} \beta_A \mathbf{s}_A = (0, \ldots, 0), \tag{9.1}$$

and the β_A's are all integers.

Clearly at least one of the β_A's is negative. Hence, if we define $M = \min\{\beta_A\}$, then $M < 0$. Now define $\mathcal{A}$ to be the collection of blocks where, for every $A \subseteq X$, $|A| = k$, A occurs exactly $\beta_A - M$ times in $\mathcal{A}$ (note that $\beta_A - M \geq 0$ for all A).

It is not difficult to see that $(X, \mathcal{A})$ is a t-(v, k, λ)-design. First, we observe that

$$\sum_{A \subseteq X, |A|=k} \mathbf{s}_A = \left(\binom{v-t}{k-t}, \ldots, \binom{v-t}{k-t}\right); \tag{9.2}$$

this follows because, as we already observed, the set of all k-subsets of a v-set is a t-$\left(v, k, \binom{v-t}{k-t}\right)$-design. Now, combining equations (9.1) and (9.2), we see that

$$\sum_{A \subseteq X, |A|=k} (\beta_A - M)\mathbf{s}_A = \left(-M\binom{v-t}{k-t}, \ldots, -M\binom{v-t}{k-t}\right). \tag{9.3}$$

Hence $(X, \mathcal{A})$ is a t-(v, k, λ)-design with $\lambda = -M\binom{v-t}{k-t}$. Finally, $(X, \mathcal{A})$ is nontrivial because $\beta_A - M = 0$ for at least one A. □

Example 9.9. We provide an illustration of Theorem 9.8 in the case $t = 2$, $k = 3$. Suppose that $v \geq 6$. Then it is easy to check that

$$\mathbf{s}_{\{1,2,3\}} + \mathbf{s}_{\{1,4,5\}} + \mathbf{s}_{\{2,4,6\}} + \mathbf{s}_{\{3,5,6\}} = \mathbf{s}_{\{1,2,4\}} + \mathbf{s}_{\{1,3,5\}} + \mathbf{s}_{\{2,3,6\}} + \mathbf{s}_{\{4,5,6\}}.$$

Therefore we have found a dependence relation with $M = -1$, and we can construct a nontrivial 2-$(v, 3, v-2)$-design for all $v \geq 6$. ∎

In the next section, we look at some specific families of t-designs with $t \geq 3$. We have already observed that 2-designs are BIBDs. Thus, there remains the case of 1-designs to be considered. However, it is not hard to show that these designs exist whenever the necessary conditions are satisfied.

Theorem 9.10. *There exists a* 1-(v,k,λ)*-design if and only if* $v\lambda \equiv 0 \bmod k$.

Proof. From Theorem 9.4, the number of blocks in a 1-(v,k,λ)-design is $b = v\lambda/k$, which must be an integer. Conversely, suppose $b = v\lambda/k$ is an integer. We will describe an easy construction for a 1-(v,k,λ)-design.

Let $u = \gcd(k,\lambda)$. Then $\lambda = u\lambda'$ and $k = uk'$, where $\gcd(\lambda',k') = 1$. Now we have $b = v\lambda/k = v\lambda'/k'$ and $\gcd(\lambda',k') = 1$, so it must be the case that $v \equiv 0 \bmod k'$. Let $v = sk'$, where s is a positive integer. Then $b = v\lambda'/k' = s\lambda'$.

Let X be a set of cardinality k', and define $Y = X \times \mathbb{Z}_s$. Then $|Y| = v$. Let $A_1, \dots, A_{\lambda'}$ be λ' arbitrary u-subsets of $\mathbb{Z}_s$. For $1 \leq i \leq \lambda'$, define $B_i = X \times A_i$. Then each B_i is a k-subset of Y. Now develop each B_i through $\mathbb{Z}_s$, obtaining a set of b blocks that contain every point in Y exactly λ times. The result is a 1-(v,k,λ)-design. □

Example 9.11. Suppose that $v = 15$, $k = 9$, and $\lambda = 6$. Then $b = 10$, $s = 5$, $k' = 3$, and $\lambda' = 2$. Suppose we take $X = \{x,y,z\}$, $A_1 = \{0,1,2\}$, and $A_2 = \{0,1,3\}$. Then

$$B_1 = \{(x,0),(y,0),(z,0),(x,1),(y,1),(z,1),(x,2),(y,2),(z,2)\}$$

and

$$B_2 = \{(x,0),(y,0),(z,0),(x,1),(y,1),(z,1),(x,3),(y,3),(z,3)\}.$$

We obtain a total of $b = 10$ blocks from B_1 and B_2 by developing the second coordinates modulo 5 (keeping the first coordinates fixed). In the resulting set of blocks, every element occurs $\lambda = 6$ times. ∎

9.2 Some Constructions for t-Designs with $t \geq 3$

We present some constructions for t-designs with $t \geq 3$ in this section. Our first result shows that certain resolvable BIBDs are automatically 3-designs.

Theorem 9.12. *A resolvable BIBD with* $v = 2k$ *is a 3-design.*

Proof. Suppose that $(X, \mathcal{A})$ is a resolvable $(2k,k,\lambda)$-BIBD. Let Π_i be the parallel classes for $1 \leq i \leq r$. Each Π_i consists of two blocks, say A_i^1 and A_i^2.

Let $x,y,z \in X$, and define a_1,a_2,a_3,a_4 as follows:

$$a_1 = |\{i : \{x,y,z\} \subseteq A_i^j, \text{ where } j = 1 \text{ or } 2\}|,$$
$$a_2 = |\{i : \{x,y\} \subseteq A_i^j \text{ and } z \notin A_i^j, \text{ where } j = 1 \text{ or } 2\}|,$$
$$a_3 = |\{i : \{x,z\} \subseteq A_i^j \text{ and } y \notin A_i^j, \text{ where } j = 1 \text{ or } 2\}|, \quad \text{and}$$
$$a_4 = |\{i : \{y,z\} \subseteq A_i^j \text{ and } x \notin A_i^j, \text{ where } j = 1 \text{ or } 2\}|.$$

Clearly

$$a_1 + a_2 + a_3 + a_4 = r$$

since each parallel class is one of the four types enumerated above. Also, looking at pairs of elements, we have that

$$a_1 + a_2 = a_1 + a_3 = a_1 + a_4 = \lambda.$$

From these equations, it follows that $a_1 = (3\lambda - r)/2$. Therefore $(X, \mathcal{A})$ is a 3-$(2k, k, (3\lambda - r)/2)$-design. □

Corollary 9.13. *If there exists a Hadamard matrix of order $4m$, then there exists a 3-$(4m, 2m, m-1)$-design.*

Proof. If there exists a Hadamard matrix of order $4m$, then there exists a resolvable $(4m, 2m, 2m-1)$-BIBD from Theorem 5.19. Apply Theorem 9.12. □

The next theorem constructs 3-designs with $k = 4$ and $\lambda = 3$.

Theorem 9.14. *For all even integers $v \geq 6$, there exists a 3-$(v, 4, 3)$-design.*

Proof. We proved in Theorem 5.2 that there exists a resolvable $(v, 2, 1)$-BIBD, say $(X, \mathcal{A})$, for all even $v \geq 4$. Suppose $v \geq 6$, and suppose $\Pi_1, \ldots, \Pi_{v-1}$ are the parallel classes in this BIBD. Define

$$\mathcal{B} = \{A_1 \cup A_2 : A_1, A_2 \in \Pi_i, A_1 \neq A_2, 1 \leq i \leq v-1\}.$$

We will show that $(X, \mathcal{B})$ is a 3-$(v, 4, 3)$-design.

Consider any three points, say x_1, x_2, x_3. Let $1 \leq i \leq 3$. There is a unique block $A_i \in \mathcal{A}$ that contains the pair $\{x_1, x_2, x_3\} \backslash \{x_i\}$. The block A_i is in a unique parallel class, say Π_{j_i}. Note that j_1, j_2, j_3 are distinct integers. Now, there is a unique block in Π_{j_i} that contains x_i, say A'_i. Then $\{x_1, x_2, x_3\} \subseteq A_i \cup A'_i$ for $1 \leq i \leq 3$. Thus we have found three blocks that contain the triple $\{x_1, x_2, x_3\}$. Clearly no other block contains this triple, so we have a 3-$(v, 4, 3)$-design, as required. □

Example 9.15. A 3-$(6, 4, 3)$-design. We begin with the resolvable $(6, 2, 1)$-BIBD presented in Example 5.3. The parallel classes of this BIBD are as follows:

$$\Pi_0 = \{\{\infty, 0\}, \{1, 4\}, \{2, 3\}\}$$
$$\Pi_1 = \{\{\infty, 1\}, \{2, 0\}, \{3, 4\}\}$$
$$\Pi_2 = \{\{\infty, 2\}, \{3, 1\}, \{4, 0\}\}$$
$$\Pi_3 = \{\{\infty, 3\}, \{4, 2\}, \{0, 1\}\}$$
$$\Pi_4 = \{\{\infty, 4\}, \{0, 3\}, \{1, 2\}\}.$$

We obtain the following fifteen blocks of a 3-$(6, 4, 3)$-design, $(\mathbb{Z}_4 \cup \{\infty\}, \mathcal{A})$:

$$\mathcal{A} = \left\{ \begin{array}{l} \infty014, \infty023, 1423, \\ \infty120, \infty134, 2034, \\ \infty231, \infty240, 3140, \\ \infty342, \infty301, 4201, \\ \infty403, \infty412, 0312 \end{array} \right\}.$$

∎

A 3-$(v,4,1)$-design is known as a *Steiner quadruple system* of *order* v and is denoted $\mathsf{SQS}(v)$. The necessary condition for the existence of an $\mathsf{SQS}(v)$ is that $v \equiv 2,4 \bmod 6$. Here is a nice doubling construction for Steiner quadruple systems.

Theorem 9.16. *If there exists an* $\mathsf{SQS}(v)$*, then there exists an* $\mathsf{SQS}(2v)$.

Proof. As in the proof of Theorem 9.14, we use the fact that when v is even, there exists a resolvable $(v,2,1)$-BIBD. Let $|X| = |Y| = v$, $X \cap Y = \emptyset$. Let $(X,\mathcal{A})$ and $(Y,\mathcal{B})$ be resolvable $(v,2,1)$-BIBDs having parallel classes $\Pi_1,\ldots,\Pi_{v-1}$ and $\Psi_1,\ldots,\Psi_{v-1}$, respectively. Also, let $(X,\mathcal{C})$ and $(Y,\mathcal{D})$ be $\mathsf{SQS}(v)$. Define

$$\mathcal{E} = \{A \cup B : A \in \Pi_i, B \in \Psi_i, 1 \le i \le v-1\}.$$

We show that $(X \cup Y, \mathcal{C} \cup \mathcal{D} \cup \mathcal{E})$ is an $\mathsf{SQS}(2v)$. Suppose that $\{z_1,z_2,z_3\} \subseteq X \cup Y$. We consider the following cases that may arise.

1. If $|\{z_1,z_2,z_3\} \cap X| = 3$, then $\{z_1,z_2,z_3\}$ is a subset of a unique block in $\mathcal{C}$, and it is not a subset of any block in $\mathcal{D} \cup \mathcal{E}$.
2. If $|\{z_1,z_2,z_3\} \cap Y| = 3$, then $\{z_1,z_2,z_3\}$ is a subset of a unique block in $\mathcal{D}$, and it is not a subset of any block in $\mathcal{C} \cup \mathcal{E}$.
3. Suppose $|\{z_1,z_2,z_3\} \cap X| = 2$, say $z_1,z_2 \in X$ and $z_3 \in Y$. There is a unique parallel class, say Π_j, such that $\{z_1,z_2\} \in \Pi_j$. There is a unique block of the form $\{z_3,z_4\} \in \Psi_j$. Then $\{z_1,z_2,z_3\} \subseteq \{z_1,z_2,z_3,z_4\} \in \mathcal{E}$, and $\{z_1,z_2,z_3\}$ is not a subset of any block in $\mathcal{C} \cup \mathcal{D}$.
4. Suppose $|\{z_1,z_2,z_3\} \cap Y| = 2$. This is similar to the previous case.

We have considered all possible cases, and the proof is complete. □

There does not exist an $\mathsf{SQS}(4)$ because of the restriction that $v > k$ in the definition of a t-(v,k,λ)-design. However, the construction presented in the proof of Theorem 9.16 can be carried out when $v = 4$, yielding an $\mathsf{SQS}(8)$, as presented in the following example.

Example 9.17. An $\mathsf{SQS}(8)$. The points are $X = \{1,2,3,4,a,b,c,d\}$ and the fourteen blocks are as follows:

$$\mathcal{A} = \left\{ \begin{array}{l} 12ab, 12cd, 34ab, 34cd, \\ 13ac, 13bd, 24ac, 24bd, \\ 14ad, 14bc, 23ad, 23bc, \\ 1234, abcd \end{array} \right\}.$$

∎

As a result of Example 9.17 and Theorem 9.16, we have the following result.

Theorem 9.18. *There exists an* $\mathsf{SQS}(2^n)$ *for all integers* $n \geq 3$.

9.2.1 Inversive Planes

In this section, we describe how certain permutation groups can be used to construct t-designs. We begin with some relevant definitions.

Definition 9.19. *Suppose that G is a subgroup of the symmetric group S_v acting on the v-set X, and suppose that $t \geq 1$ is an integer. G is* sharply t-transitive *provided that, for all choices of $2t$ elements $x_1, \ldots, x_t, y_1, \ldots, y_t \in X$ such that $x_1, \ldots, x_t$ are distinct and $y_1, \ldots, y_t$ are distinct, there is exactly one permutation $\pi \in G$ such that $\pi(x_i) = y_i$ for all i, $1 \leq i \leq t$.*

Example 9.20. Let $X = \mathbb{Z}_n$, suppose that $\alpha = (0\ 1\ \cdots\ n-1)$, and let

$$G = \{\alpha^i : 0 \leq i \leq n-1\}.$$

(Note that G is isomorphic to $(\mathbb{Z}_n, +)$.) It is easy to see that G is sharply 1-transitive. ∎

Example 9.21. Let q be prime and define $X = \mathbb{F}_q$. For $a \in \mathbb{F}_q \backslash \{0\}$ and for $b \in \mathbb{F}_q$, define $\pi_{(a,b)} : \mathbb{F}_q \rightarrow \mathbb{F}_q$ by the rule

$$\pi_{(a,b)}(x) = ax + b$$

for all $x \in \mathbb{F}_q$. It is not difficult to see that every $\pi_{(a,b)}$ is a permutation of $\mathbb{F}_q$. Define

$$G = \{\pi_{(a,b)} : a \in \mathbb{F}_q \backslash \{0\}, b \in \mathbb{F}_q\}.$$

Then it can be shown that G is a sharply 2-transitive group. This group is often denoted $\mathsf{AGL}(1, q)$. ∎

Suppose that G is a sharply t-transitive subgroup of the symmetric group S_v acting on the v-set X. Suppose that $Y \subseteq X$. Recall that the stabilizer of Y, denoted $\mathsf{stab}(Y)$, consists of all the permutations $\pi \in G$ such that $\{\pi(y) : y \in Y\} = Y$. It is not difficult to prove that $\mathsf{stab}(Y)$ is a subgroup of G. Now consider the orbit of subsets obtained by letting G act on Y, which we denote by $\mathsf{orbit}(Y)$.

We have the following result.

Theorem 9.22. *For any $Y \subseteq X$ such that $t \leq |Y| < |X|$, $(X, \mathsf{orbit}(Y))$ is a t-(v, k, λ)-design, where $v = |X|$, $k = |Y|$, and*

$$\lambda = \frac{k(k-1)\cdots(k-t+1)}{|\mathsf{stab}(Y)|}.$$

Proof. For each $\pi \in G$, let $\pi(Y) = \{\pi(y) : y \in Y\}$. Let $\mathcal{A}$ denote the multiset $\{\pi(Y) : \pi \in G\}$. Because G is sharply t-transitive, it follows that every t-subset of points occurs in exactly $k(k-1)\cdots(k-t+1)$ blocks in the collection $\mathcal{A}$. However, every block in $\mathcal{A}$ occurs exactly $|\mathsf{stab}(Y)|$ times. Therefore, if we keep only one copy of every distinct block, then we get a t-design with the stated parameters. □

As a first illustration of the application of Theorem 9.22, we show how to construct affine planes using permutation groups.

Example 9.23. Suppose that q is a prime power, and let G be the group $\mathsf{AGL}(1,q^2)$ acting on $\mathbb{F}_{q^2}$. Let $Y = \mathbb{F}_q$ (which is a subset of $X = \mathbb{F}_{q^2}$). It is not hard to see that $\mathsf{stab}(Y) = \{\pi_{(a,b)} : a, b \in \mathbb{F}_q\}$. (Note that $\mathsf{stab}(Y)$ is isomorphic to $\mathsf{AGL}(1,q)$, but it acts on the points in $\mathbb{F}_{q^2}$.) Clearly, $|\mathsf{stab}(Y)| = q(q-1)$. Then, applying Theorem 9.22, we have that $(X, \mathsf{orbit}(Y))$ is a 2-$(q^2,q,1)$-design. (This design is, in fact, isomorphic to the affine plane $\mathsf{AG}_2(q)$.) ∎

We now present a family of 3-transitive groups that can be used to construct 3-designs with $\lambda = 1$.

Let q be a prime power and suppose that $a,b,c,d \in \mathbb{F}_q$, $ad - bc \neq 0$. Let $\infty \notin \mathbb{F}_q$, and define a function $\pi_{\left(\begin{smallmatrix} a & b \\ c & d \end{smallmatrix}\right)} : (\mathbb{F}_q \cup \{\infty\}) \rightarrow (\mathbb{F}_q \cup \{\infty\})$ as follows:

$$\pi_{\left(\begin{smallmatrix} a & b \\ c & d \end{smallmatrix}\right)}(x) = \begin{cases} \frac{ax+b}{cx+d} & \text{if } x \in \mathbb{F}_q \text{ and } cx+d \neq 0 \\ \infty & \text{if } x \in \mathbb{F}_q,\ cx+d = 0, \text{ and } ax+b \neq 0 \\ \frac{a}{c} & \text{if } x = \infty \text{ and } c \neq 0 \\ \infty & \text{if } x = \infty,\ c = 0, \text{ and } a \neq 0. \end{cases}$$

We observe that the four cases enumerated above cover all the possibilities because $a = c = 0$ is not allowed, and it is impossible that $ax + b = cx + d = 0$.

The following lemma is straightforward to prove.

Lemma 9.24. *Suppose that q is a prime power, $a,b,c,d \in \mathbb{F}_q$, and $ad - bc \neq 0$. Then $\pi_{\left(\begin{smallmatrix} a & b \\ c & d \end{smallmatrix}\right)}$ is a permutation of $\mathbb{F}_q \cup \{\infty\}$.*

It is easy to see that the permutations $\pi_{\left(\begin{smallmatrix} a & b \\ c & d \end{smallmatrix}\right)}$ and $\pi_{\left(\begin{smallmatrix} ra & rb \\ rc & rd \end{smallmatrix}\right)}$ are identical if $r \neq 0$. Define $\mathsf{PGL}(2,q)$ to consist of all the distinct permutations $\pi_{\left(\begin{smallmatrix} a & b \\ c & d \end{smallmatrix}\right)}$, where $a,b,c,d \in \mathbb{F}_q$ and $ad - bc \neq 0$. Notice that there are $q-1$ identical permutations $\pi_{\left(\begin{smallmatrix} a & b \\ c & d \end{smallmatrix}\right)}$ corresponding to each permutation in $\mathsf{PGL}(2,q)$.

Lemma 9.25. $|\mathsf{PGL}(2,q)| = q^3 - q$.

Proof. There are q^4 four-tuples $(a,b,c,d) \in (\mathbb{F}_q)^4$. To compute $|\mathsf{PGL}(2,q)|$, we must subtract the number of four-tuples such that $ad = bc$ and then divide by $q-1$.

It is clear that $ad = bc$ if and only if $\det\begin{pmatrix} a & b \\ c & d \end{pmatrix} = 0$. If $(a,b) = (0,0)$, then any one of the q^2 choices for (c,d) yields a zero determinant. If $(a,b) \neq (0,0)$, then there are q scalar multiples of (a,b), each of which yields a zero determinant when it is defined to be (c,d).

Therefore we have that

$$|\mathsf{PGL}(2,q)| = \frac{q^4 - (q^2 + q(q^2-1))}{q-1} = q^3 - q.$$

□

Theorem 9.26. *$\mathsf{PGL}(2,q)$ is a sharply 3-transitive permutation group.*

Proof. First, to show that $\mathsf{PGL}(2,q)$ is a group, it is sufficient to prove that the composition of any two permutations in $\mathsf{PGL}(2,q)$ is again a permutation in $\mathsf{PGL}(2,q)$. Consider the composition of two permutations $\pi_{\begin{pmatrix} a & b \\ c & d \end{pmatrix}}$ and $\pi_{\begin{pmatrix} a' & b' \\ c' & d' \end{pmatrix}}$ in $\mathsf{PGL}(2,q)$. Using elementary algebra, it is easy to see that

$$\pi_{\begin{pmatrix} a & b \\ c & d \end{pmatrix}}\left(\pi_{\begin{pmatrix} a' & b' \\ c' & d' \end{pmatrix}}(x)\right) = \pi_{\begin{pmatrix} a'' & b'' \\ c'' & d'' \end{pmatrix}}(x)$$

for all x, where

$$\begin{pmatrix} a'' & b'' \\ c'' & d'' \end{pmatrix} = \begin{pmatrix} a & b \\ c & d \end{pmatrix}\begin{pmatrix} a' & b' \\ c' & d' \end{pmatrix}.$$

Furthermore,

$$\det\begin{pmatrix} a'' & b'' \\ c'' & d'' \end{pmatrix} = \det\begin{pmatrix} a & b \\ c & d \end{pmatrix}\det\begin{pmatrix} a' & b' \\ c' & d' \end{pmatrix} \neq 0$$

because $\det\begin{pmatrix} a & b \\ c & d \end{pmatrix} \neq 0$ and $\det\begin{pmatrix} a' & b' \\ c' & d' \end{pmatrix} \neq 0$.

We now show that $\mathsf{PGL}(2,q)$ is sharply 3-transitive. First, we prove that, for all choices of three distinct elements $r,s,t \in \mathbb{F}_q \cup \{\infty\}$, there is a permutation $\pi_{\begin{pmatrix} a & b \\ c & d \end{pmatrix}} \in \mathsf{PGL}(2,q)$ such that $\pi_{\begin{pmatrix} a & b \\ c & d \end{pmatrix}}(0) = r$, $\pi_{\begin{pmatrix} a & b \\ c & d \end{pmatrix}}(1) = s$, and $\pi_{\begin{pmatrix} a & b \\ c & d \end{pmatrix}}(\infty) = t$. Proving this assertion requires some consideration of cases. First, suppose that $r,s,t \in \mathbb{F}_q$. Then what we want is

$$\frac{b}{d} = r,$$

$$\frac{a+b}{c+d} = s, \quad \text{and}$$

$$\frac{b}{d} = t.$$

If we set $d = 1$ (which we can do without loss of generality), then we obtain

$$\begin{aligned} a &= \frac{t(s-r)}{t-s}, \\ b &= r, \\ c &= \frac{s-r}{t-s}, \quad \text{and} \\ d &= 1. \end{aligned}$$

The cases when one of $r, s, t = \infty$ can be handled by similar considerations, and in each case we find the desired permutation.

Now, suppose we choose three distinct elements $r, s, t \in \mathbb{F}_q \cup \{\infty\}$, and $r', s', t' \in \mathbb{F}_q \cup \{\infty\}$ are also distinct. We proved above that there is a permutation $\pi_{\left(\begin{smallmatrix} a & b \\ c & d \end{smallmatrix}\right)}$ that maps 0 to r, 1 to s, and ∞ to t and a permutation $\pi_{\left(\begin{smallmatrix} a' & b' \\ c' & d' \end{smallmatrix}\right)}$ that maps 0 to r', 1 to s', and ∞ to t'. Define

$$\begin{pmatrix} a'' & b'' \\ c'' & d'' \end{pmatrix} = \begin{pmatrix} a' & b' \\ c' & d' \end{pmatrix} \begin{pmatrix} a & b \\ c & d \end{pmatrix}^{-1};$$

then the permutation $\pi_{\left(\begin{smallmatrix} a'' & b'' \\ c'' & d'' \end{smallmatrix}\right)}$ maps r to r', s to s', and t to t'.

We have shown that there is at least one permutation mapping any three distinct elements r, s, t to r', s', t', respectively. However, the total number of permutations in $\mathsf{PGL}(2, q)$ is $q^3 - q$, so we conclude that there is exactly one permutation mapping any three distinct elements r, s, t to r', s', t', respectively. Therefore, we have shown that $\mathsf{PGL}(2, q)$ is sharply 3-transitive. □

Consider $\mathsf{PGL}(2, q^2)$; this is a sharply 3-transitive group acting on the set $X = \mathbb{F}_{q^2} \cup \{\infty\}$. Let $Y = \mathbb{F}_q \cup \{\infty\}$. It is not hard to prove that $\mathsf{stab}(Y)$ is isomorphic to $\mathsf{PGL}(2, q)$ (acting on X). Therefore, $|\mathsf{stab}(Y)| = q^3 - q$, and it follows from Theorem 9.22 that $(X, \mathsf{orbit}(Y))$ is a 3-$(q^2 + 1, q + 1, 1)$-design. This 3-design is called an *inversive plane*. Summarizing, we have the following result.

Theorem 9.27. *For all prime powers q, there exists a* 3-$(q^2 + 1, q + 1, 1)$*-design.*

Example 9.28. We construct a 3-$(10, 4, 1)$-design using Theorem 9.27. The design consists of 30 blocks on the points $\mathbb{F}_9 \cup \{\infty\}$ obtained by letting $\mathsf{PGL}(2, 9)$ act on the block $\mathbb{Z}_3 \cup \{\infty\}$. Using the irreducible polynomial $x^2 + 1 \in \mathbb{Z}_3[x]$, we can construct the field $\mathbb{F}_9 = \mathbb{Z}_3[x]/(x^2 + 1)$. The blocks of the resulting 3-$(10, 4, 1)$-design are listed in Figure 9.1. ∎

9.2.2 Some 5-Designs

We first present a construction (without proof) for a 5-$(12, 6, 1)$-design that uses permutation groups. For an odd prime power q, the group $\mathsf{PSL}(2, q)$ is

$$\begin{array}{lll}
\{\infty,0,1,2\}, & \{\infty,0,x,2x\}, & \{\infty,0,x+2,2x+1\}, \\
\{\infty,0,x+1,2x+2\}, & \{\infty,1,x,2x+2\}, & \{\infty,1,x+2,2x\}, \\
\{\infty,1,x+1,2x+1\}, & \{\infty,2,x,2x+1\}, & \{\infty,2,x+2,2x+2\}, \\
\{\infty,2,x+1,2x\}, & \{\infty,x,x+1,x+2\}, & \{\infty,2x,2x+1,2x+2\}, \\
\{0,1,x+2,2x+2\}, & \{0,1,x,x+1\}, & \{0,1,2x,2x+1\}, \\
\{0,2,x+1,2x+1\}, & \{0,2,x,x+2\}, & \{0,2,2x,2x+2\}, \\
\{0,x,2x+1,2x+2\}, & \{0,x+1,x+2,2x\}, & \{1,2,x,2x\}, \\
\{1,2,x+1,x+2\}, & \{1,2,2x+1,2x+2\}, & \{1,x,x+2,2x+1\}, \\
\{1,x+1,2x,2x+2\}, & \{2,x,x+1,2x+2\}, & \{2,x+2,2x,2x+1\}, \\
\{x,x+1,2x,2x+1\}, & \{x,x+2,2x,2x+2\}, & \{x+1,x+2,2x+1,2x+2\}.
\end{array}$$

Fig. 9.1. The Blocks in a 3-(10, 4, 1)-Design

the subgroup of $\mathsf{PGL}(2,q)$ consisting of all the distinct permutations $\pi_{\begin{pmatrix} a & b \\ c & d \end{pmatrix}}$ such that $ad - bc \neq 0$ is a quadratic residue in $\mathbb{F}_q$. It can be shown that $|\mathsf{PSL}(2,q)| = (q^3-1)/2$.

Example 9.29. Let $X = \mathbb{F}_{11} \cup \{\infty\}$ and let $Y = \{1,3,4,5,9\} \cup \{\infty\}$ (note that Y consists of the quadratic residues modulo 11 together with ∞). Applying the group $\mathsf{PSL}(2,11)$ to Y, it can be proven that $(X, \mathsf{orbit}(Y))$ is a 5-(12, 6, 1)-design. ∎

In the rest of this section, we will present a construction for an infinite class of 5-designs. First, we need some preliminary results on a seemingly different topic.

Let $n \geq 2$ be an integer, and let $(X, \mathcal{A})$ be a projective plane of order 2^n. A *hyperoval* in $(X, \mathcal{A})$ is a set of $2^n + 2$ points $\mathcal{O} \subseteq X$ such that $|\mathcal{O} \cap A| \in \{0,2\}$ for all $A \in \mathcal{A}$.

Theorem 9.30. *For all integers $n \geq 2$, there exists a projective plane of order 2^n containing a hyperoval.*

Proof. We construct a projective plane of order 2^n as in Section 2.3. Let V denote the three-dimensional vector space over the field $\mathbb{F}_{2^n}$. Let X consist of all the one-dimensional subspaces of V, and let $\mathcal{B}$ consist of all the two-dimensional subspaces of V. For each $B \in \mathcal{B}$, define a block

$$A_B = \{x \in X : x \subseteq B\}.$$

Finally, define

$$\mathcal{A} = \{A_B : B \in \mathcal{B}\}.$$

Then $(X, \mathcal{A})$ is a projective plane of order 2^n.

For each $x \in X$, choose a 3-tuple $(x_1, x_2, x_3) \in x$ such that $(x_1, x_2, x_3) \neq (0,0,0)$. Also, for each $(x_1, x_2, x_3) \in (\mathbb{F}_{2^n})^3$ such that $(x_1, x_2, x_3) \neq (0,0,0)$, $\mathsf{span}((x_1, x_2, x_3)) \in X$ is the unique point $x \in X$ such that $(x_1, x_2, x_3) \in x$ (i.e., it is the one-dimensional subspace generated by (x_1, x_2, x_3)).

Now, define

$$\mathcal{O} = \{x \in X : x_1x_2 + x_2x_3 + x_3x_1 = 0\} \cup \{\mathsf{span}((1,1,1))\}.$$

We will show that $\mathcal{O}$ is a hyperoval. First, we find all the points in $\mathcal{O}$. Consider the equation $ab + bc + ca = 0$, $a, b, c \in \mathbb{F}_q$. If $a = 0$, then $bc = 0$, so $b = 0$ or $c = 0$. Similarly, if $b = 0$, then $a = 0$ or $c = 0$; and if $c = 0$, then $a = 0$ or $b = 0$. This gives us three points in $\mathcal{O}$: $\mathsf{span}((0,0,1))$, $\mathsf{span}((0,1,0))$, and $\mathsf{span}((1,0,0))$.

We have considered all cases where at least one of $a, b, c = 0$. Therefore we can now assume $a, b, c \neq 0$. Since points are one-dimensional subspaces, we can assume without loss of generality that $a = 1$. Then $c = b(b+1)^{-1}$. In order for $(b+1)^{-1}$ to exist, $b \neq 1$. Therefore we obtain $q - 1$ more points in $\mathcal{O}$: $\mathsf{span}((1, b, b(b+1)^{-1}))$, where $b \neq 0, 1$. Finally, $\mathsf{span}((1,1,1)) \in \mathcal{O}$ by definition, and so we have shown that there are $q + 2$ points in $\mathcal{O}$.

To show that $\mathcal{O}$ is a hyperoval, we must show that any block in the projective plane intersects $\mathcal{O}$ in either zero or two points. A block can be defined as the solution set of a linear equation

$$B_{d,e,f} = \{\mathsf{span}((a,b,c)) : (a,b,c) \in (\mathbb{F}_q)^3, (d,e,f) \cdot (a,b,c) = 0\},$$

where $d, e, f \in \mathbb{F}_q$. If (d, e, f) and (d', e', f') are scalar multiples of each other, then they define the same block. Therefore, without loss of generality, we can take the first nonzero coefficient of (d, e, f) to be 1.

There are several cases to consider.

1. Suppose $(d, e, f) = (1, 0, 0)$. Then
$$B_{d,e,f} \cap \mathcal{O} = \{\mathsf{span}((0,1,0)), \mathsf{span}((0,0,1))\}.$$
2. Suppose $(d, e, f) = (0, 1, 0)$. Then
$$B_{d,e,f} \cap \mathcal{O} = \{\mathsf{span}((1,0,0)), \mathsf{span}((0,0,1))\}.$$
3. Suppose $(d, e, f) = (0, 0, 1)$. Then
$$B_{d,e,f} \cap \mathcal{O} = \{\mathsf{span}((1,0,0)), \mathsf{span}((0,1,0))\}.$$
4. Suppose $(d, e, f) = (1, 0, 1)$. Then
$$B_{d,e,f} \cap \mathcal{O} = \{\mathsf{span}((0,1,0)), \mathsf{span}((1,1,1))\}.$$
5. Suppose $(d, e, f) = (1, 0, f), f \neq 0, 1$. Then
$$B_{d,e,f} \cap \mathcal{O} = \{\mathsf{span}((0,1,0)), \mathsf{span}((1, (f+1)^{-1}, f^{-1}))\}.$$
6. Suppose $(d, e, f) = (1, 1, 0)$. Then
$$B_{d,e,f} \cap \mathcal{O} = \{\mathsf{span}((0,0,1)), \mathsf{span}((1,1,1))\}.$$

7. Suppose $(d,e,f) = (1,e,0)$, $e \neq 0,1$. Then
$$B_{d,e,f} \cap \mathcal{O} = \{\mathsf{span}((0,0,1)), \mathsf{span}((1,e^{-1},(1+e)^{-1}))\}.$$
8. Suppose $(d,e,f) = (0,1,1)$. Then
$$B_{d,e,f} \cap \mathcal{O} = \{\mathsf{span}((1,0,0)), \mathsf{span}((1,1,1))\}.$$
9. Suppose $(d,e,f) = (0,1,f)$, $f \neq 0,1$. Then
$$B_{d,e,f} \cap \mathcal{O} = \{\mathsf{span}((1,0,0)), \mathsf{span}((1,f^{-1},f(1+f)^{-1}))\}.$$
10. Suppose $(d,e,f) = (1,e,f)$, where $e,f \neq 0$.
    a) If $e+f = 1$, then
    $$B_{d,e,f} \cap \mathcal{O} = \{\mathsf{span}((1,1,1)), \mathsf{span}((1,r^{-1},(1+r)^{-1}))\},$$
    where r is the unique square root (in $\mathbb{F}_q$) of e (this is why we require that q be even: in a finite field of even order, every nonzero field element has a unique square root).
    b) If $e+f \neq 1$, then we form the quadratic equation
    $$eb^2 + (e+f+1)b + 1 = 0.$$
    The linear coefficient in this equation, namely $e+f+1$, is nonzero, so this equation has either zero or two roots over $\mathbb{F}_q$. The roots (if any) determine the values of b such that $\mathsf{span}((1,b,b(b+1)^{-1})) \in B_{d,e,f} \cap \mathcal{O}$.

The cases above exhaust all the possibilities, and the desired result is proven. □

Example 9.31. A hyperoval $\mathcal{O}$ in a projective plane of order 4. We begin by constructing the field $\mathbb{F}_4 = \mathbb{Z}_2[x]/(x^2+x+1)$. Then the six points in $\mathcal{O}$ are as follows:
$$\mathcal{O} = \left\{ \begin{array}{lll} \mathsf{span}((0,0,1)), & \mathsf{span}((0,1,0)), & \mathsf{span}((1,0,0)), \\ \mathsf{span}((1,x,x+1)), & \mathsf{span}((1,x+1,x)), & \mathsf{span}((1,1,1)) \end{array} \right\}.$$
■

Theorem 9.32. *For all integers $n \geq 3$, there exists a* 5-$(2^n+2,6,15)$-*design.*

Proof. Let $\mathcal{O}$ be a hyperoval in a projective plane of order 2^n, say $(X,\mathcal{A})$. For each point $x \in X\backslash\mathcal{O}$, define
$$\mathcal{P}(x) = \{A \in \mathcal{A} : x \in A \text{ and } |A \cap \mathcal{O}| = 2\}.$$
Note that each $\mathcal{P}(x)$ consists of $2^{n-1}+1$ blocks. Now, for $x \in X\backslash\mathcal{O}$, define

$$\Pi_x = \{A \cap \mathcal{O} : A \in \mathcal{P}(x)\}.$$

Using the fact that $\mathcal{O}$ is a hyperoval, it is not hard to see that each Π_x is a partition of $\mathcal{O}$ into $2^{n-1}+1$ 2-subsets. Further, given any two disjoint 2-subsets in $\mathcal{O}$, there is a unique Π_x that contains both of them (this follows because any two blocks in a projective plane intersect in a unique point).

Now, define

$$\mathcal{B} = \{A_1 \cup A_2 \cup A_3 : A_1, A_2, A_3 \in \Pi_x, A_1 \neq A_2 \neq A_3 \neq A_1, x \in X \backslash \mathcal{O}\}.$$

We claim that the pair $(\mathcal{O}, \mathcal{B})$ is a 5-$(2^n+2, 6, 15)$-design. To prove this, let x_1, x_2, x_3, x_4, x_5 be five distinct points in $\mathcal{O}$. There are $\binom{5}{2}\binom{3}{2}/2 = 15$ ways to choose two disjoint 2-subsets from $\{x_1, x_2, x_3, x_4, x_5\}$. Consider, for example, $\{x_1, x_2\}$ and $\{x_3, x_4\}$. As stated above, there is a unique x such that $\{x_1, x_2\} \in \Pi_x$ and $\{x_3, x_4\} \in \Pi_x$. Then, there is a unique x_6 such that $\{x_5, x_6\} \in \Pi_x$. This yields a block $\{x_1, x_2, x_3, x_4, x_5, x_6\}$ containing the five given points.

From this argument, it is easily seen that we have a 5-$(2^n+2, 6, 15)$-design. □

9.3 *t*-wise Balanced Designs

We begin by defining *t*-wise balanced designs, which generalize the notion of pairwise balanced designs.

Definition 9.33. *Let $t \geq 1$ be an integer. A* t-wise balanced design *is a design $(X, \mathcal{B})$ such that the following properties are satisfied.*

1. *$|B| \geq t$ for all $B \in \mathcal{B}$.*
2. *Every subset of t distinct points is contained in exactly one block.*

*Let $K \subseteq \{n \in \mathbb{Z} : n \geq t\}$. A t-wise balanced design $(X, \mathcal{B})$ is denoted as a t-(v, K)-*tBD *provided that $|X| = v$ and $|B| \in K$ for all $B \in \mathcal{B}$.*

As a first class of examples, we observe that it is easy to construct certain 3-wise balanced designs using the method of Theorem 9.16.

Theorem 9.34. *Suppose that $v \geq 2$ is an even integer. Then there exists a* 3-$(2v, \{4, v\})$-tBD.

Proof. Use the same construction as in the proof of Theorem 9.16, but retain X and Y as two blocks of size v. □

Next, we give an elegant construction for certain 5-wise balanced designs.

Theorem 9.35. *For all integers $n \geq 4$, there exists a* 5-$(2^n, \{6, 8\})$-tBD.

Proof. Let $X = (\mathbb{Z}_2)^n$. Define

$$\mathcal{A} = \left\{ Y \subseteq X : |Y| = 6 \text{ and } \sum_{x \in Y} x = (0, \ldots, 0) \right\}.$$

Let $\mathcal{B}$ consist of all three-dimensional subspaces of $(\mathbb{Z}_2)^n$ and all their additive cosets (recall that the blocks in $\mathcal{B}$ are called flats). Observe that a flat has the form $\mathbf{a} + \mathsf{span}(\mathbf{u}, \mathbf{v}, \mathbf{w})$, where $\mathbf{a}, \mathbf{u}, \mathbf{v}, \mathbf{w} \in (\mathbb{Z}_2)^n$; $\mathbf{u}, \mathbf{v}, \mathbf{w}$ are linearly independent; and $\mathsf{span}(\mathbf{u}, \mathbf{v}, \mathbf{w})$ denotes the subspace spanned by $\mathbf{u}$, $\mathbf{v}$, and $\mathbf{w}$.

We will show that $(X, \mathcal{A} \cup \mathcal{B})$ is a 5-$(2^n, \{6, 8\})$-tBD. Let $\mathbf{x}_1, \mathbf{x}_2, \mathbf{x}_3, \mathbf{x}_4, \mathbf{x}_5$ be five distinct vectors in $(\mathbb{Z}_2)^n$. Define $\mathbf{x}_6 = \mathbf{x}_1 + \mathbf{x}_2 + \mathbf{x}_3 + \mathbf{x}_4 + \mathbf{x}_5$. If $\mathbf{x}_6$ is distinct from $\mathbf{x}_1, \mathbf{x}_2, \mathbf{x}_3, \mathbf{x}_4, \mathbf{x}_5$, then $\{\mathbf{x}_1, \mathbf{x}_2, \mathbf{x}_3, \mathbf{x}_4, \mathbf{x}_5, \mathbf{x}_6\}$ is a block in $\mathcal{A}$ that contains $\mathbf{x}_1, \mathbf{x}_2, \mathbf{x}_3, \mathbf{x}_4, \mathbf{x}_5$. Suppose that $\mathbf{x}_6 \in \{\mathbf{x}_1, \mathbf{x}_2, \mathbf{x}_3, \mathbf{x}_4, \mathbf{x}_5\}$. Without loss of generality, suppose that $\mathbf{x}_5 = \mathbf{x}_6$. Then $\mathbf{x}_1 + \mathbf{x}_2 + \mathbf{x}_3 + \mathbf{x}_4 = 0$ and hence $\mathbf{x}_4 = \mathbf{x}_1 + \mathbf{x}_2 + \mathbf{x}_3$. It is not difficult to check that

$$\mathbf{x}_5 + \mathsf{span}(\mathbf{x}_1 + \mathbf{x}_5, \mathbf{x}_2 + \mathbf{x}_5, \mathbf{x}_3 + \mathbf{x}_5)$$

is a flat of dimension three (i.e., a block in $\mathcal{B}$) that contains $\mathbf{x}_1, \mathbf{x}_2, \mathbf{x}_3, \mathbf{x}_4, \mathbf{x}_5$. This is easy to verify since

$$\begin{aligned} &\mathsf{span}(\mathbf{x}_1 + \mathbf{x}_5, \mathbf{x}_2 + \mathbf{x}_5, \mathbf{x}_3 + \mathbf{x}_5) \backslash \{(0, \ldots, 0)\} \\ &= \{\mathbf{x}_1 + \mathbf{x}_5, \mathbf{x}_2 + \mathbf{x}_5, \mathbf{x}_3 + \mathbf{x}_5, \mathbf{x}_1 + \mathbf{x}_2, \mathbf{x}_2 + \mathbf{x}_3, \mathbf{x}_1 + \mathbf{x}_3, \mathbf{x}_4 + \mathbf{x}_5\}. \end{aligned}$$

At this point, we know that every set of five points occurs in at least one block. We need to check that no set of five points occurs in more than one block. Equivalently, we need to show that no two blocks intersect in more than four points.

Clearly no two blocks in $\mathcal{A}$ intersect in more than four points since two blocks in $\mathcal{A}$ that intersect in five points would be identical. The intersection of two blocks in $\mathcal{B}$ is a flat and therefore cannot contain more than four points, so we need only to consider the intersection of a block in $A \in \mathcal{A}$ with a block in $B \in \mathcal{B}$. Since B is a flat, the sum of an odd number of vectors in B is a vector in B. This means that $|A \cap B| \neq 5$ since any vector in A is the sum of the other five vectors in A. The only remaining possibility is that $A \subseteq B$. Now the sum of all the vectors in B is $(0, \ldots, 0)$, as is the sum of all the vectors in A. This means that the sum of the two vectors in $B \backslash A$ is also $(0, \ldots, 0)$, which implies that they are equal. This is a contradiction, and we conclude that we have constructed a 5-wise balanced design. □

9.3.1 Holes and Subdesigns

Definition 9.36. *Let $t \geq 1$ be an integer. An* incomplete t-wise balanced design *is a triple $(X, Y, \mathcal{B})$ such that the following properties are satisfied.*

1. *X is a set of elements called* points.
2. *$Y \subseteq X$ is called the* hole.
3. *$\mathcal{B}$ is a set of subsets of X called* blocks *such that $|B| \geq t$ for all $B \in \mathcal{B}$.*
4. *No block contains t points from Y.*
5. *Every subset of t points $Z \subseteq X$ such that $Z \not\subseteq Y$ is contained in exactly one block.*

Let $K \subseteq \{n \in \mathbb{Z} : n \geq t\}$. An incomplete t-wise balanced design $(X, Y, \mathcal{B})$ is denoted as a t-(v, h, K)-ItBD provided that $|X| = v$, $|Y| = h$, and $|B| \in K$ for all $B \in \mathcal{B}$.

Example 9.37. Corollary 7.4 shows how to construct a $(2v-1, \{3, v-1\})$-PBD having exactly one block of size $v-1$ whenever $v \geq 4$ is an even integer. If the block of size $v-1$ is taken to be the hole, then we have a 2-$(2v-1, v-1, \{3\})$-ItBD. ∎

Observe that a t-(v, h, K)-ItBD is the same thing as a t-(v, K)-tBD when $0 \leq h \leq t-1$. When $h \geq t$, we have the following result.

Lemma 9.38. *Suppose there is a t-(v, h, K)-ItBD, where $h \geq t$. Then there is a t-$(v, K \cup \{h\})$-tBD.*

Proof. Let $(X, Y, \mathcal{B})$ be a t-(v, h, K)-ItBD. Define $\mathcal{C} = \mathcal{B} \cup \{Y\}$. Then $(X, \mathcal{C})$ is a t-$(v, K \cup \{h\})$-tBD. □

Holes can sometimes be filled in with t-wise balanced designs, as is shown in the following lemma.

Lemma 9.39 (Filling in Holes). *Suppose there exists a t-(v, h, K)-ItBD and a t-(h, K)-tBD. Then there exists a t-(v, K)-tBD.*

Proof. Let $(X, Y, \mathcal{B})$ be a t-(v, h, K)-ItBD and let $(Y, \mathcal{C})$ be a t-(h, K)-tBD. Then it is easy to see that $(X, \mathcal{B} \cup \mathcal{C})$ is a t-(h, K)-tBD. □

Suppose that $(X, \mathcal{B})$ is a t-(v, K)-tBD, and suppose further that $Y \subseteq X$ and $\mathcal{C} \subseteq \mathcal{B}$. Then $(Y, \mathcal{C})$ is a *subdesign* of $(X, \mathcal{B})$ provided that $(Y, \mathcal{C})$ is itself a t-(h, K)-tBD, where $h = |Y|$. Any block of a t-wise balanced design yields a subdesign. Subdesigns consisting of more than one block are more interesting, however.

Observe that Lemma 9.39 creates an incomplete t-wise balanced design that contains a subdesign, $(Y, \mathcal{C})$. The following lemma is a type of converse result.

Lemma 9.40. *Suppose that $(X, \mathcal{B})$ is a t-(v, K)-tBD and $(Y, \mathcal{C})$ is a subdesign. Define $\mathcal{D} = \mathcal{B} \backslash \mathcal{C}$. Then $(X, Y, \mathcal{D})$ is a t-(v, h, K)-ItBD, where $h = |Y|$.*

In the rest of this section, we find some necessary conditions for the existence of certain incomplete t-wise balanced designs.

Theorem 9.41. *Suppose that t, k, h, and v are positive integers such that $2 \le t < k < h < v$. In a t-$(v, h, \{k\})$-ItBD, say $(X, Y, \mathcal{B})$, the number of blocks disjoint from the hole is exactly*

$$a(t, v, h, k) = \sum_{i=0}^{t-1} \frac{(-1)^i \binom{h}{i} \left(\binom{v-i}{t-i} - \binom{h-i}{t-i} \right)}{\binom{k-i}{t-i}}.$$

Proof. The proof is similar to the first part of the proof of Theorem 9.7. For any $y \in Y$, define

$$\mathcal{B}_y = \{B \in \mathcal{B} : y \in B\}.$$

Then, for any $Y_0 \subseteq Y$ such that $|Y_0| = i \le t$, it is easy to see that

$$\left| \bigcap_{y \in Y_0} \mathcal{B}_y \right| = \frac{\binom{v-i}{t-i} - \binom{h-i}{t-i}}{\binom{k-i}{t-i}}.$$

Applying the principle of inclusion-exclusion, as in the proof of Theorem 9.7, the desired result is obtained. □

Corollary 9.42. *Suppose that t, k, h, and v are positive integers such that $2 \le t < k < h < v$. If a t-$(v, h, \{k\})$-ItBD exists, then $a(t, v, h, k) \ge 0$.*

Corollary 9.42 can be used to prove some useful necessary conditions. The first interesting case is $t = 2$, which can easily be analyzed. We have the following:

$$\begin{aligned} a(2, v, h, k) &= \sum_{i=0}^{1} \frac{(-1)^i \binom{h}{i} \left(\binom{v-i}{2-i} - \binom{h-i}{2-i} \right)}{\binom{k-i}{2-i}} \\ &= \frac{\binom{v}{2} - \binom{h}{2}}{\binom{k}{2}} - \frac{h(v-h)}{k-1} \\ &= \frac{v-h}{k-1} \left(\frac{v+h-1}{k} - h \right). \end{aligned}$$

Therefore it follows that $a(2, v, h, k) \ge 0$ if and only if $v \ge h(k-1) + 1$, and we obtain the following well-known result by applying Corollary 9.42.

Theorem 9.43. *Suppose that k, h, and v are positive integers such that $2 < k < h < v$. If a 2-$(v, h, \{k\})$-ItBD exists, then $v \ge h(k-1) + 1$.*

We observe that the 2-$(2v - 1, v - 1, \{3\})$-ItBDs, which were constructed in Example 9.37, meet the bound of Theorem 9.43 with equality.

Another case that can be solved is when t is even and $k = t + 1$. First, we rewrite the function $a(t, v, h, k)$ and apply a certain binomial identity. For a positive integer $x > k$, define

$$b(t,k,h,x) = \sum_{i=0}^{t-1} \frac{(-1)^i \binom{h}{i} \binom{x-i}{t-i}}{\binom{k-i}{t-i}}.$$

Then it is clear that $a(t,v,h,k) = b(t,k,h,v) - b(t,k,h,h)$. We study the function b a bit further.

Using the fact that

$$\frac{\binom{x-i}{t-i}}{\binom{k-i}{t-i}} = \frac{\binom{x-i}{x-k}}{\binom{x-t}{k-t}},$$

we have that

$$b(t,k,h,x) = \sum_{i=0}^{t-1} \frac{(-1)^i \binom{h}{i} \binom{x-i}{x-k}}{\binom{x-t}{k-t}} = \frac{1}{\binom{x-t}{k-t}} \sum_{i=0}^{t-1} (-1)^i \binom{h}{i} \binom{x-i}{x-k}. \tag{9.4}$$

Now we can apply a binomial identity, which we state without proof.

$$\sum_{i=0}^{k} (-1)^i \binom{h}{i} \binom{x-i}{x-k} = \binom{x-h}{k}. \tag{9.5}$$

From (9.4) and (9.5), it follows immediately that

$$b(t,k,h,x) = \frac{1}{\binom{x-t}{k-t}} \left(\binom{x-h}{k} - \sum_{i=t}^{k} (-1)^i \binom{h}{i} \binom{x-i}{x-k} \right).$$

We apply the results above when $k = t+1$ and t is even:

$$\begin{aligned} b(t,t+1,h,x) &= \frac{1}{x-t} \left(\binom{x-h}{t+1} - \sum_{i=t}^{t+1} (-1)^i \binom{h}{i} \binom{x-i}{x-t-1} \right) \\ &= \frac{1}{x-t} \left(\binom{x-h}{t+1} - \binom{h}{t}(x-t) + \binom{h}{t+1} \right). \end{aligned}$$

Now we can compute the function $a(t,v,h,t+1)$ (when t is even) as follows:

$$\begin{aligned} a(t,v,h,t+1) &= b(t,t+1,h,v) - b(t,t+1,h,h) \\ &= \frac{1}{v-t} \left(\binom{v-h}{t+1} - \binom{h}{t}(v-t) + \binom{h}{t+1} \right) \\ &\quad - \frac{1}{h-t} \left(-\binom{h}{t}(h-t) + \binom{h}{t+1} \right) \\ &= \frac{1}{v-t} \binom{v-h}{t+1} - \binom{h}{t+1} \left(\frac{1}{v-t} - \frac{1}{h-t} \right) \\ &= \frac{1}{v-t} \left(\binom{v-h}{t+1} - \binom{h}{t+1} \frac{v-h}{h-t} \right). \end{aligned}$$

Then it is easily seen that $a(t,v,h,t+1) \geq 0$ if and only if

$$(v-h-1)\times\cdots\times(v-h-t)\geq h\times\cdots\times(h-t+1).$$

This is true if and only if $v \geq 2h+1$.

Applying Corollary 9.42, we have the following result.

Theorem 9.44. *Suppose that t, h, and v are positive integers such that $t+1 < h < v$, and suppose that t is even. If a t-$(v, h, \{t+1\})$-ItBD exists, then $v \geq 2h+1$.*

We can obtain a slightly stronger result when $t \geq 3$ is odd.

Theorem 9.45. *Suppose that t, h, and v are positive integers such that $t+1 < h < v$, and suppose that $t \geq 3$ is odd. If a t-$(v, h, \{t+1\})$-ItBD exists, then $v \geq 2h$.*

Proof. Suppose that $(X, Y, \mathcal{B})$ is a t-$(v, h, \{t+1\})$-ItBD with $t \geq 3$, t odd. Let $y \in Y$. Then

$$(X\backslash\{y\}, Y\backslash\{y\}, \{B\backslash\{y\} : y \in B \in \mathcal{B}\})$$

is a $(t-1)$-$(v-1, h-1, \{t\})$-ItBD. Applying Theorem 9.44, we have that $v - 1 \geq 2(h-1)+1$, or $v \geq 2h$. □

9.4 Notes and References

For more information on the topics described in this chapter, see Kramer [70], Kreher [72], and Colbourn and Mathon [31], all of which are sections in "The CRC Handbook of Combinatorial Designs".

Steiner quadruple systems of all possible orders were shown to exist by Hanani [56]. A very readable proof of this difficult result can be found in Chapter 7 of "Design Theory" by Lindner and Rodger [77]. Hartman and Phelps [58] is a useful survey on Steiner quadruple systems.

Theorem 9.8 is due to Wilson [120]. The proof we give is from Cameron [19]. Theorem 9.27 is due to Witt [124].

The proofs of Theorems 9.14 and 9.32 are due to Lonz and Vanstone; their techniques are discussed further by Jungnickel and Vanstone in [66] and [67].

The construction of simple t-designs with $t \geq 3$ has been a problem of ongoing interest. There are quite a number of results for $t = 3$, but relatively little is known for $t \geq 4$. It is known that such designs exist for all t; this is a famous result of Teirlinck [108]. The existence of a t-design with $t \geq 6$ and $\lambda = 1$ is currently unknown, however.

Theorem 9.35 is unpublished work due to Wilson; the construction is presented in Kramer [69]. In 1983, Kramer [69] conjectured the results that we stated as Theorems 9.44 and 9.45. These theorems were proven by Kreher and Rees in 2001 [73].

9.5 Exercises

9.1 A t-(v, k, λ)-design, say $(X, \mathcal{A})$, is said to be a *graphical t-design* if X consists of the edges of a complete graph K_n (i.e., all the 2-subsets of a v-set, where $v = \binom{n}{2}$) and $\mathcal{A}$ is formed by taking all subgraphs of K_n that are isomorphic to one of the graphs in a set $\mathcal{G}$ of specified subgraphs of K_n.

(a) Suppose that $v = 15, n = 6$, and

$$\mathcal{G} = \{\{12, 34, 56\}, \{12, 13, 23\}\}.$$

Prove that the result is a graphical 2-$(15, 3, 1)$-design (i.e., a $(15, 3, 1)$-BIBD).

(b) Suppose that $v = 10, n = 5$, and

$$\mathcal{G} = \{\{12, 13, 14, 15\}, \{12, 13, 23, 45\}, \{12, 23, 34, 14\}\}.$$

Prove that the result is a graphical 3-$(10, 4, 1)$-design.

9.2 Assuming that a 5-$(12, 6, 1)$-design exists, compute the values λ_i^j for all i, j such that $i + j \leq 5$.

9.3 Construct a 1-$(14, 6, 3)$-design.

9.4 Theorem 9.14 describes how to construct a 3-$(v, 4, 3)$-design from a resolvable $(v, 2, 1)$-BIBD, say $(X, \mathcal{A})$. Let $\Pi_1, \ldots, \Pi_{v-1}$ denote the parallel classes in this BIBD, and denote by $(X, \mathcal{B})$ the resulting 3-$(v, 4, 3)$-design. Prove the following assertions:

(a) For any $i \neq j$, $\Pi_i \cup \Pi_j$ consists of disjoint cycles that partition X.

(b) The length of any cycle in any union $\Pi_i \cup \Pi_j$ is an even integer that is ≥ 4.

(c) $(X, \mathcal{B})$ is a simple 3-$(v, 4, 3)$-design if and only if there is no cycle of length four in any union $\Pi_i \cup \Pi_j$.

9.5 Suppose there is an SQS(v). Prove that there is a 3-$(2v, \{v, 4\})$-tBD.

9.6 Suppose there is a 3-$(v, \{4, 6\})$-tBD. Prove that there is a 3-$(2v, \{4, 6\})$-tBD.

9.7 Use the existence of hyperovals to establish the following.

(a) Prove that there is a 2-$(2^{2n} - 1, \{2^n + 1, 2^n - 1\})$-tBD for all integers $n \geq 2$.

(b) For all integers $n \geq 2$ and all integers i such that $1 \leq i \leq 2^n + 1$, prove that there is a 2-$(2^{2n} + 2^n + 1 - i, \{2^n + 1, 2^n, 2^n - 1\})$-tBD.

9.8 Let $m \geq 2$ be an integer, and let $(X, \mathcal{A})$ be a projective plane of order m. Suppose that $\mathcal{O} \subseteq X$ is a set of points such that $|\mathcal{O} \cap A| \leq 2$ for all $A \in \mathcal{A}$. Prove the following.

(a) $|\mathcal{O}| \leq m + 2$.

(b) If $|\mathcal{O}| = m + 2$, then $|\mathcal{O} \cap A| \in \{0, 2\}$ for all $A \in \mathcal{A}$.

(c) If $|\mathcal{O}| = m + 2$, then m is even.

9.9 We outline a proof that the identity (9.5) holds. The proof uses the following two simpler identities:

$$\binom{n}{i} = (-1)^i \binom{i - n - 1}{i} \tag{9.6}$$

and

$$\sum_{i=0}^{j} \binom{n}{i} \binom{m}{j-i} = \binom{n+m}{j}. \tag{9.7}$$

Remark: You are not asked to prove these two identities. However, we note that (9.6) follows easily from the definition of a binomial coefficient, and (9.7) is a classical result known as the *Vandermonde convolution formula*.

Assuming that (9.6) and (9.7) hold, prove that (9.5) holds as follows:

- replace $\binom{x-i}{x-k}$ by $\binom{x-i}{k-i}$,
- apply (9.6),
- apply (9.7), and finally
- apply (9.6).

9.10 Prove that a 2-$(v, h, \{k\})$-ItBD with $v = h(k-1)+1$ exists if and only if a resolvable $(v-h, k-1, 1)$-BIBD exists.

10

Orthogonal Arrays and Codes

10.1 Orthogonal Arrays

We defined orthogonal arrays in Section 6.5. We give a more general definition now.

Definition 10.1. *Let t, v, k, and λ be positive integers such that $k \geq t \geq 2$. A t-(v,k,λ)* orthogonal array *(denoted t-(v,k,λ)-OA) is a pair (X, D) such that the following properties are satisfied.*

1. *X is a set of v elements called* points.
2. *D is a λv^t by k array whose entries are chosen from the set X.*
3. *Within any t columns of D, every t-tuple of points is contained in exactly λ rows.*

An orthogonal array (X, D) is a simple orthogonal array *if all the rows in D are different (i.e., D does not contain "repeated rows"). An orthogonal array (X, D) is a* linear orthogonal array *if $X = \mathbb{F}_q$ for some prime power q and the rows of D form a subspace (of the vector space $(\mathbb{F}_q)^k$) having dimension $\log_q |D|$. It is clear from the definitions that a linear orthogonal array is necessarily simple.*

We already defined a special type of orthogonal array in Section 6.5; an OA(k, n) (as defined in Section 6.5) is the same thing as a 2-$(v,k,1)$-OA. That is, the previous definition is just the special case $t = 2$ and $\lambda = 1$.

We have defined orthogonal arrays using array notation. Each row of an orthogonal array D is a k-tuple. It is possible to define an orthogonal array to be the collection (or multiset) of k-tuples formed from the rows of D. We will sometimes use this alternative viewpoint, particularly when we consider the connections with codes in later sections of this chapter.

We illustrate the definition above with a simple construction for certain orthogonal arrays from Hadamard matrices.

Theorem 10.2. *Suppose there exists a Hadamard matrix of order $4m$. Then there exists a 2-$(2, 4m-1, m)$-OA.*

Proof. Let H be a standardized Hadamard matrix of order $4m$ (see Section 4.1). Delete the first row of H, and then transpose this array to form a $4m$ by $4m-1$ array, D. It is easy to see that D is the desired orthogonal array using the counting arguments from the proof of Theorem 4.4. □

Example 10.3. A 2-$(2,7,2)$-OA constructed from the Hadamard matrix of order 8 presented in Example 4.6.

$$\begin{pmatrix} 1 & 1 & 1 & 1 & 1 & 1 & 1 \\ 1 & 1 & 1 & -1 & -1 & -1 & -1 \\ 1 & -1 & -1 & 1 & 1 & -1 & -1 \\ 1 & -1 & -1 & -1 & -1 & 1 & 1 \\ -1 & 1 & -1 & 1 & -1 & 1 & -1 \\ -1 & 1 & -1 & -1 & 1 & -1 & 1 \\ -1 & -1 & 1 & 1 & -1 & -1 & 1 \\ -1 & -1 & 1 & -1 & 1 & 1 & -1 \end{pmatrix}.$$

▮

The following important construction enables orthogonal arrays to be constructed for a wide variety of parameter situations.

Theorem 10.4. *Let ℓ and n be positive integers, and let q be a prime power. Let M be an ℓ by n matrix of elements from $\mathbb{F}_q$ such that every set of t columns of M is linearly independent. Define D to be the q^ℓ by n matrix whose rows consist of all the linear combinations of the rows of M. Then $(\mathbb{F}_q, D)$ is a linear t-(q,n,λ)-OA, where $\lambda = q^{\ell-t}$.*

Proof. Choose t columns of D, say the ones labeled $c_1, \ldots, c_t$. Let $(y_1, \ldots, y_t)$ be an arbitrary t-tuple of elements of $\mathbb{F}_q$. We want to determine the rows i of D such that $D(i, c_j) = y_j$ for $1 \le j \le t$.

A row of D is constructed as $\mathbf{r}M$, where $\mathbf{r} = (r_1, \ldots, r_\ell) \in (\mathbb{F}_q)^\ell$. Let $\mathbf{c}_j$ denote the jth column of M for $1 \le j \le n$. We want to determine all vectors $\mathbf{r}$ such that

$$\mathbf{r}\mathbf{c}_{i_j} = y_{i_j}, \quad 1 \le j \le t. \tag{10.1}$$

The column vectors $\mathbf{c}_{i_1}, \ldots, \mathbf{c}_{i_t}$ are linearly independent by assumption. Therefore, (10.1) is a system of t independent linear equations in ℓ unknowns, and it has a solution space of dimension $\ell - t$. The number of solutions $\mathbf{r}$ is $q^{\ell-t}$, as desired. □

We present a couple of important corollaries of Theorem 10.4.

Corollary 10.5. *Let $\ell \ge 2$ be a positive integer, and let q be a prime power. Then there exists a* 2-$(q, (q^\ell - 1)/(q-1), q^{\ell-2})$-OA.

Proof. Excluding the zero vector, there are $q^\ell - 1$ distinct ℓ-tuples of elements of $\mathbb{F}_q$. Each ℓ-tuple has $q - 1$ nonzero scalar multiples, so the $q^\ell - 1$ nonzero vectors are partitioned into $(q^\ell - 1)/(q - 1)$ subspaces each of dimension equal to one. Arbitrarily pick one vector from each subspace, and let these vectors be the columns of M. Then apply Theorem 10.4. □

When we take $\ell = 2$ in Corollary 10.5, we get a 2-$(q, q+1, 1)$-OA. This is equivalent to the projective plane $\mathrm{PG}_2(q)$.

Example 10.6. Suppose we take $q = 5$ and $\ell = 2$ in Corollary 10.5. Each pair of columns of the following 2×6 matrix is linearly independent over $\mathbb{Z}_5$:

$$\begin{pmatrix} 0\,1\,1\,1\,1\,1 \\ 1\,0\,1\,2\,3\,4 \end{pmatrix}.$$

Applying Theorem 10.4, the following 2-$(5, 6, 1)$-OA is obtained:

$$\begin{pmatrix} 0\,0\,0\,0\,0\,0 \\ 0\,1\,1\,1\,1\,1 \\ 0\,2\,2\,2\,2\,2 \\ 0\,3\,3\,3\,3\,3 \\ 0\,4\,4\,4\,4\,4 \\ 1\,0\,1\,2\,3\,4 \\ 2\,1\,2\,3\,4\,0 \\ 3\,2\,3\,4\,0\,1 \\ 4\,3\,4\,0\,1\,2 \\ 0\,4\,0\,1\,2\,3 \\ 2\,0\,2\,4\,1\,3 \\ 3\,1\,3\,0\,2\,4 \\ 4\,2\,4\,1\,3\,0 \\ 0\,3\,0\,2\,4\,1 \\ 1\,4\,1\,3\,0\,2 \\ \vdots \\ 4\,0\,4\,3\,2\,1 \\ 0\,1\,0\,4\,3\,2 \\ 1\,2\,1\,0\,4\,3 \\ 2\,3\,2\,1\,0\,4 \\ 3\,4\,3\,2\,1\,0 \end{pmatrix}.$$

Corollary 10.7. *Let $t \geq 2$ be an integer, and let q be a prime power. Then there exists a t-$(q, q, 1)$-OA.*

Proof. For every $x \in \mathbb{F}_q$, construct the vector $\mathbf{x} = (1, x, x^2, \ldots, x^{t-1}) \in (\mathbb{F}_q)^t$. Transpose these q vectors to form the columns of M. Therefore M has the following form:

$$M = \begin{pmatrix} 1 & 1 & 1 & \cdots & 1 \\ x_1 & x_2 & x_3 & \cdots & x_q \\ {x_1}^2 & {x_2}^2 & {x_3}^2 & \cdots & {x_q}^2 \\ \vdots & \vdots & \vdots & \ddots & \vdots \\ {x_1}^{t-1} & {x_2}^{t-1} & {x_3}^{t-1} & \cdots & {x_q}^{t-1} \end{pmatrix},$$

where $x_1, \ldots, x_q$ are the q distinct elements of $\mathbb{F}_q$.

In order to apply Theorem 10.4, we need to show that any t of the vectors $\mathbf{x}$ are linearly independent. Suppose that this is not the case. Then there exists a $t \times t$ submatrix of M, say M_0, whose columns are linearly dependent. M_0 has the form

$$M_0 = \begin{pmatrix} 1 & 1 & 1 & \cdots & 1 \\ y_1 & y_2 & y_3 & \cdots & y_t \\ {y_1}^2 & {y_2}^2 & {y_3}^2 & \cdots & {y_t}^2 \\ \vdots & \vdots & \vdots & \ddots & \vdots \\ {y_1}^{t-1} & {y_2}^{t-1} & {y_3}^{t-1} & \cdots & {y_t}^{t-1} \end{pmatrix},$$

where $y_1, \ldots, y_t$ are t distinct elements of $\mathbb{F}_q$.

If the columns of M_0 are linearly dependent, then the rows of M_0 are also linearly dependent. Therefore, there exist $a_1, \ldots, a_t \in \mathbb{F}_q$, not all equal to 0, such that $(a_1, \ldots, a_t)M_0 = (0, \ldots, 0)$. Define the polynomial

$$a(x) = a_1 + a_2 x + \cdots + a_t x^{t-1};$$

then $a(y_j) = 0$ for $1 \leq j \leq t$. This means that the degree $t-1$ polynomial $a(x)$ has t roots in the field $\mathbb{F}_q$, which is impossible. This contradiction establishes the desired result. □

Example 10.8. A 3-$(5,5,1)$-OA. The matrix M described in Corollary 10.7 is as follows:

$$M = \begin{pmatrix} 1\,1\,1\,1\,1 \\ 0\,1\,2\,3\,4 \\ 0\,1\,4\,4\,1 \end{pmatrix}.$$

The 125 rows that are the linear combinations (over $\mathbb{Z}_5$) of the three rows of M comprise the desired orthogonal array. ∎

The constructions above all yield linear orthogonal arrays. Here is a construction for orthogonal arrays that makes use of quadratic, instead of linear, functions.

Theorem 10.9. *Let q be an odd prime power. For $a, b \in \mathbb{F}_q$, define $f_{a,b} : \mathbb{F}_q \to \mathbb{F}_q$ by the rule*

$$f_{a,b}(x) = (x+a)^2 + b.$$

Then the q^2 by q array $D = (d_{i,j})$, where $d_{i,j} = f_{a,b}(j)$ $(i = (a,b) \in (\mathbb{F}_q)^2, j \in \mathbb{F}_q)$, is a 2-$(q,q,1)$-OA.

Proof. Let $x_1, x_2 \in \mathbb{F}_q$ (where $x_1 \neq x_2$) and let $y_1, y_2 \in \mathbb{F}_q$. We want to show that there is exactly one ordered pair $(a, b) \in (\mathbb{F}_q)^2$ such that

$$(x_1 + a)^2 + b = y_1$$

and

$$(x_2 + a)^2 + b = y_2.$$

Subtracting the two equations, we can solve uniquely for a:

$$a = \frac{y_1 - y_2}{2(x_1 - x_2)} - \frac{x_1 + x_2}{2}.$$

Then, given a, we obtain a unique solution for b. □

Example 10.10. The following 2-(3, 3, 1)-OA is constructed using Theorem 10.9.

$$\begin{array}{l} \quad\;\; 0\,1\,2 \\ \hline f_{0,0} : 0\,1\,1 \\ f_{0,1} : 1\,2\,2 \\ f_{0,2} : 2\,0\,0 \\ f_{1,0} : 1\,1\,0 \\ f_{1,1} : 2\,2\,1 \\ f_{1,2} : 0\,0\,2 \\ f_{2,0} : 1\,0\,1 \\ f_{2,1} : 2\,1\,2 \\ f_{2,2} : 0\,2\,0 \end{array} \rightarrow \begin{pmatrix} 0\,1\,1 \\ 1\,2\,2 \\ 2\,0\,0 \\ 1\,1\,0 \\ 2\,2\,1 \\ 0\,0\,2 \\ 1\,0\,1 \\ 2\,1\,2 \\ 0\,2\,0 \end{pmatrix}$$

This orthogonal array is not linear. This can be seen, for example, by observing that the sum of the first two rows (modulo 3) is $(1, 0, 0)$, which is not a row of the array. ∎

Finally, we give a powerful nonconstructive existence result for orthogonal arrays.

Theorem 10.11 (Gilbert-Varshamov Bound). *Let ℓ, t and n be positive integers such that $2 \leq t \leq \ell$, and let q be a prime power. Suppose that*

$$\sum_{i=0}^{t-1} \binom{n-1}{i} (q-1)^i < q^\ell. \tag{10.2}$$

Then there exists a linear t-(q, n, λ)-OA, where $\lambda = q^{\ell - t}$.

Proof. We will prove that there exists an ℓ by n matrix, say M, satisfying the hypotheses of Theorem 10.4 whenever (10.2) holds. Suppose that M_ℓ is the ℓ by ℓ identity matrix. It is clear that any t columns of M_ℓ are linearly independent.

Now suppose that M_j is an ℓ by j matrix having entries from $\mathbb{F}_q$ such that any subset of t columns of M_j is linearly independent. The number of linear combinations of at most $t-1$ columns of M_j is

$$\sum_{i=0}^{t-1} \binom{j}{i} (q-1)^i.$$

(Note that not all of these linear combinations necessarily yield distinct vectors.) There are q^ℓ possible column vectors of length ℓ. Therefore there is a column vector, say $\mathbf{c}$, such that $\mathbf{c}$ is not one of these linear combinations, provided that

$$\sum_{i=0}^{t-1} \binom{j}{i} (q-1)^i < q^\ell. \tag{10.3}$$

Then we can construct the matrix M_{j+1} by adjoining the column vector $\mathbf{c}$ to M_j, and M_{j+1} again satisfies the property that any subset of t columns is linearly independent.

We assumed that (10.2) holds, which implies that (10.3) is true for $j = \ell, \ell+1, \ldots, n-1$. This means that we can construct matrices $M_j, \ldots, M_n$ satisfying the required properties, and the matrix M_n is the desired matrix M.

□

10.2 Codes

Definition 10.12. *A* code *is a pair* (Q, C) *such that the following properties are satisfied.*

1. *Q is a set of elements called* symbols.
2. *C is a set of n-tuples of symbols called* codewords *(i.e., $C \subseteq Q^n$), where $n \geq 1$ is an integer.*

If $Q = \mathbb{F}_2$, then a code (Q, C) is called a binary code.

The concept of "distance" is fundamental to the study of codes. We give several relevant definitions now.

Definition 10.13. *Let (Q, C) be a code, where $C \subseteq Q^n$. For $\mathbf{x}, \mathbf{y} \in Q^n$, define the* Hamming distance *between $\mathbf{x}$ and $\mathbf{y}$ to be*

$$\mathsf{d}(\mathbf{x}, \mathbf{y}) = |\{i : x_i \neq y_i\}|,$$

where $\mathbf{x} = (x_1, \ldots, x_n)$ and $\mathbf{y} = (y_1, \ldots, y_n)$.

The distance *of the code (Q, C), denoted $\mathsf{d}(C)$, is the smallest positive integer d such that $\mathsf{d}(\mathbf{x}, \mathbf{y}) \geq d$ for all $\mathbf{x}, \mathbf{y} \in C$, $\mathbf{x} \neq \mathbf{y}$.*

*(Q, C) is an (n, M, d, q)-*code *if the following properties are satisfied:*

1. $|Q| = q$,

2. $C \subseteq Q^n$,
3. $|C| = M$, *and*
4. $\mathsf{d}(C) \geq d$.

For future reference, we record some basic facts about the Hamming distance.

Lemma 10.14. *For all* $\mathbf{x}, \mathbf{y}, \mathbf{z} \in X^n$, *the following properties hold:*

1. $\mathsf{d}(\mathbf{x}, \mathbf{y}) = \mathsf{d}(\mathbf{y}, \mathbf{x})$,
2. $\mathsf{d}(\mathbf{x}, \mathbf{y}) = 0$ *if and only if* $\mathbf{x} = \mathbf{y}$, *and*
3. $\mathsf{d}(\mathbf{x}, \mathbf{y}) \leq \mathsf{d}(\mathbf{x}, \mathbf{z}) + \mathsf{d}(\mathbf{y}, \mathbf{z})$ *(this is known as the* triangle inequality*).*

Now we define linear codes.

Definition 10.15. *A code* (Q, C) *is a* linear code *of* dimension m *if* $Q = \mathbb{F}_q$ *for some prime power* q *and* C *is an* m*-dimensional subspace of the vector space* $(\mathbb{F}_q)^n$. *The* dual code *of a linear code* (Q, C) *is the code* $(Q, C^\perp)$, *where*

$$C^\perp = \{\mathbf{y} \in (\mathbb{F}_q)^n : \mathbf{x} \cdot \mathbf{y} = 0 \text{ for all } \mathbf{x} \in C\}.$$

(As usual, "$\mathbf{x} \cdot \mathbf{y}$*" denotes the inner product over* $\mathbb{F}_q$ *of the two vectors* $\mathbf{x}$ *and* $\mathbf{y}$*. The subspaces* C *and* $C^\perp$ *are called* orthogonal complements *of each other.) Then* $(Q, C^\perp)$ *is a linear code of dimension* $n - \mathsf{dim}(C)$.

Suppose that $\mathbf{x} \in (\mathbb{F}_q)^n$. Define the *weight* of $\mathbf{x}$ to be

$$\mathsf{wt}(\mathbf{x}) = |\{i : x_i \neq 0\}|,$$

where $\mathbf{x} = (x_1, \ldots, x_n)$.

Lemma 10.16. *Suppose* $(\mathbb{F}_q, C)$ *is a linear code, where* $C \subseteq (\mathbb{F}_q)^n$*. Then*

$$\mathsf{d}(C) = \min\{\mathsf{wt}(\mathbf{x}) : \mathbf{x} \in C, \mathbf{x} \neq (0, \ldots, 0)\}.$$

Proof. Denote $\mathsf{wt}(C) = \min\{\mathsf{wt}(\mathbf{x}) : \mathbf{x} \in C, \mathbf{x} \neq (0, \ldots, 0)\}$. Let $\mathbf{x}, \mathbf{y} \in C$ be two codewords such that $\mathsf{d}(\mathbf{x}, \mathbf{y}) = \mathsf{d}(C)$. The vector $\mathbf{x} - \mathbf{y} \in C$ because C is linear, and $\mathsf{wt}(\mathbf{x} - \mathbf{y}) = \mathsf{d}(\mathbf{x}, \mathbf{y}) = \mathsf{d}(C)$. Therefore $\mathsf{wt}(C) \leq \mathsf{d}(C)$.

Conversely, let $\mathbf{x} \in C$ be a codeword such that $\mathsf{wt}(\mathbf{x}) = \mathsf{wt}(C)$. The vector $(0, \ldots, 0) \in C$ because C is linear. Then $\mathsf{d}(\mathbf{x}, (0, \ldots, 0)) = \mathsf{wt}(\mathbf{x}) = \mathsf{wt}(C)$, so $\mathsf{d}(C) \leq \mathsf{wt}(C)$. □

Theorem 10.17. *Suppose that* $C \subseteq (\mathbb{F}_q)^n$ *is a linear code of dimension* m*. Then* $(\mathbb{F}_q, C)$ *is an* (n, q^m, d, q)*-code if and only if* $C^\perp$ *is a (linear)* $(d-1)$*-*(q, n, λ)*-OA, where* $\lambda = q^{n-m-d+1}$.

Proof. Suppose that $(\mathbb{F}_q, \mathcal{C})$ is a linear (n, q^m, d, q)-code. Clearly $\mathcal{C}^\perp$ is a subspace having dimension $n - m$; we will show that it is an orthogonal array. Let D be a basis for $\mathcal{C}^\perp$, and write the vectors in D as an $n - m$ by n matrix. We will prove that D satisfies the conditions of Theorem 10.4, and hence it will follow that $\mathcal{C}^\perp$ is an orthogonal array with the stated parameters.

Suppose that there exist $e \leq d - 1$ columns of D that are linearly dependent, and therefore there exists a dependence relation of the form

$$\sum_{i=1}^{e} \alpha_i \mathbf{c}_{i_j} = (0, \ldots, 0)^T,$$

where $\mathbf{c}_1, \ldots, \mathbf{c}_n$ are the columns of D. Define a vector $\mathbf{x} = (x_1, \ldots, x_n)$ as follows:

$$x_h = \begin{cases} \alpha_h & \text{if } h = i_j \text{ for some } j \\ 0 & \text{otherwise.} \end{cases}$$

Then $\mathbf{x} \cdot \mathbf{r} = 0$ for every row $\mathbf{r}$ of D and hence $\mathbf{x} \in \mathcal{C}$. However, $\mathsf{wt}(\mathbf{x}) = e < \mathsf{d}(\mathcal{C})$, which contradicts Lemma 10.16.

Conversely, suppose that $\mathcal{C}^\perp$ is a linear $(d-1)$-(q, n, λ)-OA, where $\lambda = q^{n-m-d+1}$. This implies that $\mathcal{C}^\perp$ has dimension $n - m$, and hence $\mathcal{C}$ has dimension m. Let D be a basis for $\mathcal{C}^\perp$; then D has $n - m$ rows when it is written as an array.

We will prove that the minimum distance of $\mathcal{C}$ is at least d. If not, then there exists a vector $\mathbf{x} \in \mathcal{C}$ such that $0 \leq \mathsf{wt}(\mathbf{x}) \leq d - 1$. Suppose that the nonzero entries of $\mathbf{x}$ occur in coordinates $i_1, \ldots, i_e$, where $e = \mathsf{wt}(\mathbf{x})$. Clearly $\mathbf{x} \cdot \mathbf{y} = 0$ for every row $\mathbf{y} \in D$. When $\mathcal{C}^\perp$ is viewed as an orthogonal array, it follows that

$$\sum_{j=1}^{e} x_{i_j} y_{i_j} = 0$$

for every row $\mathbf{y}$. In other words, in every row of $\mathcal{C}^\perp$, the entries in columns $i_1, \ldots, i_e$ satisfy a linear dependence relation. This means that it is impossible that every e-tuple of symbols occurs in a row of $\mathcal{C}^\perp$ within the e columns under consideration. Therefore $\mathcal{C}^\perp$ is not a $(d-1)$-(q, n, λ)-OA, which is a contradiction. This contradiction proves that the minimum distance of $\mathcal{C}$ is at least d. □

Example 10.18. Consider the linear 3-(5, 5, 1)-OA presented in Example 10.8. The following three vectors in $(\mathbb{Z}_5)^3$ form a basis of this orthogonal array: $(1, 1, 1, 1, 1)$, $(0, 1, 2, 3, 4)$, and $(0, 1, 4, 4, 1)$. Using standard techniques from linear algebra, it is not hard to determine a basis for the orthogonal complement; the vectors $(4, 3, 2, 1, 0)$ and $(2, 3, 4, 0, 1)$ form one such basis. The code generated by these two vectors is a $(5, 25, 4, 5)$-code. ∎

10.3 Bounds on Codes and Orthogonal Arrays

In this section, we present a few bounds on codes and orthogonal arrays and give some constructions that meet these bounds.

Theorem 10.19 (Singleton Bound). *Suppose that $\mathcal{C}$ is an (n, M, d, q)-code. Then $M \leq q^{n-d+1}$.*

Proof. Suppose that $M > q^{n-d+1}$. Then, by the pigeonhole principle, there exist two codewords $\mathbf{x}, \mathbf{y} \in \mathcal{C}$ such that $x_i = y_i$ for all i such that $1 \leq i \leq n - d + 1$. Then $\mathsf{d}(\mathbf{x}, \mathbf{y}) \leq n - (n - d + 1) = d - 1$. □

Theorem 10.19 can be restated as an upper bound on the distance of a code, as follows.

Corollary 10.20. *Suppose that $\mathcal{C}$ is an (n, M, d, q)-code. Then $d \leq n + 1 - \log_q M$.*

Orthogonal arrays with $\lambda = 1$ turn out to be equivalent to codes that meet the Singleton Bound with equality.

Theorem 10.21. *An (n, M, d, q)-code in which $M = q^{n-d+1}$ is equivalent to a t-$(q, n, 1)$-OA in which $t = n - d + 1$.*

Proof. Suppose that (X, D) is any t-$(q, n, 1)$-OA. Construct a code $(X, \mathcal{C})$ by taking the q^t rows of D to be the codewords in $\mathcal{C}$. We will prove that $(X, \mathcal{C})$ is an $(n, q^t, n - t + 1, q)$-code, as follows. Suppose that $\mathsf{d}(\mathcal{C}) \leq n - t$. Then there exist two codewords $\mathbf{x}, \mathbf{y} \in \mathcal{C}$ such that the entries of $\mathbf{x}$ and $\mathbf{y}$ are the same in at least t columns. Within these t columns, the corresponding rows of D are identical, which contradicts the assumption that $\lambda = 1$ in the orthogonal array (X, D).

Conversely, suppose that $(X, \mathcal{C})$ is an (n, M, d, q)-code in which $M = q^{n-d+1}$. Construct an $M \times n$ array, D, by taking the codewords in $\mathcal{C}$ to be the rows of D. Consider the restriction of D to any subset of $n - d + 1$ columns. The q^{n-d+1} $(n - d + 1)$-tuples obtained from the rows of D in this restriction must all be different (as in the proof of Theorem 10.19). Since there are q^{n-d+1} different $(n - d + 1)$-tuples, it follows that every possible $(n - d + 1)$-tuple occurs in exactly one row of D in this restriction. Because this property holds for all possible subsets of $n - d + 1$ columns of D, it follows that D is an $(n - d + 1)$-$(q, n, 1)$-OA. □

A code in which the Singleton Bound is met with equality is called a *maximum distance separable code* (or *MDS code*). Theorem 10.21 establishes that MDS codes are equivalent to orthogonal arrays with $\lambda = 1$. Since we have already constructed various families of orthogonal arrays with $\lambda = 1$, we can translate these results into the language of codes. For example, from Corollary 10.7, we can state the following result.

Corollary 10.22. *Let $t \geq 2$ be an integer, and let q be a prime power. Then there exists an (MDS) $(q, q^t, q-t+1, q)$-code.*

The codes obtained in Corollary 10.22 are commonly known as *Reed-Solomon codes.*

Theorem 10.23 (Sphere-packing Bound). *Suppose that $(X, \mathcal{C})$ is an (n, M, d, q)-code. Then*

$$M \leq \frac{q^n}{\sum_{i=0}^{e} \binom{n}{i} (q-1)^i},$$

where $e = \lfloor \frac{d-1}{2} \rfloor$.

Proof. Suppose $\mathbf{x} \in X^n$. Define the *sphere* with center $\mathbf{x}$ and radius e to be the following set of vectors, denoted $\mathsf{S}(\mathbf{x}, e)$:

$$\mathsf{S}(\mathbf{x}, e) = \{\mathbf{y} \in X^n : \mathsf{d}(\mathbf{x}, \mathbf{y}) \leq e\}.$$

It is not hard to see that

$$|\mathsf{S}(\mathbf{x}, e)| = \sum_{i=0}^{e} \binom{n}{i} (q-1)^i. \tag{10.4}$$

We next prove that $\mathsf{S}(\mathbf{x}, e) \cap \mathsf{S}(\mathbf{y}, e) = \emptyset$ if $\mathbf{x}, \mathbf{y} \in \mathcal{C}$, $\mathbf{x} \neq \mathbf{y}$. Suppose $\mathbf{z} \in \mathsf{S}(\mathbf{x}, e) \cap \mathsf{S}(\mathbf{y}, e)$. Then $\mathsf{d}(\mathbf{x}, \mathbf{z}) \leq e$ and $\mathsf{d}(\mathbf{y}, \mathbf{z}) \leq e$. Applying the Triangle Inequality (Lemma 10.14), we see that

$$\mathsf{d}(\mathbf{x}, \mathbf{y}) \leq \mathsf{d}(\mathbf{x}, \mathbf{z}) + \mathsf{d}(\mathbf{y}, \mathbf{z}) \leq 2e < d.$$

This contradicts the fact that $\mathsf{d}(\mathcal{C}) \geq d$.

Now consider all the spheres $\mathsf{S}(\mathbf{x}, e)$, $\mathbf{x} \in \mathcal{C}$. These spheres are mutually disjoint, and all of them are contained in the set X^n, which consists of q^n vectors. Applying (10.4), the following is immediate:

$$q^n \geq M \sum_{i=0}^{e} \binom{n}{i} (q-1)^i.$$

Thus the desired result is proven. □

A code in which the Sphere-packing Bound is met with equality is known as a *perfect code*. We are easily able to construct infinite families of perfect codes with distance 3 using results we have already established.

Theorem 10.24. *Let $\ell \geq 2$ be a positive integer, and let q be a prime power. Then there is a perfect $(n, q^m, 3, q)$-code in which $n = (q^\ell - 1)/(q-1)$ and $m = n - \ell$.*

Proof. Corollary 10.5 shows there is a linear 2-$(q,(q^\ell-1)/(q-1),q^{\ell-2})$-OA for the stated values of ℓ and q. Consider the code that is the orthogonal complement of this orthogonal array, as described in Theorem 10.17. This code has distance 3, and m can be computed from the equation

$$q^m = \frac{q^n}{\lambda q^2} = q^{n-\ell}.$$

The fact that the code is perfect is a simple computation:

$$\sum_{i=0}^{1} \binom{n}{i} (q-1)^i = 1 + n(q-1) = q^\ell,$$

and hence

$$q^m \sum_{i=0}^{1} \binom{n}{i} (q-1)^i = q^{m+\ell} = q^n.$$

□

The codes constructed in Theorem 10.24 are known as *Hamming codes.* When $q = 2$, a Hamming code is a $(2^\ell-1, 2^{2^\ell-\ell-1}, 3, 2)$-code. This is the *binary Hamming code.* It turns out that the vectors of weight three in a binary Hamming code yield a Steiner triple system, which we prove now.

Theorem 10.25. *Suppose that C are the vectors in a $(2^\ell-1, 2^{2^\ell-\ell-1}, 3, 2)$-code. Form a matrix M whose columns consist of all the codewords of weight three. Then M is the incidence matrix of a $(2^\ell-1, 3, 1)$-BIBD.*

Proof. Let C_3 denote the set of codewords in C that have weight three. Each codeword in C_3 yields a column vector that corresponds to a block in the set system having incidence matrix M. This set system therefore consists of $|C_3|$ blocks, each having cardinality equal to three.

It suffices to show that every pair of points in the set system is contained in a unique block. A pair of points corresponds to a column vector, say $\mathbf{u}^T$, of weight two. The code C is perfect, so the spheres of radius 1 whose centers are the codewords in C partition the space $\{0,1\}^n$ (where $n = 2^\ell - 1$). Therefore there is a unique $\mathbf{x} \in C$ such that $\mathbf{u} \in \mathsf{S}(\mathbf{x}, 1)$. We have that $\mathsf{wt}(\mathbf{u}) = 2$, $\mathsf{wt}(\mathbf{x}) \geq 3$ or $\mathsf{wt}(\mathbf{x}) = 0$, and $\mathsf{d}(\mathbf{u},\mathbf{x}) \leq 1$. It follows that $\mathsf{wt}(\mathbf{x}) = 3$. Therefore the pair having incidence vector $\mathbf{u}$ occurs in the unique block having incidence vector $\mathbf{x}$, and the proof is complete. □

Example 10.26. Suppose we take $\ell = 3$. The corresponding Hamming code is a $(7, 16, 3, 2)$-code. This code is the dual of the linear orthogonal array obtained from the matrix

$$M = \begin{pmatrix} 1\,0\,0\,0\,1\,1\,1 \\ 0\,1\,0\,1\,1\,0\,1 \\ 0\,0\,1\,1\,0\,1\,1 \end{pmatrix}.$$

The code therefore consists of the following 16 vectors:

$$\begin{array}{ll}(0,0,0,0,0,0,0) & (0,1,1,1,0,0,0)\\(1,1,0,0,1,0,0) & (1,0,1,1,1,0,0)\\(1,0,1,0,0,1,0) & (1,1,0,1,0,1,0)\\(0,1,1,0,1,1,0) & (0,0,0,1,1,1,0)\\(1,1,1,0,0,0,1) & (1,0,0,1,0,0,1)\\(0,0,1,0,1,0,1) & (0,1,0,1,1,0,1)\\(0,1,0,0,0,1,1) & (0,0,1,1,0,1,1)\\(1,0,0,0,1,1,1) & (1,1,1,1,1,1,1).\end{array}$$

There are seven codewords having weight three. Treating these as incidence vectors of points $1,\ldots,7$, we obtain the following seven blocks:

$$\begin{array}{ll}\{2,3,4\} & \{1,2,5\}\\\{1,3,6\} & \{4,5,6\}\\\{1,4,7\} & \{3,5,7\}\\\{2,6,7\}. &\end{array}$$

These seven blocks form a $(7,3,1)$-BIBD. ∎

10.4 New Codes from Old

There are many methods of producing new codes from old ones. We describe a few useful techniques in this section.

Shortening a Code

Suppose that $(X,\mathcal{C})$ is an (n,M,d,q)-code. Let $x \in X$, and define

$$\mathcal{C}_x = \{\mathbf{y} = (y_1,\ldots,y_n) \in \mathcal{C} : y_1 = x\}.$$

Then define

$$\mathsf{short}(\mathcal{C},x) = \{(y_2,\ldots,y_n) : (y_1,\ldots,y_n) \in \mathcal{C}_x\}.$$

It is clear that $(X,\mathsf{short}(\mathcal{C},x))$ is an $(n-1,|\mathcal{C}_x|,d,q)$-code.

The following result can now be proven.

Theorem 10.27 (Shortening a Code). *Suppose there is an (n,M,d,q)-code. Then there is an $(n-1,M',d,q)$-code, where $M' \geq M/q$.*

Proof. Suppose that $(X,\mathcal{C})$ is an (n,M,d,q)-code. It is clear that the q sets of vectors $\mathcal{C}_x$ $(x \in X)$ are disjoint and partition $\mathcal{C}$. Hence, there exists some $x_0 \in X$ such that $|\mathcal{C}_{x_0}| \geq M/q$, and consequently $(X,\mathsf{short}(\mathcal{C},x_0))$ is an $(n-1,M',d,q)$-code with $M' \geq M/q$. □

Pasting Codes Together

Suppose that $(X, \mathcal{C})$ is an (n_1, M_1, d_1, q)-code and $(X, \mathcal{D})$ is an (n_2, M_2, d_2, q)-code. Without loss of generality, suppose that $M_1 \leq M_2$. Let $\mathcal{C} = \{\mathbf{x}_1, \ldots, \mathbf{x}_{M_1}\}$ and let $\mathcal{D} = \{\mathbf{y}_1, \ldots, \mathbf{y}_{M_2}\}$. Let u and v be positive integers, and define the code $(X, u\mathcal{C} \oplus v\mathcal{D})$ to consist of the following M_1 vectors:

$$\underbrace{\mathbf{x}_i \parallel \cdots \parallel \mathbf{x}_i}_{u} \parallel \underbrace{\mathbf{y}_i \parallel \cdots \parallel \mathbf{y}_i}_{v},$$

for $1 \leq i \leq M_1$. In other words, the ith codeword in $u\mathcal{C} \oplus v\mathcal{D}$ is formed by concatenating u copies of the ith codeword in $\mathcal{C}$ and v copies of the ith codeword in $\mathcal{D}$. The code $(X, u\mathcal{C} \oplus v\mathcal{D})$ is easily seen to have parameters as stated in the following theorem.

Theorem 10.28 (Pasting Codes Together). *Suppose there is an (n_1, M_1, d_1, q)-code and an (n_2, M_2, d_2, q)-code, where $M_1 \leq M_2$. Let u and v be positive integers. Then there exists a $(u\,n_1 + v\,n_2, M_1, u\,d_1 + v\,d_2, q)$-code.*

The $\mathbf{u}, \mathbf{u} + \mathbf{v}$ Construction

Suppose that $(X, \mathcal{C})$ is an $(n, M_1, d_1, 2)$-code and $(X, \mathcal{D})$ is an $(n, M_2, d_2, 2)$-code, where $X = \{0, 1\}$. We construct a code $(X, \mathcal{E})$ by taking all vectors formed as follows:

$$\mathcal{E} = \{\mathbf{u} \parallel \mathbf{u} + \mathbf{v} : \mathbf{u} \in \mathcal{C}, \mathbf{v} \in \mathcal{D}\}.$$

Here, addition denotes addition of vectors modulo 2, as usual. Clearly every vector in $\mathcal{E}$ has length $2n$, and there are $M_1 M_2$ vectors in $\mathcal{E}$. We compute a lower bound on the distance of $(X, \mathcal{E})$ as follows.

First, suppose that $\mathbf{u} \neq \mathbf{u}'$. Then

$$\mathsf{d}(\mathbf{u} \parallel \mathbf{u} + \mathbf{v}, \mathbf{u}' \parallel \mathbf{u}' + \mathbf{v}) = 2\,\mathsf{d}(\mathbf{u}, \mathbf{u}') \geq 2d_1.$$

Next, suppose that $\mathbf{v} \neq \mathbf{v}'$. To handle this case, we use the following lemma.

Lemma 10.29. *Suppose that $\mathbf{u}, \mathbf{u}', \mathbf{v}, \mathbf{v}' \in (\mathbb{Z}_2)^n$. Then*

$$\mathsf{d}(\mathbf{v}, \mathbf{v}') \leq \mathsf{d}(\mathbf{u}, \mathbf{u}') + \mathsf{d}(\mathbf{u} + \mathbf{v}, \mathbf{u}' + \mathbf{v}').$$

Proof. Let $\mathbf{u} = (u_1, \ldots, u_n)$, $\mathbf{u}' = (u'_1, \ldots, u'_n)$, $\mathbf{v} = (v_1, \ldots, v_n)$, and $\mathbf{v}' = (v'_1, \ldots, v'_n)$. Define the following subsets of $\{1, \ldots, n\}$:

$$\begin{aligned} A &= \{i : u_i = u'_i \text{ and } v_i = v'_i\}, \\ B &= \{i : u_i = u'_i \text{ and } v_i \neq v'_i\}, \\ C &= \{i : u_i \neq u'_i \text{ and } v_i = v'_i\}, \quad \text{and} \\ D &= \{i : u_i \neq u'_i \text{ and } v_i \neq v'_i\}. \end{aligned}$$

It is not hard to see that

$$\begin{aligned} \mathsf{d}(\mathbf{u},\mathbf{u}') &= |C| + |D|, \\ \mathsf{d}(\mathbf{v},\mathbf{v}') &= |B| + |D|, \quad \text{and} \\ \mathsf{d}(\mathbf{u}+\mathbf{v},\mathbf{u}'+\mathbf{v}') &= |B| + |C|. \end{aligned}$$

Hence,

$$\mathsf{d}(\mathbf{u},\mathbf{u}') + \mathsf{d}(\mathbf{u}+\mathbf{v},\mathbf{u}'+\mathbf{v}') = |B| + 2|C| + |D| \geq |B| + |D| = \mathsf{d}(\mathbf{v},\mathbf{v}')$$

because $|C| \geq 0$. □

Now, assuming that $\mathbf{v} \neq \mathbf{v}'$ and applying Lemma 10.29, we have that

$$\begin{aligned} \mathsf{d}(\mathbf{u} \parallel \mathbf{u}+\mathbf{v}, \mathbf{u}' \parallel \mathbf{u}'+\mathbf{v}') &= \mathsf{d}(\mathbf{u},\mathbf{u}') + \mathsf{d}(\mathbf{u}+\mathbf{v},\mathbf{u}'+\mathbf{v}') \\ &\geq \mathsf{d}(\mathbf{u},\mathbf{u}') + \mathsf{d}(\mathbf{v},\mathbf{v}') - \mathsf{d}(\mathbf{u},\mathbf{u}') \\ &= \mathsf{d}(\mathbf{v},\mathbf{v}') \\ &\geq d_2. \end{aligned}$$

Summarizing the above, we obtain the following result.

Theorem 10.30 ($\mathbf{u}, \mathbf{u}+\mathbf{v}$ Construction). *Suppose there exists an $(n, M_1, d_1, 2)$-code and an $(n, M_2, d_2, 2)$-code. Then there exists a $(2n, M_1M_2, d, 2)$-code, where $d = \min\{2d_1, d_2\}$.*

Example 10.31. Suppose that $\mathcal{C}$ consists of the following vectors:

$$\begin{array}{cc} (0,0,0,0) & (0,0,1,1) \\ (0,1,0,1) & (0,1,1,0) \\ (1,0,0,1) & (1,0,1,0) \\ (1,1,0,0) & (1,1,1,1), \end{array}$$

and suppose that $\mathcal{D}$ is as follows:

$$(0,0,0,0) \ (1,1,1,1).$$

$(\mathbb{Z}_2, \mathcal{C})$ is a $(4,8,2,2)$-code and $(\mathbb{Z}_2, \mathcal{D})$ is a $(4,2,4,2)$-code. Applying Theorem 10.30, we get the following $(4,16,4,2)$-code:

$$\begin{array}{cc} (0,0,0,0,0,0,0,0) & (0,0,1,1,0,0,1,1) \\ (0,1,0,1,0,1,0,1) & (0,1,1,0,0,1,1,0) \\ (1,0,0,1,1,0,0,1) & (1,0,1,0,1,0,1,0) \\ (1,1,0,0,1,1,0,0) & (1,1,1,1,1,1,1,1) \\ (0,0,0,0,1,1,1,1) & (0,0,1,1,1,1,0,0) \\ (0,1,0,1,1,0,1,0) & (0,1,1,0,1,0,0,1) \\ (1,0,0,1,0,1,1,0) & (1,0,1,0,0,1,0,1) \\ (1,1,0,0,0,0,1,1) & (1,1,1,1,0,0,0,0). \end{array}$$

■

10.5 Binary Codes

10.5.1 The Plotkin Bound and Hadamard Codes

Recall that a binary code is one in which the alphabet is $\mathbb{F}_2 = \{0,1\}$. In this section, we prove some results on binary codes. We begin by stating and proving a bound for binary codes having a "large" Hamming distance.

Theorem 10.32 (Plotkin Bound). *Suppose that $(\{0,1\}, \mathcal{C})$ is an $(n, M, d, 2)$-code, and suppose that $d > n/2$. Then*

$$M \leq 2\left\lfloor \frac{d}{2d-n} \right\rfloor.$$

Proof. Let the codewords in $\mathcal{C}$ be named $\mathbf{x}_i$, $1 \leq i \leq M$, and construct an $M \times n$ matrix, say N, whose rows are the codewords. Define

$$S = \sum_{i=1}^{M}\sum_{j=1}^{M} \mathsf{d}(\mathbf{x}_i, \mathbf{x}_j).$$

This sum contains $M(M-1)$ terms that are each at least d, and M terms equal to 0. Hence, we have that

$$S \geq M(M-1)d. \tag{10.5}$$

We now determine an upper bound on S. Suppose that column c contains t_c "1"s and $M - t_c$ "0"s $(1 \leq c \leq n)$. Then

$$S = \sum_{c=1}^{n} 2t_c(M - t_c).$$

Suppose that M is even. Then the maximum value of $t(M-t)$ (for $0 \leq t \leq M$) occurs when $t = M/2$, and hence $t(M-t) \leq M^2/4$. Therefore, it holds that

$$S \leq \frac{nM^2}{2}. \tag{10.6}$$

Now, combining (10.5) and (10.6), we see that

$$M(M-1)d \leq \frac{nM^2}{2}$$

or

$$M(2d-n) \leq 2d.$$

Because $2d > n$, it follows that

$$M \leq \frac{2d}{2d-n}.$$

Because M is an even integer, it follows that

$$M \le 2 \left\lfloor \frac{d}{2d-n} \right\rfloor.$$

Now suppose that M is odd. In this case, the maximum value of $t(M-t)$ (for $0 \le t \le M$, t an integer) occurs when $t = (M+1)/2$ or $t = (M-1)/2$, and hence $t(M-t) \le (M^2-1)/4$. Therefore, it holds that

$$S \le \frac{n(M^2-1)}{2}. \tag{10.7}$$

Now, combining (10.5) and (10.6), we see that

$$M(M-1)d \le \frac{n(M^2-1)}{2}$$

or

$$M(2d-n) \le n.$$

Because $2d > n$ and M is an integer, it follows that

$$M \le \left\lfloor \frac{n}{2d-n} \right\rfloor = \left\lfloor \frac{2d}{2d-n} \right\rfloor - 1.$$

For any real number $\epsilon > 0$, it holds that $\lfloor 2\epsilon \rfloor \le 2\lfloor \epsilon \rfloor + 1$. Hence, taking $\epsilon = 2/(2d-n)$, it follows that

$$M \le 2 \left\lfloor \frac{d}{2d-n} \right\rfloor.$$

This completes the proof. □

Codes meeting the Plotkin Bound with equality can be constructed provided that certain Hadamard matrices exist. Naturally enough, they are known as *Hadamard codes*. We first establish a preliminary result.

Lemma 10.33. *Suppose there is a Hadamard matrix of order n. Then there exists an* $(n-1, n, \frac{n}{2}, 2)$*-code and an* $(n-2, \frac{n}{2}, \frac{n}{2}, 2)$*-code.*

Proof. Let H be a standardized Hadamard matrix of order n. Delete the first column of H, and take the rows of the resulting matrix to be codewords of a code. This yields the $(n-1, n, \frac{n}{2}, 2)$-code (see the proof of Theorem 4.4). Then apply Theorem 10.27 to obtain the second code. □

Now, suppose that d and n are both even and $d > n/2$. Define $k = \left\lfloor \frac{d}{2d-n} \right\rfloor$. Then, define

$$u = \frac{d(2k+1) - n(k+1)}{2} \tag{10.8}$$

and

$$v = \frac{nk - d(2k-1)}{2}. \tag{10.9}$$

Using a bit of arithmetic, we can show that u and v are integers such that $u > 0$ and $v \geq 0$. First, it is clear that u and v are integers because n and d are even. We have upper and lower bounds on k:

$$\frac{d}{2d-n} - 1 < k \leq \frac{d}{2d-n}.$$

We will use these bounds on k to prove lower bounds on u and v.

First, we have that $u > 0$ if and only if $d(2k+1) - n(k+1) > 0$. But

$$\begin{aligned} d(2k+1) - n(k+1) &= k(2d-n) + d - n \\ &> \left(\frac{d}{2d-n} - 1\right)(2d-n) + d - n \\ &= d - (2d-n) + d - n \\ &= 0; \end{aligned}$$

hence $u > 0$.

Similarly, $v \geq 0$ if and only if $nk - d(2k-1) \geq 0$. But

$$\begin{aligned} nk - d(2k-1) &= d - k(2d-n) \\ &\geq d - \left(\frac{d}{2d-n}\right)(2d-n) \\ &= d - d \\ &= 0; \end{aligned}$$

hence $v \geq 0$.

Suppose that Hadamard matrices of orders $4k$ and $4k+4$ both exist. Let $\mathcal{C}$ be the codewords of a $(4k-2, 2k, 2k, 2)$-code and let $\mathcal{D}$ be the codewords of a $(4k+2, 2k+2, 2k+2, 2)$-code (these are constructed using Theorem 10.33 with $n = 4k$ and $n = 4k+4$, respectively). Now construct the code having codewords $u\mathcal{C} \oplus v\mathcal{D}$. Using the formulas in Theorem 10.28, the resulting code is seen to meet the Plotkin Bound, and we obtain the following result.

Theorem 10.34 (Levenshtein's Theorem). *Suppose that n and d are even positive integers such that $2d > n$. Define $k = \left\lfloor \frac{d}{2d-n} \right\rfloor$, and suppose that Hadamard matrices of order $4k$ and $4k+4$ exist. Then there exists an $(n, 2k, d, 2)$-code, which meets the Plotkin Bound with equality.*

Proof. We need only to check that the constructed code has n and d as stated. The code is formed by pasting together u copies of a $(4k-2, 2k, 2k, 2)$-code and v copies of a $(4k+2, 2k+2, 2k+2, 2)$-code, where u and v are defined in (10.8) and (10.9), respectively. The resulting code is a

$$(u(4k-2) + v(4k+2), 2k, u(2k) + v(2k+2), 2)\text{-code.}$$

However,

$$\begin{aligned} &u(4k-2)+v(4k+2) \\ &= \left(\frac{d(2k+1)-n(k+1)}{2}\right)(4k-2)+\left(\frac{nk-d(2k-1)}{2}\right)(4k+2) \\ &= n, \end{aligned}$$

and

$$\begin{aligned} &u(2k)+v(2k+2) \\ &= \left(\frac{d(2k+1)-n(k+1)}{2}\right)(2k)+\left(\frac{nk-d(2k-1)}{2}\right)(2k+2) \\ &= d. \end{aligned}$$

Finally, notice that the number of codewords in this code is $2k = 2\left\lfloor \frac{d}{2d-n} \right\rfloor$. □

Example 10.35. Suppose we take $n = 24$ and $d = 14$. Then $k = 3$, and $u = v = 1$. We can use Hadamard matrices of orders 12 and 16 to construct a $(10,6,6,2)$-code and a $(14,8,8,2)$-code, respectively. Pasting these two codes together, we would obtain a $(24,6,14,2)$-code, which meets the Plotkin Bound with equality. ▮

10.5.2 Reed-Muller Codes

Reed-Muller codes are closely connected to Boolean functions. We first review some notions and notation from Section 4.8. Recall that a Boolean function of n variables is any function $f : (\mathbb{Z}_2)^n \rightarrow \mathbb{Z}_2$; and $\mathcal{B}_n$ denotes the set of all 2^{2^n} Boolean functions of n variables. For a function $f \in \mathcal{B}_n$, $\phi(f) \in (\mathbb{Z}_2)^{2^n}$ is the vector formed by evaluating f at all $\mathbf{x} \in (\mathbb{Z}_2)^n$.

Recall also that the affine functions in $\mathcal{B}_n$ are the 2^{n+1} functions $f \in \mathcal{B}_n$ having the form

$$f(\mathbf{x}) = a_0 + a_1x_1 + \cdots + a_nx_n \bmod 2,$$

where $\mathbf{x} = (x_1, \ldots, x_n)$ and $a_0, a_1, \ldots, a_n \in \mathbb{Z}_2$.

The *first-order Reed-Muller code*, denoted $\mathcal{R}(1,n)$, is the code whose codewords are all the vectors $\phi(f)$, where $f \in \mathcal{B}_n$ is an affine function. First, we show that $\mathcal{R}(1,n)$ is a linear code, as follows. The sum of any two affine functions, say f_1 and f_2, is again an affine function. Furthermore, the modulo 2 sum of the corresponding codewords, $\phi(f_1)$ and $\phi(f_2)$, is another codeword because $\phi(f_1) + \phi(f_2) = \phi(f_1 + f_2)$.

Recall that the distance of a linear code equals the minimum weight of a nonzero codeword (Lemma 10.16). Therefore we can determine the distance of $\mathcal{R}(1,n)$ if we know the weights of the codewords in $\mathcal{R}(1,n)$. We prove the following simple lemma concerning these weights.

Lemma 10.36. *Suppose that $f \in \mathcal{B}_n$ is an affine function. Then*

$$\mathrm{wt}(\phi(f)) = \begin{cases} 0 & \text{if } f = 0 \\ 2^n & \text{if } f = 1 \\ 2^{n-1} & \text{otherwise.} \end{cases}$$

Proof. Clearly $\mathrm{wt}(\phi(0)) = 0$ and $\mathrm{wt}(\phi(1)) = 2^n$. Now, suppose that $f \neq 0, 1$ is an affine function, say $f(\mathbf{x}) = a_0 + a_1x_1 + \cdots + a_nx_n \bmod 2$. Because $f \neq 0, 1$, there exists an integer $i \geq 1$ such that $a_i = 1$. Suppose that arbitrary values for $x_1, \ldots, x_{i-1}, x_{i+1}, \ldots, x_n \in \mathbb{Z}_2$ have been chosen, and denote

$$A = a_0 + \sum_{1 \leq j \leq n, j \neq i} a_jx_j \bmod 2.$$

Then $f(\mathbf{x}) = 0$ if $x_i = A$, and $f(\mathbf{x}) = 1$ if $x_i \neq A$.

Summing over all 2^{n-1} choices for $x_1, \ldots, x_{i-1}, x_{i+1}, \ldots, x_n$, we find that there are exactly 2^{n-1} vectors $\mathbf{x}$ such that $f(\mathbf{x}) = 0$ (and there are also 2^{n-1} vectors such that $f(\mathbf{x}) = 1$). □

Corollary 10.37. *For any integer $n \geq 2$, $\mathcal{R}(1, n)$ is a linear $(2^n, 2^{n+1}, 2^{n-1}, 2)$-code.*

Example 10.38. We construct the code $\mathcal{R}(1, 2)$. The eight affine functions $f \in \mathcal{B}_2$ yield codewords $\phi(f)$, where $\phi(f) = (f(0,0), f(0,1), f(1,0), f(1,1))$, as follows:

f	$\phi(f)$	$\mathrm{wt}(f)$
0	$(0,0,0,0)$	0
1	$(1,1,1,1)$	4
x_2	$(0,1,0,1)$	2
$1 + x_2$	$(1,0,1,0)$	2
x_1	$(0,0,1,1)$	2
$1 + x_1$	$(1,1,0,0)$	2
$x_1 + x_2$	$(0,1,1,0)$	2
$1 + x_1 + x_2$	$(1,0,0,1)$	2

(Compare this to Example 4.40.) ∎

Reed-Muller codes of order $r > 1$ are constructed by generalizing the approach above. Instead of using affine functions, which can be thought of as polynomials of degree at most one, we use polynomials of degree at most r. We begin this discussion by establishing some basic results about Boolean polynomials.

Let $x_1, \ldots, x_n$ be indeterminates taking on values in $\mathbb{Z}_2$. Then, for $1 \leq i \leq n$, x_i and $x_i{}^2$ are equivalent polynomials because $0^2 = 0$ and $1^2 = 1$ in $\mathbb{Z}_2$. Therefore, in our consideration of Boolean polynomials, we can assume that there are no occurrences of any terms of the form $x_i{}^j$, where $j > 1$.

A *Boolean monomial* of degree r is a polynomial of the form $x_{i_1}x_{i_2}\dots x_{i_r}$, where $1 \le i_1 < i_2 < \cdots < i_r \le n$. (The Boolean monomials of degree zero are 0 and 1.) A *Boolean polynomial* is a modulo 2 sum of one or more different Boolean monomials. Let $\mathcal{P}_n$ denote the set of all Boolean polynomials in n indeterminates. The degree of a Boolean polynomial $p \in \mathcal{P}_n$ is the maximum degree of any monomial that occurs in the representation of p as a sum of monomials.

The number of different monomials is 2^n because there is a monomial associated with every possible subset of $\{1,\dots,n\}$ and there are 2^n subsets of $\{1,\dots,n\}$. The number of Boolean polynomials, $|\mathcal{P}_n|$, is therefore equal to 2^{2^n} because a Boolean polynomial is expressed as a sum of a subset of the 2^n possible monomials.

Recall that there are 2^{2^n} different Boolean functions on n variables. It is not hard to see that the 2^{2^n} Boolean polynomials are distinct (when considered as functions) and there is a natural bijection between the set $\mathcal{P}_n$ and the set $\mathcal{B}_n$. This is proven in the following lemma.

Lemma 10.39. *For every Boolean function $f \in \mathcal{B}_n$, there is a unique polynomial $p_f \in \mathcal{P}_n$ such that $f(\mathbf{x}) = p_f(\mathbf{x})$ for all $\mathbf{x} \in (\mathbb{Z}_2)^n$.*

Proof. For any $\mathbf{z} \in (\mathbb{Z}_2)^n$, define $T_\mathbf{z} \in \mathcal{P}_n$ as follows:

$$T_\mathbf{z} = \prod_{\{i: z_i=0\}} (1+x_i) \prod_{\{i: z_i=1\}} x_i,$$

where $\mathbf{z} = (z_1,\dots,z_n)$. Then it is clear that $T_\mathbf{z}(\mathbf{x}) = 1$ if and only if $\mathbf{x} = \mathbf{z}$.

Now, for any $f \in \mathcal{B}_n$, define $p_f \in \mathcal{P}_n$ by the following formula:

$$p_f = \sum_{\{\mathbf{z}\in(\mathbb{Z}_2)^n : f(\mathbf{z})=1\}} T_\mathbf{z}. \tag{10.10}$$

It is easy to verify that $f(\mathbf{x}) = p_f(\mathbf{x})$ for all $\mathbf{x} \in (\mathbb{Z}_2)^n$. This proves that there is at least one polynomial with the stated property for every $f \in \mathcal{B}_n$. However, there are the same number of functions as polynomials (i.e., $|\mathcal{B}_n| = |\mathcal{P}_n|$), so there must be exactly one polynomial with the stated property for every $f \in \mathcal{B}_n$. □

Example 10.40. Let's first do a specific example computation of a polynomial p_f. Suppose that $f(0,0) = f(1,0) = f(1,1) = 1$ and $f(0,1) = 0$. Then, applying (10.10), we have that

$$\begin{aligned} p_f &= x_1x_2 + x_1(1+x_2) + (1+x_1)(1+x_2) \bmod 2 \\ &= x_1x_2 + x_1 + x_1x_2 + 1 + x_1 + x_2 + x_1x_2 \bmod 2 \\ &= 1 + x_2 + x_1x_2. \end{aligned}$$

By doing similar computations, it is possible to tabulate all $2^{2^2} = 16$ Boolean functions of two variables x_1 and x_2 and their (simplified) representations as polynomials. These are presented in Table 10.1.

f	$f(0,0)$	$f(0,1)$	$f(1,0)$	$f(1,1)$	p_f	$\deg(p_f)$
f_0	0	0	0	0	0	0
f_1	0	0	0	1	x_1x_2	2
f_2	0	0	1	0	$x_1 + x_1x_2$	2
f_3	0	0	1	1	x_1	1
f_4	0	1	0	0	$x_2 + x_1x_2$	2
f_5	0	1	0	1	x_2	1
f_6	0	1	1	0	$x_1 + x_2$	1
f_7	0	1	1	1	$x_1 + x_2 + x_1x_2$	2
f_8	1	0	0	0	$1 + x_1 + x_2 + x_1x_2$	2
f_9	1	0	0	1	$1 + x_1 + x_2$	1
f_{10}	1	0	1	0	$1 + x_2$	1
f_{11}	1	0	1	1	$1 + x_2 + x_1x_2$	2
f_{12}	1	1	0	0	$1 + x_1$	1
f_{13}	1	1	0	1	$1 + x_1 + x_1x_2$	2
f_{14}	1	1	1	0	$1 + x_1x_2$	2
f_{15}	1	1	1	1	1	0

Table 10.1. Boolean Functions of Two Variables

Note that the function f_{11} (in the Table 10.1) is the function f considered initially. ∎

Let $0 \le r \le n$. The *rth-order Reed-Muller code*, denoted $\mathcal{R}(r,n)$, is the code whose codewords are all the vectors $\phi(f)$, where $f \in \mathcal{P}_n$ is a Boolean polynomial of degree less than or equal to r. It is not hard to see that $\mathcal{R}(r,n)$ is a linear code.

The number of monomials of degree i is $\binom{n}{i}$. Therefore, the number of monomials of degree at most r is

$$m = \sum_{i=0}^{r} \binom{n}{i}.$$

These monomials form a basis for $\mathcal{R}(r,n)$, and hence the number of codewords in $\mathcal{R}(r,n)$ is 2^m.

We now consider the distance of the code $\mathcal{R}(r,n)$.This can be determined fairly easily by showing how to construct Reed-Muller codes using the **u**, **u** + **v** construction. The argument we use will be inductive, and we will use the codes $\mathcal{R}(0,n)$ and $\mathcal{R}(n,n)$ as base cases. These base cases are easily analyzed as follows.

Lemma 10.41. *For all integers $n \ge 1$, $\mathcal{R}(0,n)$ is a $(2^n, 2, 2^n, 2)$-code, and $\mathcal{R}(n,n)$ is a $(2^n, 2^{2^n}, 1, 2)$-code.*

Proof. It is easy to see that $\mathcal{R}(0,n)$ consists of the two vectors $(0,\ldots,0)$ and $(1,\ldots,1)$, and $\mathcal{R}(n,n)$ consists of all 2^{2^n} vectors in $(\mathbb{Z}_2)^{2^n}$. □

Lemma 10.42. *Suppose that* $0 < r < n$. *Then the code* $\mathcal{R}(r,n)$ *can be constructed by applying the* $\mathbf{u}, \mathbf{u}+\mathbf{v}$ *construction to the codes* $\mathcal{R}(r,n-1)$ *and* $\mathcal{R}(r-1,n-1)$.

Proof. Let C be the codewords of the code that is constructed by applying the $\mathbf{u}, \mathbf{u}+\mathbf{v}$ construction to $\mathcal{R}(r,n-1)$ and $\mathcal{R}(r-1,n-1)$. We first prove that $\mathcal{R}(r,n) \subseteq C$. A codeword in $\mathcal{R}(r,n)$ has the form $\phi(f)$, where $f \in \mathcal{P}_n$ has degree at most r. We can write the polynomial f in the form

$$f = x_1 f_1 + f_2,$$

where f_1 and f_2 are polynomials in the $n-1$ indeterminates $x_2, \ldots, x_n$. Also, the degree of f_1 is at most $r-1$ and the degree of f_2 is at most r, so $\phi(f_1) \in \mathcal{R}(r-1,n-1)$ and $\phi(f_2) \in \mathcal{R}(r,n-1)$. Now, it is not difficult to see that

$$\phi(f) = \phi(f_2) \parallel \phi(f_1) + \phi(f_2).$$

This is because the first 2^{n-1} binary n-tuples in lexicographic order have $x_1 = 0$, and the last 2^{n-1} binary n-tuples have $x_1 = 1$.

By Theorem 10.30, we have that

$$|C| = |\mathcal{R}(r,n-1)| \times |\mathcal{R}(r-1,n-1)|.$$

If we can show that

$$|\mathcal{R}(r,n-1)| \times \mathcal{R}(r-1,n-1)| = |\mathcal{R}(r,n)|, \tag{10.11}$$

then we will be finished because $\mathcal{R}(r,n) \subseteq C$. Proving (10.11) is a straightforward computation involving binomial coefficients, which we leave for the reader to do. □

Example 10.43. Consider the Boolean polynomial $f = x_1 + x_2 + x_1x_3 + x_2x_3$. We can write $f = x_1(1+x_3) + x_2 + x_2x_3$, so f_1 and f_2 (as defined in the proof of Lemma 10.42) are computed to be $f_1(x_2,x_3) = 1 + x_3$ and $f_2(x_2,x_3) = x_2 + x_2x_3$. It is easy to verify that

$$\begin{aligned}
\phi(f_1) &= (f_1(0,0), f_1(0,1), f_1(1,0), f_1(1,1)) \\
&= (1,0,1,0), \\
\phi(f_2) &= (f_2(0,0), f_2(0,1), f_2(1,0), f_2(1,1)) \\
&= (0,0,1,0), \quad \text{and} \\
\phi(f) &= (f(0,0,0), f(0,0,1), \ldots, f(1,1,0), f(1,1,1)) \\
&= (0,0,1,0,1,0,0,0) \\
&= (0,0,1,0) \parallel (0,0,1,0) + (1,0,1,0),
\end{aligned}$$

as shown in Lemma 10.42. ∎

Now it is a simple matter to determine the minimum distance of $\mathcal{R}(r,n)$ using Theorem 10.30.

Lemma 10.44. *The minimum distance of $\mathcal{R}(r,n)$ is 2^{n-r} for all $0 \le r \le n$.*

Proof. The assertion is true for the "base cases" $r = 0$ and $r = n$ by Lemma 10.41. We proceed by induction on n, assuming that $n \ge 2$. We proved that $\mathcal{R}(r,n)$ is constructed from $\mathcal{R}(r,n-1)$ and $\mathcal{R}(r-1,n-1)$ using the $\mathbf{u}, \mathbf{u}+\mathbf{v}$ construction. By induction or by a base case, it holds that

$$d(\mathcal{R}(r,n-1)) = 2^{n-r-1}$$

and

$$d(\mathcal{R}(r-1,n-1)) = 2^{n-r}.$$

From Theorem 10.30, we have that

$$d(\mathcal{R}(r,n-1)) \ge \min\{2 \times 2^{n-r-1}, 2^{n-r}\} = 2^{n-r}.$$

The minimum distance is seen to be equal to 2^{n-r} by exhibiting a codeword in $\mathcal{R}(r,n)$ having weight 2^{n-r}. Let $\mathbf{u} \in \mathcal{R}(r,n-1)$ have weight 2^{n-r-1} and let $\mathbf{v} = (0,\ldots,0)$. Then the codeword $\mathbf{u} \parallel \mathbf{u}+\mathbf{v} = \mathbf{u} \parallel \mathbf{u}$ has weight $2 \times 2^{n-r-1} = 2^{n-r}$. Since the distance of a linear code is the same as the minimum weight of a nonzero codeword, the desired result follows. □

Summarizing the results above, we have the following.

Theorem 10.45 (Reed-Muller Codes). *Suppose that r and n are integers such that $0 \le r \le n$. Then the Reed-Muller code $\mathcal{R}(r,n)$ is a linear $(2^n, 2^m, 2^{n-r}, 2)$-code, where*

$$m = \sum_{i=0}^{r} \binom{n}{i}.$$

Example 10.46. We present a basis for the code $\mathcal{R}(2,4)$ that consists of the codewords generated by the 11 monomials $f \in \mathcal{P}_4$ of degree at most two:

f	$\phi(f)$
1	1111 1111 1111 1111
x_1	0000 0000 1111 1111
x_2	0000 1111 0000 1111
x_3	0011 0011 0011 0011
x_4	0101 0101 0101 0101
x_1x_2	0000 0000 0000 1111
x_1x_3	0000 0000 0011 0011
x_1x_4	0000 0000 0101 0101
x_2x_3	0000 0011 0000 0011
x_2x_4	0000 0101 0000 0101
x_3x_4	0001 0001 0001 0001

■

Reed-Muller codes turn out to be closely related to affine geometries over $\mathbb{Z}_2$. We discuss some of these connections now. Suppose that F is an $(n-k)$-flat in $\mathrm{AG}_n(2)$. Then F is the solution set of a system of k linear equations in n indeterminates over $\mathbb{Z}_2$, which can be written in the following form:

$$\begin{aligned} a_{11}x_1 + \cdots + a_{1n}x_n &= c_1 \\ a_{21}x_1 + \cdots + a_{2n}x_n &= c_2 \\ \vdots \quad & \vdots \quad \vdots \\ a_{k1}x_1 + \cdots + a_{kn}x_n &= c_k. \end{aligned}$$

This can be expressed in an equivalent way, in the form of a single equation, as follows:

$$\prod_{i=1}^{k}(a_{i1}x_1 + \cdots + a_{in}x_n + c_i + 1) = 1.$$

The polynomial

$$p_F(\mathbf{x}) = \prod_{i}(a_{i1}x_1 + \cdots + a_{in}x_n + c_i + 1)$$

is a Boolean polynomial of degree k. Hence, $\phi(p_F)$ is a codeword in $\mathcal{R}(r,n)$ provided that $r \geq k$.

Given the flat F, we can form an incidence vector $\mathbf{s}_F \in \{0,1\}^{2^n}$ in the usual way, where the coordinates of the vector $\mathbf{s}_F \in \{0,1\}^{2^n}$ are all the points in $(\mathbb{Z}_2)^n$ in lexicographic order. The incidence vector $\mathbf{s}_F$ records which points $(x_1, \ldots, x_n)$ are in the flat F. Then it is easy to see that $\mathbf{s}_F = \phi(p_F)$, and we have the following.

Lemma 10.47. *Let F be an $(n-k)$-flat in $\mathrm{AG}_n(2)$. Then the incidence vector of F is a codeword in $\mathcal{R}(r,n)$ whenever $r \geq k$.*

The lemma above shows that every flat in $\mathrm{AG}_n(2)$ yields a codeword in a Reed-Muller code. Not every codeword can be formed in this manner, however. For example, $\mathcal{R}(2,4)$ contains codewords of weight six, and there are no flats in $\mathrm{AG}_2(4)$ containing exactly six points. However, we will show that the codewords $\phi(f)$, where f is a monomial, all correspond to flats. This is not hard to see: a monomial of degree k, say $x_{i_1}x_{i_2}\cdots x_{i_k}$, takes on the value 1 if and only if

$$x_{i_1} = x_{i_2} = \cdots = x_{i_k} = 1.$$

This is equivalent to the following system of k linear equations:

$$x_{i_1} = 1, \quad x_{i_2} = 1, \quad \ldots, \quad x_{i_k} = 1,$$

which is a flat of dimension $n-k$.

Because the monomials of degree at most r form a basis for the code $\mathcal{R}(r,n)$, this means that the codewords corresponding to flats of dimension at least $n-r$ generate this code. Expressed mathematically, we have the following theorem.

Theorem 10.48.

$$\text{span}(\mathbf{s}_F : F \textit{ is a d-flat in } \mathrm{AG}_n(2) \textit{ and } d \geq n - r) = \mathcal{R}(r, n).$$

Example 10.49. Consider the code $\mathcal{R}(2,4)$. The results proven above establish that every 2-flat, 3-flat, and 4-flat in $\mathrm{AG}_4(2)$ yields a codeword in $\mathcal{R}(2,4)$; and conversely, every codeword in $\mathcal{R}(2,4)$ is a sum of codewords corresponding to flats of dimension 2, 3, or 4.

Consider the 2-flat $F = \{1010, 1100, 1111, 1001\}$. It is not hard to see that F is the solution set of the following system of two linear equations:

$$\begin{aligned} x_1 + x_2 + x_3 + x_4 &= 0 \\ x_1 &= 1. \end{aligned}$$

Therefore the corresponding polynomial $p_F(\mathbf{x})$ is the following:

$$\begin{aligned} p_F(\mathbf{x}) &= x_1(x_1 + x_2 + x_3 + x_4 + 1) \\ &= {x_1}^2 + x_1x_2 + x_1x_3 + x_1x_4 + x_1 \\ &= x_1x_2 + x_1x_3 + x_1x_4. \end{aligned}$$

(Notice that we simplified p_F using the fact that ${x_1}^2 + x_1 = x_1 + x_1 = 0$.) The codeword in $\mathcal{R}(2,4)$ associated with the flat F is

$$\mathbf{s}_F = \phi(p_F) = 0000\,0000\,0110\,1001.$$

Conversely, suppose we start with a codeword in $\mathcal{R}(2,4)$, say

$$1110\,0001\,1110\,0001.$$

This codeword is derived from the following sum of three monomials: $1 + x_2 + x_3x_4$. The three monomials 1, x_2, and x_3x_4 correspond to flats of dimensions 4, 3, and 2, respectively. ∎

10.6 Resilient Functions

In this section, we consider Boolean functions of the form $f : (\mathbb{Z}_2)^n \to (\mathbb{Z}_2)^m$ (the Boolean functions we studied previously were the special case $m = 1$). We write $(y_1, \ldots, y_m) = f(x_1, \ldots, x_n)$, where $x_1, \ldots, x_n$ are the n *input variables* and $y_1, \ldots, y_m$ are the m *output variables*. The set of all such functions is denoted $\mathcal{B}_{n,m}$.

Definition 10.50. *Let t, m, and n be positive integers such that $t < n$, and suppose that $f \in \mathcal{B}_{n,m}$. Suppose that the values of t of the n input variables are fixed, and the remaining $n - t$ input variables are chosen independently and uniformly at random. Then f is said to be a* t-resilient function *provided that every possible vector of* *output variables is equally likely to occur. More formally, the property can be stated as follows: For every t-subset* $\{i_1, \ldots, i_t\} \subseteq \{1, \ldots, n\}$, *for every choice of* $z_j \in \mathbb{Z}_2$ $(1 \leq j \leq t)$, *and for every* $(y_1, \ldots, y_m) \in (\mathbb{Z}_2)^m$, *we have that*

$$\mathsf{Pr}[f(x_1, \ldots, x_n) = (y_1, \ldots, y_m) | x_{i_j} = z_j, 1 \leq j \leq t] = 2^{-m}.$$

We will refer to such a function f *as an* (n, m, t)-resilient function.

Example 10.51. Let $m = 1$ and $t = n - 1$. Define

$$f(x_1, \ldots, x_n) = x_1 + \cdots + x_n \bmod 2.$$

Then f is an $(n, 1, n-1)$-resilient function. ∎

Example 10.52. Let $m = n - 1$ and $t = 1$. Define

$$f(x_1, \ldots, x_n) = (x_1 + x_2 \bmod 2, x_2 + x_3 \bmod 2, \ldots, x_{n-1} + x_n \bmod 2).$$

Then f is an $(n, n-1, 1)$-resilient function. ∎

Example 10.53. Let $m = 2$, $n = 3h$, and $t = 2h - 1$. Define

$$f(x_1, \ldots, x_n) = (x_1 + \cdots + x_{2h} \bmod 2, x_{h+1} + \cdots + x_{3h} \bmod 2).$$

Then f is an $(n, 2, 2n/3 - 1)$-resilient function. ∎

Resilient functions are closely related to certain collections of orthogonal arrays, which we define now. A *large set of* t-(v, k, λ)-*orthogonal arrays*, denoted t-(v, k, λ)-LOA, is defined to be a set of v^{k-t}/λ simple t-(v, k, λ)-OAs such that every possible k-tuple of symbols occurs in exactly one of the orthogonal arrays in the set. (Equivalently, the union of the orthogonal arrays forms a k-$(v, k, 1)$-OA.)

Theorem 10.54. *An* (n, m, t)*-resilient function is equivalent to a* t-$(2, n, 2^{n-m-t})$-LOA.

Proof. First, suppose $f : (\mathbb{Z}_2)^n \rightarrow (\mathbb{Z}_2)^m$ is an (n, m, t)-resilient function. For any $\mathbf{y} \in (\mathbb{Z}_2)^m$, form an array $A_\mathbf{y}$ whose rows are the vectors in the inverse image $f^{-1}(\mathbf{y})$. $A_\mathbf{y}$ is an $|f^{-1}(\mathbf{y})| \times n$ binary array. It is clear that the 2^m arrays $A_\mathbf{y}$ together contain every possible n-tuple as a row, so if each $A_\mathbf{y}$ is a t-$(2, n, 2^{n-m-t})$-OA, then we automatically get a t-$(2, n, 2^{n-m-t})$-LOA.

Let $\{i_1, \ldots, i_t\} \subseteq \{1, \ldots, n\}$ be a t-subset, and let $z_j \in \mathbb{Z}_2$ $(1 \leq j \leq t)$. For every $\mathbf{y} \in (\mathbb{Z}_2)^m$, let $\lambda(\mathbf{y})$ denote the number of rows in $A_\mathbf{y}$ in which z_j occurs in column i_j for all j, $1 \leq j \leq t$. It is easy to see that

$$\sum_{\mathbf{y} \in (\mathbb{Z}_2)^m} \lambda(\mathbf{y}) = 2^{n-t}.$$

This is because the total number of possible n-tuples satisfying the conditions that z_j occurs in position i_j for all j, $1 \le j \le t$, is 2^{n-t}.

Now, it is clear that

$$\Pr[f(x_1, \ldots, x_n) = (y_1, \ldots, y_m) | x_{i_j} = z_j, 1 \le j \le t] = \frac{\lambda(\mathbf{y})}{2^{n-t}}. \quad (10.12)$$

Since f is t-resilient, we get

$$\frac{\lambda(\mathbf{y})}{2^{n-t}} = 2^{-m},$$

or $\lambda(\mathbf{y}) = 2^{n-m-t}$. Since $\{i_1, \ldots, i_t\}$ and z_j $(1 \le j \le t)$ are arbitrary, we have shown that each $A_{\mathbf{y}}$ is a t-$(2, n, 2^{n-m-t})$-OA, as desired.

Conversely, suppose we start with a t-$(2, n, 2^{n-m-t})$-LOA. There are 2^m arrays in the large set; arbitrarily name them $A_{\mathbf{y}}$, $\mathbf{y} \in (\mathbb{Z}_2)^m$. Then define a function $f : (\mathbb{Z}_2)^n \to (\mathbb{Z}_2)^m$ by the rule

$$f(x_1, \ldots, x_n) = (y_1, \ldots, y_m) \Leftrightarrow (x_1, \ldots, x_n) \in A_{(y_1, \ldots, y_m)}.$$

Using (10.12), it is easy to see that the function f is t-resilient. □

Example 10.55. Consider Example 10.53 with $h = 2$:

$$f(x_1, x_2, x_3, x_4, x_5, x_6) = (x_1 + x_2 + x_3 + x_4 \bmod 2, x_3 + x_4 + x_5 + x_6 \bmod 2).$$

This is a $(6, 2, 3)$-resilient function, and by Theorem 10.54, it is equivalent to a 3-$(2, 6, 2)$-LOA. There are four orthogonal arrays in the large set, one of which is obtained from $f^{-1}(0, 0)$:

$$\begin{pmatrix} 0&0&0&0&0&0 \\ 0&1&0&1&0&1 \\ 0&0&0&0&1&1 \\ 0&1&0&1&1&0 \\ 1&1&0&0&0&0 \\ 1&0&0&1&0&1 \\ 1&1&0&0&1&1 \\ 1&0&0&1&1&0 \\ 0&0&1&1&0&0 \\ 0&1&1&0&0&1 \\ 0&0&1&1&1&1 \\ 0&1&1&0&1&0 \\ 1&1&1&1&0&0 \\ 1&0&1&0&0&1 \\ 1&1&1&1&1&1 \\ 1&0&1&0&1&0 \end{pmatrix}.$$

The other three orthogonal arrays in the large set are constructed easily as well. ∎

A resilient function is a *linear resilient function* if every output variable is a linear function of the input variables. All of the examples of resilient function considered above are linear. The following theorem gives a characterization of linear resilient functions in terms of linear codes.

Theorem 10.56. *There is a linear $(n, 2^m, d, 2)$-code if and only if there is a linear $(n, m, d-1)$-resilient function.*

Proof. Let G be an $m \times n$ matrix whose rows form a basis for a linear $(n, 2^m, d, 2)$-code, say $\mathcal{C}$. Define the function $f : (\mathbb{Z}_2)^n \to (\mathbb{Z}_2)^m$ by the rule

$$f(x_1, \ldots, x_n) = (x_1, \ldots, x_n)G^T,$$

where all arithmetic is modulo 2. Clearly f is linear; we will establish that f is an $(n, m, d-1)$-resilient function with the aid of Theorem 10.54.

It is easy to see that the inverse image $f^{-1}(0, \ldots, 0)$ is in fact the dual code $\mathcal{C}^\perp$. Theorem 10.17 asserts that $\mathcal{C}^\perp$ is a $(d-1)$-$(2, n, 2^{n-m-d+1})$-OA. Now, any other inverse image $f^{-1}(\mathbf{y})$ ($\mathbf{y} \in (\mathbb{Z}_2)^m$) is an additive coset of $\mathcal{C}^\perp$, and thus it is also a $(d-1)$-$(2, n, 2^{n-m-d+1})$-OA. Hence we obtain 2^m orthogonal arrays that form a large set. By Theorem 10.54, f is an $(n, m, d-1)$-resilient function.

Conversely, suppose that f is a linear $(n, m, d-1)$-resilient function. Because f is linear, it can be written in the form $f(\mathbf{x}) = \mathbf{x}G^T$, where G is an $m \times n$ matrix. The proof of Theorem 10.54 shows that $f^{-1}(0, \ldots, 0)$ is a $(d-1)$-$(2, n, 2^{n-m-d+1})$-OA. Clearly this orthogonal array is linear, so Theorem 10.17 can be applied. This theorem shows that the dual of the orthogonal array is a linear $(n, 2^m, d, 2)$-code (the rows of G are actually a basis for this code). □

We illustrate the application of Theorem 10.56 in an example.

Example 10.57. From Corollary 10.37, a first-order Reed-Muller code, $\mathcal{R}(1, n)$, is a linear $(2^n, 2^{n+1}, 2^{n-1}, 2)$-code. Therefore there exists a $(2^n, n+1, 2^{n-1}-1)$-resilient function for all positive integers n. When $n = 2$, the code $\mathcal{R}(1, n)$ has basis $(1, 1, 1, 1)$, $(0, 1, 0, 1)$, and $(0, 0, 1, 1)$. The matrix G, described in the proof of Theorem 10.56, is as follows:

$$G = \begin{pmatrix} 1\,1\,1\,1 \\ 0\,1\,0\,1 \\ 0\,0\,1\,1 \end{pmatrix},$$

and the resulting $(4, 3, 1)$-resilient function is defined to be

$$f(x_1, x_2, x_3, x_4) = (x_1 + x_2 + x_3 + x_4 \bmod 2, x_2 + x_4 \bmod 2, x_3 + x_4 \bmod 2).$$

▮

Theorem 10.56 can also be used to verify the resiliency of linear functions. Basically, all that is required is to write down the matrix G and determine the distance of the resulting linear code. We illustrate this process now.

Example 10.58. Consider the resilient function described in Example 10.53. The matrix G is as follows:

$$G = \begin{pmatrix} 1 \cdots 1\,1 \cdots 1\,0 \cdots 0 \\ 0 \cdots 0\,1 \cdots 1\,1 \cdots 1 \end{pmatrix},$$

and the code C consists of the following four codewords:

$$(\underbrace{0,\ldots,0}_{h},\underbrace{0,\ldots,0}_{h},\underbrace{0,\ldots,0}_{h})$$
$$(\underbrace{1,\ldots,1}_{h},\underbrace{1,\ldots,1}_{h},\underbrace{0,\ldots,0}_{h})$$
$$(\underbrace{0,\ldots,0}_{h},\underbrace{1,\ldots,1}_{h},\underbrace{1,\ldots,1}_{h})$$
$$(\underbrace{1,\ldots,1}_{h},\underbrace{0,\ldots,0}_{h},\underbrace{1,\ldots,1}_{h}).$$

The distance of the code C is equal to $2h$, and hence f is a $(3h, 2, 2h-1)$-resilient function. ∎

10.7 Notes and References

Coding theory is an enormous topic in its own right. We have just mentioned a few results that are closely connected to combinatorial designs in general and orthogonal arrays in particular. Most of the results on codes are "classical" and can be found in standard reference works and textbooks.

Useful books on coding theory include "Introduction to Coding Theory" by van Lint [78], "The Theory of Error-correcting Codes" by MacWilliams and Sloane [80], and "Coding and Information Theory" by Roman [87]. Two books that describe connections between designs and codes are "Designs and Their Codes" by Assmus and Key [3] and "Designs, Codes, Graphs and Their Links" by Cameron and van Lint [20]. See also the survey on codes by Tonchev [111].

"Orthogonal Arrays, Theory and Applications", by Hedayat, Sloane, and Stufken [59], is a recent book devoted specifically to orthogonal arrays.

Resilient functions were invented by Bennett, Brassard, and Robert [7] and independently by Chor et al. [22]. These functions have interesting applications in cryptography. Section 10.6 is based on Stinson [104]. For additional information on resilient functions, see Bierbrauer, Gopalakrishnan, and Stinson [11].

10.8 Exercises

10.1 Assuming there is a t-(v_1, k, λ_1)-OA and a t-(v_2, k, λ_2)-OA, prove that there is a t-$(v_1v_2, k, \lambda_1\lambda_2)$-OA.

10.2 Use the Gilbert-Varshamov Bound to prove that the following orthogonal arrays exist:

(a) a 4-(2,9,8)-OA;
(b) a 4-(2,12,16)-OA;
(c) a 3-(3,10,9)-OA.

10.3 Prove that a $(23, 2^{12}, 7, 2)$-code and an $(11, 729, 5, 3)$-code are both perfect codes.

Remark: These codes exist, and they are known as the binary and ternary Golay codes, respectively.

10.4 Suppose that $(\mathbb{F}_2, \mathcal{C})$ is an $(n, M, d, 2)$-code in which d is odd. For all codewords $(x_1, \ldots, x_n) \in \mathcal{C}$, define

$$x_{n+1} = \begin{cases} 0 & \text{if } \mathsf{wt}(x_1, \ldots, x_n) \text{ is even} \\ 1 & \text{if } \mathsf{wt}(x_1, \ldots, x_n) \text{ is odd.} \end{cases}$$

Then define

$$\mathcal{D} = \{(x_1, \ldots, x_{n+1}) : (x_1, \ldots, x_n) \in \mathcal{C}\}.$$

Prove that $(\mathbb{F}_2, \mathcal{D})$ is an $(n+1, M, d+1, 2)$-code.

Remark: This process is called *extending a code.*

10.5 Construct the $(24, 6, 14, 2)$-code that is described in Example 10.35.

10.6 Suppose n, d, k, u, and v are defined as in the proof of Theorem 10.34. Suppose also that $2d > n$, d is even, n is odd, and k is even.

(a) Prove that $2u$ and v are both nonnegative integers.
(b) Prove that the code that is formed by pasting together $2u$ copies of a $(2k-2, k, k, 2)$-code and v copies of a $(4k+2, 2k+2, 2k+2, 2)$-code meets the Plotkin Bound with equality.

10.7 An (n, M, d, q)-code, say $(Q, \mathcal{C})$, is an *equidistant code* if $\mathsf{d}(\mathbf{x}, \mathbf{y}) = d$ for all $\mathbf{x}, \mathbf{y} \in \mathcal{C}, \mathbf{x} \neq \mathbf{y}$.

(a) Suppose there is a resolvable $(v, b, r, k, 1)$-BIBD. Prove that there is an equidistant (n, M, d, q)-code, where $n = r$, $M = v$, $d = r - \lambda$, and $q = v/k$.
(b) If an equidistant (n, M, d, q)-code exists and $d > (q-1)n/q$, then it is known that

$$M \geq \frac{qd}{qd - (q-1)n}.$$

Prove that the code constructed in part (a) meets this bound with equality.

Remark: This bound is a q-ary analogue of the Plotkin Bound.

10.8 Suppose we first choose 2^n codewords from the code $\mathcal{R}(1, n)$, then we form a square matrix whose rows are the 2^n chosen codewords, and then we replace every entry "0" by "1" and every entry "1" by "-1".

(a) Determine the conditions under which the resulting matrix is a Hadamard matrix of order n.
(b) Determine the conditions under which the resulting matrix is the Sylvester matrix S_n.

10.9 A binary code $(\mathbb{F}_2, \mathcal{C})$ is a *constant-weight code* if there exists a positive integer w such that $\text{wt}(\mathbf{x}) = w$ for all $\mathbf{x} \in \mathcal{C}$. A (v,k,t)-*packing* is a design $(X, \mathcal{A})$ in which $|X| = v$, every block $A \in \mathcal{A}$ has size k, and no t-subset of points is contained in more than one block.

(a) Suppose M is the incidence matrix of a (v,b,r,k,λ)-BIBD. Define a binary code $\mathcal{C}$ whose codewords are the rows of M. Prove that $(\mathbb{F}_2, \mathcal{C})$ is a $(b, v, 2(r-\lambda), 2)$-code having constant weight r.

(b) Suppose M is the incidence matrix of a $(v,b,r,k,1)$-BIBD. Define a binary code $\mathcal{D}$ whose codewords are the columns of M. Prove that $(\mathbb{F}_2, \mathcal{D})$ is a $(v, b, 2(k-1), 2)$-code having constant weight k.

(c) Prove that a (v,k,t)-packing having b blocks exists if and only if there exists a $(v, b, 2(k-t+1), 2)$-code having constant weight k.

(d) Let $D(v,k,t)$ denote the maximum number of blocks in any (v,k,t)-packing. Prove the following assertions.

i. $D(v,k,1) \leq \lfloor \frac{v}{k} \rfloor$.

ii. $D(v,k,t) \leq \lfloor \frac{v}{k} D(v-1, k-1, t-1) \rfloor$.

iii. $D(v,k,t) \leq \lfloor \frac{v}{k} \lfloor \frac{v-1}{k-1} \cdots \lfloor \frac{v-t+1}{k-t+1} \rfloor \cdots \rfloor \rfloor$.

Remark: This bound is known as the *Johnson Bound*.

10.10 Let G be the following matrix:

$$\begin{pmatrix} 0\,0\,0\,1\,1\,1\,1 \\ 0\,1\,1\,0\,0\,1\,1 \\ 1\,0\,1\,0\,1\,0\,1 \end{pmatrix}.$$

(a) Prove that the linear code whose basis consists of the rows of G is a $(7, 8, 4, 2)$-code.

(b) Describe how to construct a $(7h, 3, 4h-1)$-resilient function for all integers $h \geq 1$.

11

Selected Applications of Combinatorial Designs

There are many interesting and important applications of combinatorial designs to areas including computer networks, design and analysis of algorithms, cryptography, design and analysis of experiments, and tournament scheduling. In this chapter, we present four applications of combinatorial designs. The four applications are authentication codes, threshold schemes, group testing algorithms, and the two-point sampling technique. These applications consist of two from the field of cryptography, one from experimental design, and one from algorithm design. They should just be considered as a sample or an appetizer; we do not even begin to cover the range of the many ingenious and diverse applications of designs that have been discovered.

11.1 Authentication Codes

The eminent cryptologist Gustavus Simmons has referred to cryptology as "the science of information integrity". Most people are familiar with the idea of encryption, which is used to keep the contents of a message secret from an eavesdropper. However, as suggested by the term "integrity", there are, in addition, other objectives in providing secure communications over an insecure network. One of the most important is the question of authenticity. When Alice sends a message to Bob (encrypted or not), how can Bob be sure that it was Alice who sent the message, and how does he know that the message was not altered by someone else during its transmission?

One elegant way to solve this problem is to use an authentication code. We will discuss authentication codes, and a construction for them that uses combinatorial designs, in this section.

Here is the mathematical setting in which we study the problem. There are three participants: Alice, Bob, and Oscar. Alice and Bob want to communicate over an insecure channel (e.g., by e-mail, fax, or cell-phone). Oscar (the "bad guy") has the ability to introduce his own messages into the channel and/or to modify existing messages. We consider two types of attacks by Oscar. When Oscar places a (new) message m' into the channel, it is called *impersonation*. When Oscar sees a message m and changes it to a (different) message $m' \neq m$, it is called *substitution*.

As an example, suppose that Bob is Alice's stockbroker. When Alice sends a message to Bob, such as "buy 100 shares of Acme stock", she would not be very happy if Oscar changed "buy" to "sell"!

The goal of an authentication code is to allow Bob to detect with high probability when such an attack has taken place. Here is a formal mathematical definition of an authentication code.

Definition 11.1. *An* authentication code *is a four-tuple* $(\mathcal{S}, \mathcal{A}, \mathcal{K}, \mathcal{E})$, *where the following conditions are satisfied.*

1. $\mathcal{S}$ *is a finite set of* source states.
2. $\mathcal{A}$ *is a finite set of* authenticators.
3. $\mathcal{K}$ *is a finite set of* keys.
4. *For each* $K \in \mathcal{K}$, *there is an* authentication rule $e_K \in \mathcal{E}$, *where* $e_K : \mathcal{S} \rightarrow \mathcal{A}$.

Here is how an authentication code works. Alice and Bob jointly choose a secret key $K \in \mathcal{K}$ at random. They do this "ahead of time", either when they are together in the same place or when they have access to a secure channel. A source state is just the information that Alice wants to communicate to Bob (e.g., "buy 100 shares ..."). When Alice wants to communicate the source state $s \in \mathcal{S}$ to Bob, she uses the authentication rule e_K to construct the authenticator $a = e_K(s)$. The *message* m is formed by concatenating s and a, i.e., $m = (s, a)$. The message m is then sent over the channel. When Bob receives m, he verifies that $a = e_K(s)$ to authenticate the source state s. If $a \neq e_K(s)$, then Bob is able to detect that an attack has taken place.

An authentication code can be represented by the $|\mathcal{K}| \times |\mathcal{S}|$ *authentication matrix* in which the rows are indexed by the keys, the columns are indexed by source states, and the entry in row K and column s of the matrix is $e_K(s)$.

When Oscar performs impersonation or substitution, his goal is to have his bogus message $m' = (s', a')$ accepted as authentic by Bob, thus misleading Bob as to the state of the source. That is, if K is the secret key (the value of which is not known to Oscar), then Oscar is hoping that $a' = e_K(s')$.

The strength of an authentication code is measured by the *deception probabilities* P_0 and P_1, which represent the probability that Oscar can deceive Bob by impersonation and substitution, respectively. In computing the deception probabilities, it is assumed that Oscar is using an optimal strategy. When Alice and Bob use an authentication code, they want P_0 and P_1 to be small (so Oscar has only a small possibility of carrying out a successful attack). They also want $|\mathcal{K}|$ (the number of possible keys) to be small because the key must be stored securely by both Alice and Bob until the time that Alice sends a message to Bob and he authenticates it.

11.1.1 A Construction from Orthogonal Arrays

Orthogonal arrays provide a nice way of constructing authentication codes. Suppose that B is an $\mathsf{OA}(m,n)$ on symbol set $\{1,\ldots,n\}$. We define $\mathcal{S} = \{1,\ldots,m\}$, $\mathcal{A} = \{1,\ldots,n\}$, and $\mathcal{K} = \{1,\ldots,n^2\}$. The rows of B are indexed by $\mathcal{K}$ and the columns are indexed by $\mathcal{S}$. For $1 \leq K \leq n^2$, the authentication rule e_K is defined as

$$e_K(s) = B(K,s)$$

for $1 \leq s \leq m$. In other words, the orthogonal array B is used as the authentication matrix of our code.

Let's analyze the deception probabilities of this authentication code. In computing the deception probabilities, we assume that the authentication matrix is known to Oscar. The only information that Oscar does not know is the particular key (i.e., the row of the orthogonal array) that is being used by Alice and Bob.

P_0 is quite simple to compute. Suppose that Oscar places any message $m = (s,a)$ into the channel. Then m is accepted as authentic if and only if $e_K(s) = a$, which happens if and only if $B(K,s) = a$. Here K is a random row of the orthogonal array B, and the value of K is known by Alice and Bob but not by Oscar.

Let $\mathcal{L}(s,a) = \{L : B(L,s) = a\}$. Then it is not difficult to see that $|\mathcal{L}(s,a)| = n$, and Oscar's deception will succeed if and only if $K \in \mathcal{L}(s,a)$. Since $|\mathcal{K}| = n^2$, it follows that the attack succeeds with probability

$$\frac{|\mathcal{L}(s,a)|}{|\mathcal{K}|} = \frac{1}{n}.$$

Since this probability is independent of the message (s,a) that Oscar inserts into the channel, we see that $P_0 = 1/n$ for this code.

We now turn to the analysis of P_1. Here, we suppose that Oscar sees a valid message $m = (s,a)$, and he replaces it with a bogus message $m' = (s',a')$, where $s \neq s'$. If we again define $\mathcal{L}(s,a) = \{L : B(L,s) = a\}$, then observation of the message m allows Oscar to conclude that $K \in \mathcal{L}(s,a)$. In other words, the number of "possible keys" is reduced from n^2 to n (however, we will see that this does not increase Oscar's probability of a successful deception).

Now, Oscar's deception will succeed if and only if $K \in \mathcal{L}(s',a')$. However, since it is known that $K \in \mathcal{L}(s,a)$, it must be the case that $K \in \mathcal{L}(s,a) \cap \mathcal{L}(s',a')$. Now, we use the fact that B is an $\mathsf{OA}(m,n)$ to observe that $|\mathcal{L}(s,a) \cap \mathcal{L}(s',a')| = 1$. Since it is known that $K \in \mathcal{L}(s,a)$, and the deception succeeds if and only if $K \in \mathcal{L}(s,a) \cap \mathcal{L}(s',a')$, the success probability of this substitution attack is

$$\frac{|\mathcal{L}(s,a) \cap \mathcal{L}(s',a')|}{|\mathcal{L}(s,a)|} = \frac{1}{n}.$$

Since this probability is independent of the original message (s,a) and the bogus message (s',a') that Oscar inserts into the channel, we see that $P_1 = 1/n$.

Summarizing, we have the following theorem.

Theorem 11.2. *Suppose there is an* $\mathsf{OA}(m,n)$*. Then there is an authentication code for m source states, having n authenticators and* n^2 *keys, in which* $P_0 = P_1 = 1/n$*.*

Example 11.3. As above, suppose that Alice owns 100 shares of Acme stock. For $0 \leq i \leq 99$, we will let source state i correspond to the order "sell $i+1$ shares"; and for $100 \leq i \leq 199$, we will let source state i correspond to the order "buy $i-99$ shares". Thus we desire a code with (at least) 200 source states, so we need an $\mathsf{OA}(m,n)$ with $m \geq 200$.

Now suppose that Alice and Bob want a security level of $1/1000$; i.e., they want a code with $P_0 \leq 1/1000$ and $P_1 \leq 1/1000$. This means that they will use an $\mathsf{OA}(m,n)$ with $n \geq 1000$.

The simplest way to accommodate these requirements is to take n to be the smallest prime exceeding 1000, i.e., $n = 1009$. Then they construct an $\mathsf{OA}(200,1009)$. This can easily be done using Theorem 6.39. To be specific, let $\mathcal{S} = \{0,\ldots,199\}$, $\mathcal{A} = \mathbb{Z}_{1009}$, and $\mathcal{K} = \mathbb{Z}_{1009} \times \mathbb{Z}_{1009}$. For $K = (i,j)$, where $i,j \in \mathbb{Z}_{1009}$, the authentication rule $e_{(i,j)}$ is defined as

$$e_{(i,j)}(s) = i + sj \bmod 1009$$

for $0 \leq s \leq 199$.

Suppose that the key is $K = (427,886)$. If Alice wants to buy 50 shares of Acme stock, then the source state is $s = 149$. She computes the authenticator to be

$$a = e_{(427,886)}(149) = 427 + 886 \times 149 \bmod 1009 = 262.$$

Then the message she transmits to Bob is $m = (149,262)$. When Bob receives this message, he recomputes the authenticator using the authentication rule $e_{(427,886)}$ to verify the authenticity of the message. ∎

When constructing an authentication code using an $\mathsf{OA}(m,n)$, the parameter n relates to the security of the code, while the parameter m determines the number of source states. Furthermore, in order for an $\mathsf{OA}(m,n)$ to exist, we have that $m \leq n+1$ by Theorem 6.29 and Theorem 6.38. These facts must be taken into account when constructing an authentication code.

Another observation about this orthogonal array code is that it is a one-time code: a key should be used to authenticate only one source state. This is seen as follows. Suppose that Alice uses the same key K to authenticate two different source states, s and s'. Thus she transmits two messages, (s,a) and (s',a'), where $a = e_K(s)$ and $a' = e_K(s')$. Because the authentication matrix B is an orthogonal array, there is a unique row of B in which a appears in column s and a' appears in column s'. This row, K, is the key, and it can easily be computed by Oscar after observation of the two messages. Once Oscar knows the key, he can determine the correct authenticator for any source state and perform successful deceptions (as long as the key is not changed).

11.2 Threshold Schemes

Suppose that a bank has a vault that must be opened every day. The bank employs three senior tellers, but they do not want to trust any individual with the combination. Hence, they would like to devise a system that enables any two of the three senior tellers to gain access to the vault. This problem can be solved by means of threshold schemes. Here is an informal definition.

Definition 11.4. *Suppose that t and w are integers such that $2 \leq t \leq w$. A* perfect (t, w)-threshold scheme *is a method of sharing a secret value K among a finite set $\mathcal{P} = \{P_1, \ldots, P_w\}$ of w* participants *in such a way that any t participants can compute the value of K but no group of $t-1$ (or fewer) participants can compute any information about the value of K from the information they hold collectively.*

The value of K is chosen from a specified set of *secrets*, denoted $\mathcal{K}$, by a special player, the *dealer*. The dealer is denoted by D, and it is assumed that $D \notin \mathcal{P}$. When D wants to share the secret K among the participants in $\mathcal{P}$, he gives each participant some partial information called a *share*. Each share is chosen from a specified *share set*, denoted by $\mathcal{S}$. The shares should be distributed in a secure manner, so no participant knows the share given to another participant.

At a later time, a subset of participants $B \subseteq \mathcal{P}$ pool their shares in an attempt to compute the secret K. If $|B| \geq t$, then they should be able to compute the value of K as a function of the shares they jointly hold; if $|B| < t$, then they should not be able to compute K. In the "bank" example described above, we are asking for a $(2,3)$-threshold scheme.

11.2.1 A Construction from Orthogonal Arrays

It is easy to obtain a (t, w)-threshold scheme from any t-$(v, w+1, 1)$-OA. Suppose that this orthogonal array, A, is defined on symbol set X, the columns are labeled $1, \ldots, w+1$, and the rows are labeled $1, \ldots, v^t$. The scheme will have $\mathcal{K} = \mathcal{S} = X$, so it accommodates v possible secrets. Associate the first w columns of the array with the w participants and the last column with the secret. For every $K \in X$, define $R_K = \{r : A(r, w+1) = K\}$. In other words, R_K is the set of rows of A having the element K in the last column. Now, when D wants to share the secret $K \in X$, he chooses a random row $r \in R_K$. Then D gives the share $A(r, i)$ to participant P_i for $1 \leq i \leq w$.

Suppose that t participants, say $P_{i_1}, \ldots, P_{i_t}$, wish to determine the secret. Note that the orthogonal array A is known to all the participants in $\mathcal{P}$. Let s_j be P_{i_j}'s share, $1 \leq j \leq t$. Because A is a t-$(v, w+1, 1)$-OA, there is a unique row r such that $A(r, i_j) = s_j$, $1 \leq j \leq t$. It is a simple matter for the t given participants to determine r and then to compute $K = A(r, w+1)$.

To prove that the scheme is secure, we show that knowledge of any $t-1$ shares leaves the secret completely undetermined. This implies that no subset of $t-1$ participants can determine anything about the value of K (except that $K \in X$, of course). Suppose that P_{i_j} has share s_j, $1 \leq j \leq t-1$. For any $L \in X$, there is a unique row r_L of A such that $A(r_L, i_j) = s_j$ for $1 \leq j \leq t-1$, and $A(r_L, w+1) = L$ (again, this follows because A is a t-$(v, w+1, 1)$-OA). In other words, for any possible value L of the secret, there is exactly one row $r_L \in R_L$ such that the share s_j is given to P_{i_j}, $1 \leq j \leq t-1$. The given subset of $t-1$ participants has no way of knowing which of these v possible rows was actually used by D to compute the shares, and hence any possible value for the secret is consistent with the given subset of $t-1$ participants holding the specified $t-1$ shares.

We summarize the above as follows.

Theorem 11.5. *Suppose that there exists a t-$(v, w+1, 1)$-OA. Then there exists a perfect (t, w)-threshold scheme with $|\mathcal{S}| = |\mathcal{K}| = v$.*

Example 11.6. Suppose we want a perfect $(2, 10)$-threshold scheme with $|\mathcal{S}| = |\mathcal{K}| = 101$. We can use an OA$(11, 101)$ to do this. Because 101 is prime, Theorem 6.39 can be applied. The rows of the orthogonal array are indexed by $\mathbb{Z}_{101} \times \mathbb{Z}_{101}$ and the columns are named $0, \ldots, 10$. The entries in the orthogonal array A are defined by the formula

$$A((i,j),c) = i + jc \bmod 101,$$

$i, j \in \mathbb{Z}_{101}, 0 \leq c \leq 10$. Suppose we relabel column 0 as column 11 (this is the column of the orthogonal array that corresponds to the secret). Then, observe that $R_K = \{K\} \times \mathbb{Z}_{101}$ for $0 \leq K \leq 100$.

Suppose that D wishes to share the secret $K = 55$. He chooses a random row in R_{55}, say $(55, 17)$. This row determines the shares $s_1, \ldots, s_{10}$ to be distributed to $P_1, \ldots, P_{10}$, respectively. These shares are computed as follows:

$$\begin{aligned}
s_1 &= 55 + 17 \times 1 \bmod 101 &= 72 \\
s_2 &= 55 + 17 \times 2 \bmod 101 &= 89 \\
s_3 &= 55 + 17 \times 3 \bmod 101 &= 5 \\
s_4 &= 55 + 17 \times 4 \bmod 101 &= 22 \\
s_5 &= 55 + 17 \times 5 \bmod 101 &= 39 \\
s_6 &= 55 + 17 \times 6 \bmod 101 &= 56 \\
s_7 &= 55 + 17 \times 7 \bmod 101 &= 73 \\
s_8 &= 55 + 17 \times 8 \bmod 101 &= 90 \\
s_9 &= 55 + 17 \times 9 \bmod 101 &= 6 \\
s_{10} &= 55 + 17 \times 10 \bmod 101 &= 23.
\end{aligned}$$

Now, suppose that P_2 and P_9 want to compute K. Their shares provide two equations in two unknowns, i and j (where (i,j) is the row of the orthogonal array that D used to generate the shares):

$$\begin{aligned}
i + 2j &\equiv 89 \bmod 101 \\
i + 9j &\equiv 6 \bmod 101.
\end{aligned}$$

Subtracting the first congruence from the second, we get

$$7j \equiv 18 \bmod 101.$$

To solve this congruence, we compute $7^{-1} \bmod 101 = 29$. Then,

$$j \equiv 29 \times 18 \bmod 101 = 17.$$

Having determined that $j = 17$, it is a simple matter to substitute back into the first congruence to obtain

$$i = 89 - 2 \times 17 \bmod 101 = 55.$$

Then the secret is seen to be $K = i = 55$. ∎

11.2.2 Anonymous Threshold Schemes

A perfect (t, w)-threshold scheme is an *anonymous threshold scheme* if the following two properties are satisfied:

1. the w participants receive w distinct shares,
2. the secret can be computed solely as a function of t shares, without the knowledge of which participant holds which share.

Observe that the threshold schemes, constructed in Theorem 11.5 from orthogonal arrays, are not anonymous.

In an anonymous scheme, the computation of the secret can be performed by a black box that is given t shares and does not know the identities of the participants holding those shares. This could allow a secret to be reused many times without constructing new shares.

Resolvable $(v, w, 1)$-BIBDs provide a nice way to construct anonymous $(2, w)$-threshold schemes. Suppose that $(X, \mathcal{A})$ is a resolvable $(v, w, 1)$-BIBD. There are $r = (v-1)/(w-1)$ parallel classes in this BIBD, which we name $\Pi_i, \ldots, \Pi_r$. The scheme we construct will have $\mathcal{K} = \{1, \ldots, r\}$ and $\mathcal{S} = X$ (i.e., we have r possible secrets, and the share set has cardinality v).

Suppose that D wants to share the secret K, where $1 \leq K \leq r$. Then D chooses a random block $A \in \Pi_K$, and he gives the w points in A to the w participants (i.e., one point is given to each of the w participants).

Suppose that two participants wish to determine the secret. The design $(X, \mathcal{A})$ and its resolution are known to all the participants in $\mathcal{P}$. Let s and t be the shares held by any two participants. Since $(X, \mathcal{A})$ is a BIBD with $\lambda = 1$, there is a unique block A such that $\{s, t\} \subseteq A$. Then the two participants can find the parallel class Π_K that contains the block A, and the secret is revealed as K. Note that this computation depends only on the values of the two shares and not on the identities of the participants holding them. Thus the scheme is anonymous.

Now we show that the scheme is secure (i.e., that knowledge of any one share leaves the secret completely undetermined). Suppose a participant has share s. For any L such that $1 \leq L \leq r$, there is a unique block $A_L \in \Pi_L$ such that $s \in A_L$ (this follows from the fact that each Π_L is a parallel class). Hence, for any possible value L of the secret, there is exactly one block $A_L \in \Pi_L$ such that the share $s \in A_L$. Any of these r possible blocks could have been used by D to distribute shares to the participants in $\mathcal{P}$, and hence any possible value for the secret is consistent with any given share $s \in X$.

We summarize the above as follows.

Theorem 11.7. *Suppose there is a resolvable* $(v, w, 1)$-BIBD. *Then there exists an anonymous perfect* $(2, w)$*-threshold scheme with* $|\mathcal{S}| = v$ *and* $|\mathcal{K}| = (v-1)/(w-1)$.

Example 11.8. We will use a resolvable $(15, 3, 1)$-BIBD to construct an anonymous perfect $(2, 3)$-threshold scheme with $|\mathcal{S}| = 15$ and $|\mathcal{K}| = 7$. We present an example of a resolvable $(15, 3, 1)$-BIBD. The BIBD has point set

$$X = \{a, b, c, d, e, f, g, h, i, j, k, l, m, n, o\}.$$

The 35 blocks are arranged into seven parallel classes, named $\Pi_1, \ldots \Pi_7$, as follows:

Π_1	Π_2	Π_3	Π_4	Π_5	Π_6	Π_7
abc	*ahi*	*ajk*	*ade*	*afg*	*alm*	*ano*
djn	*beg*	*bmo*	*bln*	*bhj*	*bik*	*bdf*
ehm	*cmn*	*cef*	*cij*	*clo*	*cdg*	*chk*
fio	*dko*	*dhl*	*fkm*	*dim*	*ejo*	*eil*
gkl	*fjl*	*gin*	*gho*	*ekn*	*fhn*	*gjm*

Suppose that D wants to share the secret 4. He picks a random block in Π_4, say *cij*. The shares c, i, and j are given to the three participants.

At a later time, any two of these shares can be used to reveal the secret. For example, given the shares c and i, we find that the unique block containing c and i is *cij*. Then we determine that the parallel class that contains this block is Π_4, so the secret is $K = 4$. ∎

11.3 Group Testing Algorithms

Suppose that a large number of blood samples need to be tested for the presence of a rare disease. If each test is expensive, it might be more efficient to combine several samples before testing them. Such a scheme is called a *group testing algorithm*. Then a negative result to a test ensures that none of the samples are positive (assuming, for simplicity, that the tests always give the correct answer). On the other hand, a positive result would reveal only the fact that at least one of the samples in the test is positive. Further tests would be required to reveal which particular samples are in fact positive.

In general, we might set up a procedure where we perform a sequence of group tests in which the samples used in later tests depend on the outcomes of earlier tests. For example, as mentioned above, if a particular test

T is negative, then there is no need to retest any of the samples in T. However, in many applications of group testing, there are some practical benefits to a special type of group testing called nonadaptive testing. In nonadaptive group testing, a predetermined set of group tests is performed. This has several potential advantages, three of which are as follows.

- There is less probability of error in the testing procedure (i.e., testing the wrong samples) since exactly the same tests are done each time the group testing algorithm is carried out.
- There is potentially less overhead, due to the fact that the tests are known ahead of time and can be organized in a convenient manner.
- The tests can be performed in parallel to any desired degree. This is extremely important if it takes a long time to set up and/or carry out an individual test.

A *nonadaptive group testing algorithm* can be modeled or defined as a design in a straightforward way. Let X be a set of m elements called *samples*, and let $\mathcal{A}$ be a set of n subsets of X called *tests*. We will refer to the pair $(X, \mathcal{A})$ as an (m, n)-NAGTA. In general, the tests can be of different sizes if desired, and we are not assuming any kind of balance property. At this point, all we have is a set X and a set $\mathcal{A}$ of subsets of X.

Suppose that we define $\mathcal{A} = \{\{x\} : x \in X\}$. Then $(X, \mathcal{A})$ is a (trivial) (m, m)-NAGTA. Since we want to minimize n (the number of tests), we are interested primarily in (m, n)-NAGTAs with $n < m$.

The objective of a group testing algorithm will be to identify the subset $U \subseteq X$ of positive samples, which we call the *positive subset*. This will be done by using a *test function* $f : 2^X \to \{0, 1\}$, which works as follows:

$$f(Y) = \begin{cases} 1 & \text{if } Y \cap U \neq \emptyset \\ 0 & \text{if } Y \cap U = \emptyset \end{cases}$$

for any $Y \subseteq X$ (where 2^X denotes the set of all subsets of X). Of course the test function f depends on U.

The *result vector* of the (m, n)-NAGTA $(X, \mathcal{A})$, given the positive subset U, will be the binary n-tuple $R(U) = (f(A) : A \in \mathcal{A})$. In other words, we apply the test function to every test $A \in \mathcal{A}$. We will say that $(X, \mathcal{A})$ *identifies* the positive subset U if U is determined uniquely as a function of $R(U)$. Equivalently, this can be stated as the requirement that $R(U) \neq R(V)$ if $U \neq V$.

Often we may begin with an a priori guarantee or assumption that $|U| \leq s$, where $s \leq m$ is a specified integer. We will say that $(X, \mathcal{A})$ is (m, n)-NAGTA with *threshold* s if $R(U) \neq R(V)$ whenever $U, V \subseteq X$, $|U| \leq s$, $|V| \leq s$, and $U \neq V$.

Example 11.9. Suppose that $X = \{1, 2, 3, 4, 5, 6\}$ and

$$\mathcal{A} = \{\{1, 2, 3\}, \{1, 4, 5\}, \{2, 4, 6\}, \{3, 5, 6\}\}.$$

We tabulate the results of the (6,4)-NAGTA $(X, \mathcal{A})$ for all possible positive subsets U with $|U| \leq 2$, as follows:

U	$R(U)$	U	$R(U)$
$\emptyset$	0000	$\{1,6\}$	1111
$\{1\}$	1100	$\{2,3\}$	1011
$\{2\}$	1010	$\{2,4\}$	1110
$\{3\}$	1001	$\{2,5\}$	1111
$\{4\}$	0110	$\{2,6\}$	1011
$\{5\}$	0101	$\{3,4\}$	1111
$\{6\}$	0011	$\{3,5\}$	1101
$\{1,2\}$	1110	$\{3,6\}$	1011
$\{1,3\}$	1101	$\{4,5\}$	0111
$\{1,4\}$	1110	$\{4,6\}$	0111
$\{1,5\}$	1101	$\{5,6\}$	0111

From the tabulation above, we see that $(X, \mathcal{A})$ has (maximum) threshold $s = 1$. The fact that $s \geq 1$ follows because the seven vectors $R(U)$, where $|U| \leq 1$, are distinct. However, for sets of cardinality two, the result vectors are not always different (for example, $R(\{1,2\}) = R(\{1,4\})$). Thus $s = 1$. ∎

11.3.1 A Construction from BIBDs

Suppose that $(Y, \mathcal{B})$ is a $(v, b, r, k, 1)$-BIBD, and let $(X, \mathcal{A})$ be the dual incidence structure, as defined in Section 1.3. (In other words, $(X, \mathcal{A})$ is the design whose incidence matrix is the transpose of the incidence matrix of $(Y, \mathcal{B})$.)

We will use $(X, \mathcal{A})$ as a (b, v)-NAGTA. Recall from Theorem 1.17 that $(X, \mathcal{A})$ satisfies the following properties:

1. each sample occurs in exactly k tests,
2. each test contains exactly r samples,
3. every pair of distinct samples is contained in at most one test.

We will show that $(X, \mathcal{A})$ has threshold $k-1$. To accomplish this, we will describe a simple algorithm to identify the positive subset U, given the result vector $R(U)$ and assuming that $|U| \leq k-1$. The algorithm depends on the fundamental observation we made earlier that $U \cap Y = \emptyset$ if $f(Y) = 0$. From this observation, it follows immediately that

$$U \subseteq X \setminus \bigcup_{\{A \in \mathcal{A} : f(A) = 0\}} A$$

for any nonadaptive group testing algorithm and for any subset $U \subseteq X$.

For a NAGTA that is the dual of a BIBD with $\lambda = 1$, we will show that

$$U = X \setminus \bigcup_{\{A \in \mathcal{A} : f(A) = 0\}} A$$

if $|U| \le k-1$. Otherwise, there exists an $x \notin U$ such that

$$x \notin \bigcup_{\{A \in \mathcal{A}: f(A)=0\}} A.$$

This is equivalent to saying that $x \notin U$ and $f(A) = 1$ for every $A \in \mathcal{A}$ such that $x \in A$.

Now, the sample x occurs in k tests, each of which must contain a sample in U. Property 3 ensures that no sample in U occurs in more than one test with x, so it must be the case that $|U| \ge k$. This contradicts the assumption $|U| \le k-1$, and thus we have proved the following.

Theorem 11.10. *If there exists a* $(v, b, r, k, 1)$-BIBD, *then there exists a* (b, v)-NAGTA *with threshold* $k-1$.

Theorem 11.10 says that the positive set U can be identified if it has cardinality at most $k-1$. What happens if $|U| \ge k$? Since

$$U \subseteq X \setminus \bigcup_{\{A \in \mathcal{A}: f(A)=0\}} A,$$

it follows that

$$k \le |U| \le \left| X \setminus \bigcup_{\{A \in \mathcal{A}: f(A)=0\}} A \right|$$

in this case. Hence, even though we may not be able to identify U when $|U| \ge k$, we can always recognize when $|U| \ge k$.

Suppose that $(X, \mathcal{A})$ is the (b, v)-NAGTA of Theorem 11.10, where $X = \{1, \ldots, b\}$ and $\mathcal{A} = \{A_j : 1 \le j \le v\}$. Given the result vector $R(U) = (f(A_1), \ldots, f(A_v))$, the algorithm IDENTIFY will identify U if $|U| \le k-1$ and report that $|U| \ge k$ otherwise.

Algorithm: IDENTIFY($R(U)$)

```
U ← ∅
for i ← 1 to b
  do M[i] ← 1
for j ← 1 to v
       ⎧ if f(A_j) = 0
  do   ⎨  then for each x ∈ A_j
       ⎩   do M[x] ← 0
for i ← 1 to b
  do ⎧ if M[i] = 1
     ⎩  then U ← U ∪ {i}
if |U| ≤ k − 1
  then return (U)
  else return ("the positive subset has size at least k")
```

We present an example to illustrate this.

Example 11.11. A $(9,3,1)$-BIBD is presented in Example 1.4. The blocks of the dual incidence structure are as follows:

$$\begin{array}{l} A_1 = \{1,4,7,10\},\ A_2 = \{1,5,8,11\},\ A_3 = \{1,6,9,12\}, \\ A_4 = \{2,4,9,11\},\ A_5 = \{2,5,7,12\},\ A_6 = \{2,6,8,10\}, \\ A_7 = \{3,4,8,12\},\ A_8 = \{3,5,9,10\},\ A_9 = \{3,6,7,11\}. \end{array}$$

Suppose we obtain the following result vector:

$$R(U) = (0,1,0,0,1,0,1,1,1).$$

When we execute the algorithm IDENTIFY with input $R(U)$, we compute the following:

j	M											
	1	1	1	1	1	1	1	1	1	1	1	1
1	$\boxed{0}$	1	1	$\boxed{0}$	1	1	$\boxed{0}$	1	1	$\boxed{0}$	1	1
3	0	1	1	0	1	$\boxed{0}$	0	1	$\boxed{0}$	0	1	$\boxed{0}$
4	0	$\boxed{0}$	1	0	1	0	0	1	0	0	$\boxed{0}$	0
6	0	0	1	0	1	0	0	$\boxed{0}$	0	0	0	0

(Note that boxed entries are used to indicate when a "1" is changed to a "0".) The positive set U is thus $U = \{3,5\}$. ∎

When we use a $(v,k,1)$-BIBD to construct an (m,n)-NAGTA, we get $m = (n^2 - n)/(k^2 - k)$. For fixed k, we have that n is $O(k\sqrt{m})$.

11.4 Two-Point Sampling

11.4.1 Monte Carlo Algorithms

There are many problems for which no fast deterministic algorithm is known but that can be solved efficiently using randomized algorithms. One such problem is primality testing, where we are given an integer $n \geq 2$ and are required to answer the question "is n composite?". Primality testing is often done by means of a Monte Carlo algorithm. In general, Monte Carlo algorithms are used for decision problems, in which the objective is to correctly answer a yes-no question.

Definition 11.12. *A* yes-biased Monte Carlo algorithm, A, *is an algorithm for a decision problem that satisfies the following properties:*

1. A *is a* randomized algorithm *(i.e., it makes random choices during its execution);*
2. *for any problem instance I,* A *always gives an answer "yes" or "no";*

3. *if the instance I is a no-instance, then* A *answers "no";*
4. *if the instance I is a yes-instance, then the probability that* A *answers "yes" is at least $1 - \epsilon$, where $\epsilon \geq 0$ is some fixed constant (independent of I).*

The value ϵ is called the error probability *of the algorithm* A.

Observe that, if A answers "yes", then we know that the answer is correct. However, if A answers "no", then there is the possibility that the answer may be incorrect.

A yes-biased Monte Carlo algorithm, A, can be viewed as a two-stage procedure. In the first stage, a *sample point* x is chosen at random from a specified finite universe $U = U(I)$, where, in general, U depends on the instance I. In the second stage, a deterministic algorithm is applied to the given sample point x and instance I. The deterministic algorithm computes a yes-no valued function $f(I, x)$, which is taken to be the output of A. In order that A has error probability ϵ, the function f should satisfy the following properties for all problem instances I:

1. if I is a no-instance, then $f(I, x) = 0$ for all $x \in U(I)$;
2. if I is a yes-instance, then

$$|\{x \in U(I) : f(I, x) = 1\}| \geq (1 - \epsilon)|U(I)|.$$

Example 11.13. Primality testing is a decision problem for which Monte Carlo algorithms are often used in practice. The question to be answered is "is n composite?". This means that the instance I is just the integer n.

The well-known Miller-Rabin algorithm is a yes-biased Monte Carlo algorithm for primality testing in which $U(I) = \{0, \ldots, n - 1\}$. It has been proven that the resulting error probability of this algorithm, ϵ, is at most $1/4$. ∎

The main reason that Monte Carlo algorithms are so useful is that the error probability can be made as small as desired by repeated application of the algorithm. Assume that A is a yes-biased Monte Carlo algorithm with error probability ϵ. Suppose we are given an instance I, and we run A on I k times using k independent random sample points $x \in U(I)$ in the k trials of the algorithm. If we get at least one "yes" answer, then the instance I must be a yes-instance. On the other hand, if I is a yes-instance, then the probability of getting k "no" answers in k trials is at most ϵ^k, which approaches 0 exponentially quickly as a function of k.

This analysis is based on the assumption that the sample points used in the successive trials are chosen independently at random from $U(I)$. When a Monte Carlo algorithm is implemented in actual practice, however, one always uses a pseudo-random number generator, which is a deterministic algorithm that produces a sequence of sample points from $U(I)$ given a truly random starting point called a "seed". This means that the analysis given above does not apply. In general, analysis of the error probability will depend on the particular pseudo-random number generator that is used.

11.4.2 Orthogonal Arrays and Two-Point Sampling

Orthogonal arrays provide a convenient method of obtaining a sequence of pseudo-random sample points. Suppose that I is an instance, and let $U = U(I)$ be the universe of sample points for the instance I as before. Suppose that A is an orthogonal array $\mathsf{OA}(k,n)$ on symbol set U, where $|U| = n$. Recall that there are n^2 rows in A.

The method of *two-point sampling* proceeds as follows.

1. Let r be a random row in A.
2. Use the k values $A(r,1),\ldots,A(r,k)$ as the k sample points (note that these k sample points are not necessarily all distinct).

If the rows of A are indexed by $U \times U$, then a random row of A is specified by choosing two points independently at random from U. (The two points are not required to be distinct.) This is the reason for the term "two-point sampling".

We now present an elementary combinatorial analysis of the two-point sampling technique that allows us to calculate a bound on the resulting error probability. Suppose that I is a yes-instance, and define

$$S = \{x \in U : f(I,x) = 1\}.$$

We call S the set of *witnesses* (note that we do not know the set S explicitly). We have $|S| = m$, where $m = (1-\epsilon)n$.

Let a_i denote the number of rows of A in which there are exactly i occurrences of elements from S. Call a row of the matrix a *bad row* if none of the elements in the row is a witness. Then the error probability is simply the probability that a randomly selected row of the orthogonal array is a bad row. Hence, the error probability, when we run the algorithm A using k sample points chosen from a random row of A, is seen to be

$$\mathsf{err}(S) = \frac{a_0}{n^2}. \tag{11.1}$$

As mentioned above, we do not know the set S explicitly, but we have an upper bound on $|S|$. An upper bound on the error probability of two-point sampling can be obtained by computing

$$\mathsf{err} = \max\{\mathsf{err}(S) : S \subseteq U, |S| = m\}.$$

We first derive three simple equations using elementary properties of orthogonal arrays. Since an $\mathsf{OA}(k,n)$ has n^2 rows, we have

$$\sum_{i=0}^{n} a_i = n^2. \tag{11.2}$$

Next, we count the number of occurrences of witnesses in A in two ways. There are exactly a_i rows in which there are i occurrences of witnesses. In any column of A, each point occurs exactly n times so the number of occurrences of witnesses in a given column is nm. Since there are k columns in A, the total number of occurrences of witnesses in A is knm. This yields the following equation:

$$\sum_{i=0}^{n} ia_i = knm. \tag{11.3}$$

Similarly, we can count the number of occurrences of pairs of witnesses occurring in the same row in two ways. In any row in which there are i occurrences of witnesses, there will be $i(i-1)$ occurrences of pairs of witnesses. On the other hand, if we look at any two columns of A, the number of occurrences of pairs of witnesses in the same row is m^2. (This is because any particular pair of witnesses occurs exactly once in any given pair of columns.) Two columns can be selected in $k(k-1)$ ways, and so the total number of occurrences is $k(k-1)m^2$. This yields the following equation:

$$\sum_{i=0}^{n} i(i-1)a_i = k(k-1)m^2. \tag{11.4}$$

Let z be any real number. Then we have

$$\begin{aligned} 0 &\leq \sum_{i=1}^{n} (i-z)^2 a_i \\ &= \sum_{i=1}^{n} (i^2 - 2zi + z^2) a_i \\ &= \sum_{i=1}^{n} i^2 a_i - 2z \sum_{i=1}^{n} ia_i + z^2 \sum_{i=1}^{n} a_i \\ &= k(k-1)m^2 + knm - 2zknm + z^2 \sum_{i=1}^{n} a_i \end{aligned}$$

from equations (11.2), (11.3), and (11.4). It follows that

$$\sum_{i=1}^{n} a_i \geq \frac{2knmz - knm - k(k-1)m^2}{z^2}. \tag{11.5}$$

Elementary calculus shows that the right-hand side of (11.5) is maximized when we choose

$$z = \frac{n + (k-1)m}{n}.$$

Hence, we get

$$\sum_{i=1}^{n} a_i \geq \frac{kmn^2}{n + (k-1)m}. \tag{11.6}$$

Now, from (11.2) and (11.6), we have

$$a_0 \le n^2 - \frac{kmn^2}{n+(k-1)m}.$$

Finally, we get the following bound on the error probability from (11.1):

$$\mathsf{err}(S) \le 1 - \frac{km}{n+(k-1)m}.$$

Since $m = n(1-\epsilon)$, we have that

$$\mathsf{err}(S) \le \frac{\epsilon}{1+(k-1)(1-\epsilon)}. \tag{11.7}$$

Because (11.7) is true for any set $S \subseteq U$ of cardinality m, we have the following theorem.

Theorem 11.14. *If* err *denotes the error probability of the two-point sampling technique for a universe U of size n, using as sample points the k elements in a random row of an orthogonal array* $\mathsf{OA}(k,n)$, *then*

$$\mathsf{err} \le \frac{\epsilon}{1+(k-1)(1-\epsilon)}. \tag{11.8}$$

Note that this bound on the error probability approaches 0 only linearly quickly as a function of k.

We give a small, toy example, which actually meets the bound proved in Theorem 11.14.

Example 11.15. The following is an $\mathsf{OA}(3,4)$.

0	0	2
0	1	3
0	2	0
0	3	1
1	0	3
1	1	2
1	2	1
1	3	0
2	0	0
2	1	1
2	2	2
2	3	3
3	0	1
3	1	0
3	2	3
3	3	2

If the set of witnesses for the universe $U = \{0,1,2,3\}$ is $S = \{0,1\}$, then $n = 4$ and $\epsilon = 1/2$. It is easy to check that $a_0 = 4$ and $a_2 = 12$ for the OA(3,4) presented above, and hence

$$\mathrm{err}(S) = \frac{4}{16} = \frac{1}{4}.$$

On the other hand, since $k = 3$, we have

$$\frac{\epsilon}{1 + (k-1)(1-\epsilon)} = \frac{1}{4},$$

so the bound (11.8) is met with equality. ∎

11.5 Notes and References

Some interesting surveys on the applications of combinatorial designs to computer science include Colbourn and van Oorschot [33], Stinson [105], Gopalakrishnan and Stinson [49], and Colbourn, Dinitz, and Stinson [30].

Authentication codes were invented by Gilbert, MacWilliams, and Sloane [48]. They have been extensively studied in cryptography; Simmons [95] is a good survey. Combinatorial aspects of authentication codes are considered in various papers, such as Stinson [103].

The idea of threshold schemes is due to Blakley [12] and Shamir [93]. Connections between orthogonal arrays and threshold schemes are discussed in Dawson, Mahmoodian, and Rahilly [37]. Theorem 11.7 is from Stinson and Vanstone [106], where the idea of anonymous schemes is introduced.

Much information about group testing can be found in the book "Combinatorial Group Testing and Its Applications" by Du and Hwang [43]. Theorem 11.10 can be derived as a consequence of [43, Corollary 7.4.4].

Two-point sampling was invented by Chor and Goldreich [21]. Section 11.4 is based on Gopalakrishnan and Stinson [51]. For a brief discussion of the applications of combinatorial designs to derandomization, see Gopalakrishnan and Stinson [50].

11.6 Exercises

11.1 Suppose that $(\mathcal{S}, \mathcal{A}, \mathcal{K}, \mathcal{E})$ is an authentication code in which $\mathcal{S} = \{0,\ldots,8\}$, $\mathcal{A} = \mathbb{Z}_{11}$, and $\mathcal{K} = \mathbb{Z}_{11} \times \mathbb{Z}_{11}$. For $K = (i,j)$, where $i,j \in \mathbb{Z}_{11}$, the authentication rule $e_{(i,j)}$ is defined as

$$e_{(i,j)}(s) = i + sj \bmod 11$$

for $0 \le s \le 8$.

(a) Suppose that Oscar observes the message $(5,4)$ in the channel. Determine the set of possible keys being used by Alice and Bob.

(b) Suppose Oscar substitutes the message $(4,1)$. If this message is accepted by Bob, what must the key be?

11.2 We investigate a slightly modified model for authentication in this question. Suppose we have a four-tuple $(\mathcal{S}, \mathcal{M}, \mathcal{K}, \mathcal{E})$, where the following conditions are satisfied.

1. $\mathcal{S}$ is a finite set of *source states*.
2. $\mathcal{M}$ is a finite set of *messages*.
3. $\mathcal{K}$ is a finite set of *keys*.
4. For each $K \in \mathcal{K}$, there is an *encoding rule* $e_K \in \mathcal{E}$, where $e_K : \mathcal{S} \rightarrow \mathcal{A}$ is an injective function.

Bob will accept a message $m \in \mathcal{M}$ as authentic if there exists $s \in \mathcal{S}$ such that $e_K(s) = m$ (note that there exists at most one such s (given m) because the encoding rules are injective).

Let $(X, \mathcal{A})$ be a (v, b, r, k, λ)-BIBD. Define $\mathcal{S} = \{1, \dots, k\}$, $\mathcal{M} = X$, and $\mathcal{K} = \mathcal{A}$. For every block $A \in \mathcal{A}$, define an encoding rule e_A so that

$$\{e_A(s) : s \in \mathcal{S}\} = A.$$

(There are $k!$ possible ways to define each encoding rule e_A so that this condition is satisfied.)

Prove that this authentication code has $P_0 = k/v$ and $P_1 = (k-1)/(v-1)$.

11.3 A 3-$(17,6,1)$-OA can be used to construct a perfect $(3,5)$-threshold scheme. The entries in the orthogonal array are defined by the formula

$$A((i_0, i_1, i_2), c) = i_0 + i_1 c + i_2 c^2 \bmod 17,$$

where $(i_0, i_1, i_2) \in (\mathbb{Z}_{17})^3$ and $1 \leq c \leq 5$. The secret is

$$K = A((i_0, i_1, i_2), 0) = i_0,$$

and the shares for $P_1, \dots, P_5$ are $A((i_0, i_1, i_2), 1), \dots, A((i_0, i_1, i_2), 5)$, respectively.

Suppose that the shares given to P_1, P_3, and P_5 are 8, 10, and 11, respectively. Determine the secret.

11.4 Generalizing Exercise 11.3, we can use orthogonal arrays based on Corollary 10.7 to construct threshold schemes. The resulting threshold schemes are known as *Shamir threshold schemes*. Here is how a Shamir (t, w)-threshold scheme is constructed over $\mathbb{Z}_p$, where p is a prime.

1. D chooses w distinct, nonzero elements of $\mathbb{Z}_p$, denoted x_i, $1 \leq i \leq w$ (this is where we require $p \geq w+1$). For $1 \leq i \leq w$, D gives the value x_i to P_i. The values x_i are public.
2. Suppose D wants to share a key $K \in \mathbb{Z}_p$. D secretly chooses (independently at random) $t-1$ elements of $\mathbb{Z}_p$, which are denoted $a_1, \dots, a_{t-1}$.
3. For $1 \leq i \leq w$, D computes $y_i = a(x_i)$, where

$$a(x) = K + \sum_{j=1}^{t-1} a_j x^j \bmod p.$$

4. For $1 \leq i \leq w$, D gives the share y_i to P_i.

In summary, the dealer constructs a random polynomial $a(x)$ of degree at most $t-1$ in which the constant term is the key, K. Every participant P_i obtains a point (x_i, y_i) on this polynomial.

Suppose that participants $P_{i_1}, \ldots, P_{i_t}$ want to determine K. They know that $y_{i_j} = a(x_{i_j})$, $1 \leq j \leq t$, where $a(x) \in \mathbb{Z}_p[x]$ is the (secret) polynomial chosen by D. K can be determined by first computing

$$f(x) = \sum_{j=1}^{t} y_{i_j} \prod_{1 \leq k \leq t, k \neq j} \frac{x - x_{i_k}}{x_{i_j} - x_{i_k}} \tag{11.9}$$

and then setting $K = f(0)$.

Remark: Equation (11.9) is known as the *Lagrange Interpolation Formula*.

(a) Prove that $f(x_{i_j}) = y_{i_j}$ for $1 \leq j \leq t$.

(b) Prove that the polynomial $f(x) = a(x)$.
Hint: The polynomial $f(x) - a(x)$ has at least t roots.

(c) Prove that $K = f(0)$.

11.5 (a) Suppose that the following are the nine shares in a Shamir $(6,9)$-threshold scheme (as described in Exercise 11.3) implemented in $\mathbb{Z}_{1993}$:

i	x_i	y_i
1	1	187
2	2	1547
3	3	498
4	4	1407
5	5	1564
6	6	1176
7	7	795
8	8	185
9	9	603

Exactly one of these shares is defective (i.e., incorrect). Your task is to determine which share is defective and then figure out its correct value as well as the value of the secret. The "primitive operations" in your algorithm are polynomial interpolations (using (11.9)) and polynomial evaluations. Try to minimize the number of polynomial interpolations you perform.
Hint: The question can be solved using at most three polynomial interpolations.

(b) Suppose that a Shamir (t,w)-threshold scheme has exactly one defective share, and suppose that $w \geq 2t$. Describe how it is possible to determine which share is defective using only two polynomial interpolations.

(c) More generally, suppose that a Shamir (t,w)-threshold scheme has exactly τ defective shares, and suppose that $t \geq (\tau+1)w$.

Describe how it is possible to determine which shares are defective using only $\tau + 1$ polynomial interpolations.

11.6 Suppose an affine plane of order 7 is used to set up an anonymous (2,7) threshold scheme with $|\mathcal{K}| = 8$ and $|\mathcal{S}| = 49$. The affine plane $(X, \mathcal{A})$ can be constructed in the usual way as follows. $X = \mathbb{Z}_7 \times \mathbb{Z}_7$. For any $a, b \in \mathbb{Z}_7$, define a block

$$A_{a,b} = \{(x,y) \in X : y \equiv ax + b \bmod 7\}.$$

For any $c \in \mathbb{Z}_7$, define

$$A_{\infty,c} = \{(c,y) : c \in \mathbb{Z}_7\}.$$

Then, define

$$\mathcal{A} = \{A_{a,b} : a, b \in \mathbb{Z}_7\} \cup \{A_{\infty,c} : c \in \mathbb{Z}_7\}.$$

(a) Suppose that the secret is $K = 5$. Compute the shares to be distributed to the seven participants if the block $A_{5,3}$ is chosen by the dealer.

(b) Compute the secret if two of the shares are $(3,5)$ and $(6,2)$.

11.7 Suppose that $(X, \mathcal{A})$ is a design with m points and n blocks, and let $M = (m_{i,j})$ be its incidence matrix. Let the rows of M be labeled by the elements in the set R, and let the columns of M be labeled by the elements in the set C. M is said to be *s-disjunct* provided that for every row $r \in R$ and for all sets of rows $\{r_1, \ldots, r_s\} \subseteq R \backslash \{r\}$, there exists a column $c \in C$ such that $m_{r,c} = 1$ and $m_{r_1,c} = \cdots = m_{r_s,c} = 0$.

(a) Prove that $(X, \mathcal{A})$ is an (m,n)-NAGTA with threshold s if the incidence matrix M is s-disjunct.

(b) Suppose that a (binary) $(n, m, d, 2)$-code has constant weight w. Prove that the $m \times n$ matrix whose rows are the m codewords is s-disjunct provided that $s(w - d/2) < w$.

11.8 The blocks of the dual incidence structure of a $(9,3,1)$-BIBD are as follows:

$$\begin{aligned} A_1 &= \{1,4,7,10\},\ A_2 = \{1,5,8,11\},\ A_3 = \{1,6,9,12\},\\ A_4 &= \{2,4,9,11\},\ A_5 = \{2,5,7,12\},\ A_6 = \{2,6,8,10\},\\ A_7 &= \{3,4,8,12\},\ A_8 = \{3,5,9,10\},\ A_9 = \{3,6,7,11\}. \end{aligned}$$

Suppose that these blocks are used as tests in a nonadaptive group testing algorithm and the result vector is

$$R(U) = (1,0,1,1,0,0,1,1,0).$$

Identify the positive set U, if possible. Show all your work.

11.9 Prove that equality occurs in Theorem 11.14 if and only if there exists an OA(k,n) and a subset S of $m = (1-\epsilon)n$ symbols such that every row of this orthogonal array either contains 0 or z symbols from S, where $z = 1 + (k-1)(1-\epsilon)$.

11.10 Let $p \geq 3$ be a prime, and suppose we construct an OA$(3,p)$, say A, by the method described in Theorem 6.39. To be specific, let a_1, a_2, a_3 be three distinct elements of $\mathbb{Z}_p$. Then define the entry in row (i,j) and column c to be $A((i,j),c) = i + ja_c \bmod p$ for all $i,j \in \mathbb{Z}_p$, $c = 1,2,3$. For such an orthogonal array, and for $m = 1,2,3$, determine the exact value of

$$\max\{\mathrm{err}(S) : S \subseteq \mathbb{Z}_p, |S| = m\}.$$

A

Small Symmetric BIBDs and Abelian Difference Sets

We provide a summary of known existence and nonexistence results for "small" symmetric BIBDs and Abelian difference sets. In Table A.1, we list all parameter triples (v, k, λ) in which $\lambda(v-1) = k(k-1)$, $v/2 \geq k \geq 3$, and $3 \leq k \leq 15$ (if $k > v/2$, then apply block complementation, which was presented as Theorem 1.32, and/or Exercise 3.1).

We use the following abbreviations in Table A.1.

- "Singer" denotes a Singer difference set (Theorem 3.28).
- "QR" denotes a quadratic residue difference set (Theorem 3.21).
- "H" denotes a $(4m-1, 2m-1, m-1)$-BIBD constructed from a Hadamard matrix of order $4m$ via Theorem 4.5.
- "$PG_d(q)$" denotes a projective geometry (Theorem 2.14).
- "BRC" denotes the Bruck-Ryser-Chowla Theorems (Theorems 2.16 and 2.19).
- "MT" denotes the Multiplier Theorem (Theorem 3.33).

For existence of certain symmetric BIBDs and for the nonexistence of certain difference sets, we refer to external sources. Note also that existence of a difference set implies the existence of the corresponding symmetric BIBD, and nonexistence of a symmetric BIBD implies nonexistence of a difference set with the same parameters in any (Abelian or non-Abelian) group.

k	v	λ	SBIBD	notes	difference set	notes
3	7	1	yes	$PG_2(2)$	yes	Singer
4	13	1	yes	$PG_2(3)$	yes	Singer
5	21	1	yes	$PG_2(4)$	yes	Singer
5	11	2	yes	H	yes	QR
6	31	1	yes	$PG_2(5)$	yes	Singer
6	16	2	yes		yes	Example 3.4
7	43	1	no	BRC	no	
7	22	2	no	BRC	no	
7	15	3	yes	$PG_3(2)$, H	yes	Singer
8	57	1	yes	$PG_2(7)$	yes	Singer
8	29	2	no	BRC	no	
9	73	1	yes	$PG_2(8)$	yes	Singer
9	37	2	yes		yes	Example 3.24
9	25	3	yes	[113, Table 5.25]	no	[10, Table A.3.1]
9	19	4	yes	H	yes	QR
10	91	1	yes	$PG_2(9)$	yes	Singer
10	46	2	no	BRC	no	
10	31	3	yes		no	MT, $p = 7$
11	111	1	no	[74]	no	MT, $p = 2, 5$
11	56	2	yes	[113, Table 5.25]	no	[10, Table A.3.1]
11	23	5	yes	H	yes	QR
12	133	1	yes	$PG_2(11)$	yes	Singer
12	67	2	no	BRC	no	
12	45	3	yes		yes	Example 3.5
12	34	4	no	BRC	no	
13	157	1	unknown		no	Example 3.38
13	79	2	yes	[113, Table 5.25]	no	MT, $p = 11$
13	53	3	no	BRC	no	
13	40	4	yes	$PG_3(3)$	yes	Singer
13	27	6	yes	H	yes	QR
14	183	1	yes	$PG_2(13)$	yes	Singer
14	92	2	no	BRC	no	
15	211	1	no	BRC	no	
15	106	2	no	BRC	no	
15	71	3	yes	[113, Table 5.25]	no	[10, Table A.3.1]
15	43	5	no	BRC	no	
15	36	6	yes		yes	Example 3.6
15	31	7	yes	$PG_5(2)$, H	yes	Singer, QR

Table A.1. Small Symmetric BIBDs and Abelian Difference Sets

B

Finite Fields

In this appendix, we give a brief summary of basic facts concerning finite fields. We provide definitions of the main concepts, several illustrative examples, and statements of some important theorems, but no proofs. A reader wanting to study finite fields in more detail can consult a suitable algebra textbook.

Definition B.1. *A* finite field *is a triple* $(X, \times, +)$ *such that* X *is a finite set with* $|X| \geq 2$ *and "*$\times$*" and "*$+$*" are binary operations on* X *such that the following conditions are satisfied:*

1. *addition is* closed; *i.e., for any* $a, b \in X$, $a + b \in X$;
2. *addition is* commutative; *i.e., for any* $a, b \in X$, $a + b = b + a$;
3. *addition is* associative; *i.e., for any* $a, b, c \in X$, $(a + b) + c = a + (b + c)$;
4. 0 *is an* additive identity; *i.e., for any* $a \in X$, $a + 0 = 0 + a = a$;
5. *for any* $a \in X$, *there exists an* additive inverse *of* a, *denoted* $-a$, *such that* $a + (-a) = (-a) + a = 0$;
6. *multiplication is* closed; *i.e., for any* $a, b \in X$, $a \times b \in X$;
7. *multiplication is* commutative; *i.e., for any* $a, b \in X$, $a \times b = b \times a$;
8. *multiplication is* associative; *i.e., for any* $a, b, c \in X$, $(a \times b) \times c = a \times (b \times c)$;
9. 1 *is a* multiplicative identity; *i.e., for any* $a \in X$, $a \times 1 = 1 \times a = a$;
10. *for any* $a \in X\backslash\{0\}$, *there exists a* multiplicative inverse *of* a, *denoted* a^{-1}, *such that* $a \times a^{-1} = a^{-1} \times a = 1$;
11. *the* distributive property *is satisfied; i.e., for any* $a, b, c \in X$, $(a + b) \times c = (a \times c) + (b \times c)$, *and* $a \times (b + c) = (a \times b) + (a \times c)$.

The order *of the finite field* $(X, \times, +)$ *is the integer* $|X|$.

Suppose that $(X, \times, +)$ is a finite field. Properties 1–5 establish that $(X, +)$ is an Abelian group, and properties 6–10 show that $(X\backslash\{0\}, \times)$ is an Abelian group.

Here are some familiar examples of fields.

Example B.2. $(\mathbb{R}, \times, +)$ and $(\mathbb{Q}, \times, +)$ are both (infinite) fields. ∎

Example B.3. If p is prime, then every nonzero element of $\mathbb{Z}_p$ has a multiplicative inverse, and $(\mathbb{Z}_p, \times, +)$ is a finite field of order p. ∎

A *finite ring* is a triple $(X, \times, +)$ that satisfies every property of a finite field except for property 10.

Example B.4. If $m \geq 2$ is an integer, then $(\mathbb{Z}_m, \times, +)$ is a finite ring. If m is composite, then it is easy to see that $(\mathbb{Z}_m, \times, +)$ is not a field as follows. Suppose that d is a divisor of m, where $1 < d < n$. Then d does not have a multiplicative inverse modulo m, so property 10 is violated. ∎

There exist finite fields that are not of prime order. In fact, there is a finite field with q elements whenever $q = p^n$, p is prime, and $n \geq 1$ is an integer. We will now describe very briefly how to construct such a field when $n > 1$. First, we need several definitions.

Definition B.5. *Suppose p is prime. Define $\mathbb{Z}_p[x]$ to be the set of all polynomials in the indeterminate x in which the coefficients are elements of $\mathbb{Z}_p$. ($(\mathbb{Z}_p[x], \times, +)$ is a ring, where multiplication and addition of polynomials are defined in the usual way except that all coefficients are reduced modulo p.)*

1. *For $f(x), g(x) \in \mathbb{Z}_p[x]$, we say that $f(x)$* divides *$g(x)$ (notation: $f(x) \mid g(x)$) if there exists $q(x) \in \mathbb{Z}_p[x]$ such that*
$$g(x) = q(x)f(x).$$
2. *For $f(x) \in \mathbb{Z}_p[x]$, define* $\deg(f)$, *the* degree *of f, to be the highest exponent in a term of f.*
3. *Suppose $f(x), g(x), h(x) \in \mathbb{Z}_p[x]$, and $\deg(f) = n \geq 1$. We define*
$$g(x) \equiv h(x) \pmod{f(x)}$$
if
$$f(x) \mid (g(x) - h(x)).$$

Notice the resemblance of the definition of congruence of polynomials to that of congruence of integers.

We are now going to define a finite ring of polynomials "modulo $f(x)$", which we denote by $\mathbb{Z}_p[x]/(f(x))$. The construction of $\mathbb{Z}_p[x]/(f(x))$ from $\mathbb{Z}_p[x]$ is based on the idea of congruences modulo $f(x)$ and is analogous to the construction of $\mathbb{Z}_m$ from $\mathbb{Z}$.

Suppose $\deg(f) = n$. If we divide $g(x)$ by $f(x)$, we obtain a (unique) *quotient* $q(x)$ and *remainder* $r(x)$, where
$$g(x) = q(x)f(x) + r(x)$$
and
$$\deg(r) < n.$$

This can be done by the usual long division of polynomials. It follows that any polynomial in $\mathbb{Z}_p[x]$ is congruent modulo $f(x)$ to a unique polynomial of degree at most $n-1$.

Now we define the elements of $\mathbb{Z}_p[x]/(f(x))$ to be the p^n polynomials in $\mathbb{Z}_p[x]$ of degree at most $n-1$. Addition and multiplication in $\mathbb{Z}_p[x]/(f(x))$ are defined as in $\mathbb{Z}_p[x]$, followed by a reduction modulo $f(x)$. Equipped with these operations, $\mathbb{Z}_p[x]/(f(x))$ is a finite ring.

Recall that $\mathbb{Z}_m$ is a field if and only if m is prime. A similar situation holds for $\mathbb{Z}_p[x]/(f(x))$. The analog of primality for polynomials is irreducibility, which we define as follows.

Definition B.6. *A polynomial $f(x) \in \mathbb{Z}_p[x]$ is said to be an* irreducible polynomial *if there do not exist polynomials $f_1(x), f_2(x) \in \mathbb{Z}_p[x]$ such that*

$$f(x) = f_1(x)f_2(x),$$

where $\deg(f_1) > 0$ *and* $\deg(f_2) > 0$.

Irreducible polynomials of all possible orders exist. More precisely, we have the following theorem.

Theorem B.7. *For any prime p and for any integer $n \geq 1$, there exists an irreducible polynomial $f(x) \in \mathbb{Z}_p[x]$ having degree n.*

The relevance of irreducible polynomials to the construction of finite fields is as follows.

Theorem B.8. *Suppose p is prime and $f(x) \in \mathbb{Z}_p[x]$. Then $\mathbb{Z}_p[x]/(f(x))$ is a (finite) field if and only if $f(x)$ is irreducible.*

Here is an example to illustrate the concepts described above.

Example B.9. Let's construct a finite field having eight elements. This can be done by finding an irreducible polynomial of degree three in $\mathbb{Z}_2[x]$. It is sufficient to consider the polynomials having constant term equal to 1 since any polynomial with constant term 0 is divisible by x and hence is reducible. There are four such polynomials:

$$\begin{aligned} f_1(x) &= x^3 + 1 \\ f_2(x) &= x^3 + x + 1 \\ f_3(x) &= x^3 + x^2 + 1 \\ f_4(x) &= x^3 + x^2 + x + 1. \end{aligned}$$

Now, $f_1(x)$ is reducible because

$$x^3 + 1 = (x+1)(x^2 + x + 1)$$

(remember that all coefficients are to be reduced modulo 2). Also, f_4 is reducible because

$$x^3 + x^2 + x + 1 = (x+1)(x^2+1).$$

However, $f_2(x)$ and $f_3(x)$ are both irreducible, and either one can be used to construct a field having eight elements.

Let us use $f_2(x)$, and thus construct the field $\mathbb{Z}_2[x]/(x^3+x+1)$. The eight field elements are the eight polynomials $0, 1, x, x+1, x^2, x^2+1, x^2+x$, and x^2+x+1.

To compute a product of two field elements, we multiply the two polynomials together and reduce modulo x^3+x+1 (i.e., divide by x^3+x+1 and find the remainder polynomial). Since we are dividing by a polynomial of degree three, the remainder will have degree at most two and hence is an element of the field.

For example, to compute $(x^2+1)(x^2+x+1)$ in $\mathbb{Z}_2[x]/(x^3+x+1)$, we first compute the product in $\mathbb{Z}_2[x]$, which is x^4+x^3+x+1. Then we divide by x^3+x+1, obtaining the expression

$$x^4+x^3+x+1 = (x+1)(x^3+x+1) + x^2 + x.$$

Hence, in the field $\mathbb{Z}_2[x]/(x^3+x+1)$, we have that

$$(x^2+1)(x^2+x+1) = x^2+x.$$

Below, we present a complete multiplication table for the nonzero field elements. To save space, we write a polynomial $a_2x^2 + a_1x + a_0$ as the ordered triple $a_2a_1a_0$.

×	001	010	011	100	101	110	111
001	001	010	011	100	101	110	111
010	010	100	110	011	001	111	101
011	011	110	101	111	100	001	010
100	100	011	111	110	010	101	001
101	101	001	100	010	111	011	110
110	110	111	001	101	011	010	100
111	111	101	010	001	110	100	011

∎

We have described how to construct finite fields whose orders are primes or the power of a prime. There are no other orders for which finite fields exist.

Theorem B.10. *There exists a finite field of order q if and only if $q = p^n$, where p is prime and $n \geq 1$.*

It is natural to ask if finite fields of the same order that are constructed from different irreducible polynomials are "different". In fact, the resulting fields turn out to be isomorphic. (Two fields $(X, \times, +)$ and $(Y, \times, +)$ are *isomorphic finite fields* if there exists a bijection $\phi : X \to Y$ such that

$$\phi(x + x') = \phi(x) + \phi(x') \quad \text{and} \quad \phi(x \times x') = \phi(x) \times \phi(x')$$

for all $x, x' \in X$.)

Theorem B.11. *Suppose that $(X, \times, +)$ and $(Y, \times, +)$ are finite fields of order q. Then these two fields are isomorphic.*

We denote the (unique) finite field of order q (where $q = p^n$, p is prime, and $n \geq 1$) using the notation $\mathbb{F}_q$.

Theorem B.12. *Suppose that $\mathbb{F}_q$ is a finite field. Then $(\mathbb{F}_q \backslash \{0\}, \times)$ is a cyclic group.*

Theorem B.12 states that the nonzero elements of a finite field can be generated as powers of a single element. Such a generator is called a *primitive element* of the finite field.

Example B.13. The finite field $\mathbb{F}_8$ was constructed as $\mathbb{Z}_2[x]/(x^3 + x + 1)$ in Example B.9. The multiplicative group $(\mathbb{F}_8 \backslash \{0\}, \times)$ has order 7. Since 7 is prime, it follows that any nonzero field element is a primitive element.

For example, if we compute the powers of x, we obtain

$$\begin{aligned}
x^1 &= x \\
x^2 &= x^2 \\
x^3 &= x + 1 \\
x^4 &= x^2 + x \\
x^5 &= x^2 + x + 1 \\
x^6 &= x^2 + 1 \\
x^7 &= 1,
\end{aligned}$$

which comprise all the nonzero field elements. ∎

Theorem B.14. *Suppose that $q = p^n$, where p is prime and $n \geq 1$. Suppose also that $q - 1 \equiv 0 \pmod{r}$. Then there is a unique subgroup $(H, \times)$ of $(\mathbb{F}_q \backslash \{0\}, \times)$ having order r. Furthermore, $H = \{\alpha^{(q-1)i/r} : 0 \leq i \leq r - 1\}$, where α is a primitive element of $\mathbb{F}_q$.*

Example B.15. Suppose that $q = 81 = 3^4$. We can construct $\mathbb{F}_{81}$ by first finding an irreducible polynomial $f(x) \in \mathbb{Z}_3[x]$ having degree four. $f(x) = x^4 + x^3 + 2$ is one such polynomial, so we can take $\mathbb{F}_{81} = \mathbb{Z}_3[x]/(x^4 + x^3 + 2)$. In this field, it turns out that x is a primitive element. The multiplicative subgroup of order 8 is

$$H = \{1, x^{10}, x^{20}, \ldots, x^{70}\}.$$

Writing the elements in H as polynomials of degree at most three, it can be shown that

$$H = \{\pm 1, \pm(x^3 + 2x^2 + 1), \pm(x^3 + 2x^2 + 2), \pm(x^3 + 2x^2)\}.$$

∎

Suppose that $(X, \times, +)$ is a finite field, and let $Y \subseteq X$. We say that $(Y, \times, +)$ is a *subfield* of $(X, \times, +)$ provided that $(Y, \times, +)$ is itself a finite field.

Theorem B.16. *Suppose that $q = p^n$, where p is prime and $n \geq 1$. Then every subfield of $\mathbb{F}_q$ has order p^m, where m is a divisor of n. Conversely, for every positive integer m dividing n, there is a unique subfield of $\mathbb{F}_{p^n}$ isomorphic to $\mathbb{F}_{p^m}$.*

The subfields of $\mathbb{F}_q$ are easily constructed. $\mathbb{F}_{p^m} \backslash \{0\}$ is the unique subgroup H of $\mathbb{F}_{p^n} \backslash \{0\}$ having order $p^m - 1$ (note that $p^m - 1$ is a divisor of $p^n - 1$ whenever m is a divisor of n). Then $\mathbb{F}_{p^m} = H \cup \{0\}$.

Example B.17. $\mathbb{F}_9$ is a subfield of $\mathbb{F}_{81}$ because $81 = 3^4$, $9 = 3^2$, and 2 divides 4. $\mathbb{F}_9$ consists of $\{0, 1, \alpha^{10}, \alpha^{20}, \ldots, \alpha^{70}\}$, where α is a primitive element of $\mathbb{F}_{81}$.
∎

We now discuss the existence of square roots in finite fields. Let q be an odd prime power. Define

$$\mathsf{QR}(q) = \{z^2 : z \in \mathbb{F}_q, z \neq 0\}$$

and

$$\mathsf{QNR}(q) = \mathbb{F}_q \backslash (\mathsf{QR}(q) \cup \{0\}).$$

We have the following.

Theorem B.18. *Let q be an odd prime power. Then $|\mathsf{QR}(q)| = (q-1)/2$ and $|\mathsf{QNR}(q)| = (q-1)/2$. Furthermore, the following results hold.*

1. *If $x \in \mathsf{QR}(q)$, then there are exactly two elements $y \in \mathbb{F}_q$ such that $y^2 = x$, and these two elements sum to 0.*
2. *If $x \in \mathsf{QNR}(q)$, then there are no elements $y \in \mathbb{F}_q$ such that $y^2 = x$.*
3. *If $x = 0$, then there is exactly one element $y \in \mathbb{F}_q$ such that $y^2 = x$, namely $y = 0$.*

For even prime powers, the situation is completely different.

Theorem B.19. *Let $q = 2^n$. For every $x \in \mathbb{F}_q$, there is a unique $y \in \mathbb{F}_{2^n}$ such that $y^2 = x$.*

Notes and References

McEliece [81] is an excellent textbook on finite fields; Lidl and Niederreiter [76] is an important reference book that contains a huge amount of useful information on this subject.

References

1. R. J. R. ABEL, A. E. BROUWER, C. J. COLBOURN, AND J. H. DINITZ. Mutually orthogonal Latin squares (MOLS). In *The CRC Handbook of Combinatorial Designs* (C. J. Colbourn and J. H. Dinitz, eds.), CRC Press, Boca Raton, 1996, pp. 111–142.
2. I. ANDERSON. *Combinatorial Designs and Tournaments*, Oxford University Press, Oxford, 1997.
3. E. F. ASSMUS, JR. AND J. D. KEY. *Designs and Their Codes*, Cambridge University Press, Cambridge, 1992.
4. L. M. BATTEN AND A. BEUTELSPACHER. *The Theory of Finite Linear Spaces*. Cambridge University Press, Cambridge, 1993.
5. L. D. BAUMERT. *Cyclic Difference Sets*, Springer, Berlin, 1971.
6. L. D. BAUMERT, S. W. GOLOMB, AND M. HALL, JR. Discovery of an Hadamard matrix of order 92. *Bulletin of the American Mathematical Society* **68** (1962), 237–238.
7. C. H. BENNETT, G. BRASSARD, AND J.-M. ROBERT. Privacy amplification by public discussion. *SIAM Journal on Computing* **17** (1988), 210–229.
8. M. R. BEST. The excess of a Hadamard matrix. *Indagationes Mathematicae* **39** (1977), 357–361.
9. T. BETH, D. JUNGNICKEL, AND H. LENZ. *Design Theory, Volume 1 (Second Edition)*, Cambridge University Press, Cambridge, 1999.
10. T. BETH, D. JUNGNICKEL, AND H. LENZ. *Design Theory, Volume 2 (Second Edition)*, Cambridge University Press, Cambridge, 1999.
11. J. BIERBRAUER, K. GOPALAKRISHNAN, AND D. R. STINSON. Orthogonal arrays, resilient functions, error-correcting codes and linear programming bounds. *SIAM Journal on Discrete Mathematics* **9** (1996), 424–452.
12. G. R. BLAKLEY. Safeguarding cryptographic keys. *American Federation of Information Processing Societies – Conference Proceedings* **48** (1979), 313–317.
13. R. C. BOSE. On the construction of balanced incomplete block designs. *Annals of Eugenics* **9** (1939), 353–399.
14. R. C. BOSE. A note on the resolvability of balanced incomplete designs. *Sankhyā* **6** (1942), 105–110.
15. R. C. BOSE AND S. S. SHRIKHANDE. On the construction of sets of mutually orthogonal Latin squares and the falsity of a conjecture of Euler. *Transactions of the American Mathematical Society* **95** (1960), 191–209.

16. R. C. Bose, S. S. Shrikhande, and E. T. Parker. Further results on the construction of mutually orthogonal Latin squares and the falsity of Euler's conjecture. *Canadian Journal of Mathematics* **12** (1960), 189–203.
17. R. H. Bruck. Difference sets in a finite group. *Transactions of the American Mathematical Society* **78** (1955), 464–481.
18. R. H. Bruck and H. J. Ryser. The nonexistence of certain finite projective planes. *Canadian Journal of Mathematics* **1** (1949), 88–93.
19. P. J. Cameron. *Combinatorics: Topics, Techniques, Algorithms*, Cambridge University Press, Cambridge, 1994.
20. P. J. Cameron and J. H. van Lint. *Designs, Codes, Graphs and Their Links*, Cambridge University Press, Cambridge, 1991.
21. B. Chor and O. Goldreich. On the power of two-point based sampling. *Journal of Complexity* **5** (1989), 96–106.
22. B. Chor, O. Goldreich, J. Håstad, J. Friedman, S. Rudich, and R. Smolensky. The bit extraction problem or t-resilient functions. In *26th IEEE Symposium on Foundations of Computer Science*, IEEE Press, Washington, 1985, pp. 396–407.
23. S. Chowla and H. J. Ryser. Combinatorial problems. *Canadian Journal of Mathematics* **2** (1950), 93–99.
24. C. J. Colbourn. Construction techniques for mutually orthogonal Latin squares. In *Combinatorics Advances* (C. J. Colbourn and E. S. Mahmoodian, eds.), Kluwer Academic Publishers, Amsterdam, 1995, pp. 27–48.
25. C. J. Colbourn. Computer science: selected applications. In *The CRC Handbook of Combinatorial Designs* (C. J. Colbourn and J. H. Dinitz, eds.), CRC Press, Boca Raton, 1996, pp. 543–549.
26. C. J. Colbourn. Group testing. In *The CRC Handbook of Combinatorial Designs* (C. J. Colbourn and J. H. Dinitz, eds.), CRC Press, Boca Raton, 1996, pp. 564–565.
27. C. J. Colbourn and J. H. Dinitz, eds. *The CRC Handbook of Combinatorial Designs*, CRC Press, Boca Raton, 1996.
28. C. J. Colbourn and J. H. Dinitz. Making the MOLS table. In *Computational and Constructive Design Theory* (W. D. Wallis, ed.), Kluwer Academic Publishers, Amsterdam, 1996, pp. 67–134.
29. C. J. Colbourn and J. H. Dinitz. Mutually orthogonal Latin squares: a brief survey of constructions. *Journal of Statistical Planning and Inference* **95** (2001), 9–48.
30. C. J. Colbourn, J. H. Dinitz, and D. R. Stinson. Applications of combinatorial designs to communications, cryptography and networking. In *Surveys in Combinatorics, 1999* (J. D. Lamb and D. A. Preece, eds.), Cambridge University Press, Cambridge, 1999, pp. 37–100.
31. C. J. Colbourn and R. Mathon. Steiner systems. In *The CRC Handbook of Combinatorial Designs* (C. J. Colbourn and J. H. Dinitz, eds.), CRC Press, Boca Raton, 1996, pp. 67–75.
32. C. J. Colbourn and A. Rosa. *Triple Systems*, Oxford University Press, Oxford, 1999.
33. C. J. Colbourn and P. C. van Oorschot. Applications of combinatorial designs in computer science. *ACM Computing Surveys* **21** (1989), 223–250.
34. R. Craigen. Signed groups, sequences, and the asymptotic existence of Hadamard matrices. *Journal of Combinatorial Theory, Series A* **71** (1995), 241–254.
35. R. Craigen and H. Kharaghani. On the existence of regular Hadamard matrices. *Congressus Numerantium* **99** (1994), 277–283.

36. R. Craigen and W. D. Wallis. Hadamard matrices: 1893–1993, *Congressus Numerantium* **97** (1993), 99–129.
37. E. Dawson, E. S. Mahmoodian, and A. Rahilly. Orthogonal arrays and ordered threshold schemes, *Australasian Journal of Combinatorics* **8** (1993), 27–44.
38. N. G. de Bruijn and P. Erdös. On a combinatorial problem. *Indagationes Mathematicae* **10** (1948), 421–423.
39. P. Dembowski. *Finite Geometries*, Springer, Berlin, 1968.
40. J. F. Dillon. *Elementary Hadamard Difference Sets*, Ph.D. thesis, University of Maryland, College Park, 1974.
41. J. H. Dinitz and D. R. Stinson, eds. *Contemporary Design Theory, A Collection of Surveys*, John Wiley & Sons, New York, 1992.
42. J. H. Dinitz and D. R. Stinson. Room squares and related designs. In *Contemporary Design Theory, A Collection of Surveys* (J. H. Dinitz and D. R. Stinson, eds.), John Wiley & Sons, New York, 1992, pp. 137–204.
43. D.-Z. Du and F. K. Hwang. *Combinatorial Group Testing and Its Applications, Second Edition*. World Scientific Publishing Company, Inc., Singapore, 2000.
44. P. Erdös, R. C. Mullin, V. Sós, and D. R. Stinson. Finite linear spaces and projective planes. *Discrete Mathematics* **47** (1983), 49–62.
45. R. A. Fisher. An examination of the different possible solutions of a problem in incomplete blocks. *Annals of Eugenics* **10** (1940), 52–75.
46. S. Furino, Y. Miao, and J. Yin. *Frames and Resolvable Designs: Uses, Constructions, and Existence*. CRC Press, Boca Raton, 1996.
47. P. G. Gibbons. Computational methods in design theory. In *The CRC Handbook of Combinatorial Designs* (C. J. Colbourn and J. H. Dinitz, eds.), CRC Press, Boca Raton, 1996, pp. 718–740.
48. E. N. Gilbert, F. J. MacWilliams, and N. J. A. Sloane. Codes which detect deception. *Bell System Technical Journal* **53** (1974), 405–424.
49. K. Gopalakrishnan and D. R. Stinson. Applications of designs to cryptography. In *The CRC Handbook of Combinatorial Designs* (C. J. Colbourn and J. H. Dinitz, eds.), CRC Press, Boca Raton, 1996, pp. 549–557.
50. K. Gopalakrishnan and D. R. Stinson. Derandomization. In *The CRC Handbook of Combinatorial Designs* (C. J. Colbourn and J. H. Dinitz, eds.), CRC Press, Boca Raton, 1996, pp. 558–560.
51. K. Gopalakrishnan and D. R. Stinson. A simple analysis of the error probability of two-point based sampling. *Information Processing Letters* **60** (1996), 91–96.
52. M. Hall, Jr. Cyclic projective planes. *Duke Mathematical Journal* **14** (1947), 1079–1090.
53. M. Hall, Jr. *Combinatorial Theory (Second Edition)*, John Wiley & Sons, New York, 1986.
54. M. Hall, Jr. and W. S. Connor. An embedding theorem for balanced incomplete block designs. *Canadian Journal of Mathematics* **6** (1954), 35–41.
55. M. Hall, Jr. and H. J. Ryser. Cyclic incidence matrices. *Canadian Journal of Mathematics* **3** (1951), 495–502.
56. H. Hanani. On quadruple systems. *Canadian Journal of Mathematics* **12** (1960), 145–157.
57. H. Hanani. The existence and construction of balanced incomplete block designs. *Annals of Mathematical Statistics* **32** (1961), 361–386.

58. A. HARTMAN AND K. T. PHELPS. Steiner quadruple systems. In *Contemporary Design Theory, A Collection of Surveys* (J. H. Dinitz and D. R. Stinson, eds.), John Wiley & Sons, New York, 1992, pp. 205–240.
59. A. S. HEDAYAT, N. J. A. SLOANE, AND J. STUFKEN. *Orthogonal Arrays, Theory and Applications*, Springer, New York, 1999.
60. D. R. HUGHES AND F. C. PIPER. *Projective Planes*, Springer, Berlin, 1973.
61. D. R. HUGHES AND F. C. PIPER. *Design Theory*, Cambridge University Press, Cambridge, 1985.
62. Y. J. IONIN AND K. MACKENZIE-FLEMING. A technique for constructing non-embeddable quasi-residual designs. *Journal of Combinatorial Designs* **10** (2002), 160–172.
63. Y. J. IONIN AND M. S. SHRIKHANDE. On the λ-design conjecture. *Journal of Combinatorial Theory, Series A* **74** (1996), 100–114.
64. Y. J. IONIN AND M. S. SHRIKHANDE. λ-designs on $4p+1$ points. *Journal of Combinatorial Mathematics and Combinatorial Computing* **22** (1996), 135–142.
65. D. JUNGNICKEL. Difference sets. In *Contemporary Design Theory, A Collection of Surveys* (J. H. Dinitz and D. R. Stinson, eds.), John Wiley & Sons, New York, 1992, pp. 241–324.
66. D. JUNGNICKEL AND S. A. VANSTONE. On resolvable designs $S_3(3;4,v)$. *Journal of Combinatorial Theory, Series A* **43** (1986), 334–337.
67. D. JUNGNICKEL AND S. A. VANSTONE. Hyperfactorizations of graphs and 5-designs. *The Journal of the University of Kuwait. Science* **14** (1987), 213–224.
68. C. KOUKOUVINOS AND J. SEBERRY. New weighing matrices and orthogonal designs constructed using two sequences with zero autocorrelation function — a review. *Journal of Statistical Planning and Inference* **81** (1999), 153–182.
69. E. S. KRAMER. Some results on t-wise balanced designs. *Ars Combinatoria* **15** (1983), 179–192.
70. E. S. KRAMER. t-wise balanced designs. In *The CRC Handbook of Combinatorial Designs* (C. J. Colbourn and J. H. Dinitz, eds.), CRC Press, Boca Raton, 1996, pp. 484–490.
71. E. S. KRAMER AND D. M. MESNER. t-designs on hypergraphs. *Discrete Mathematics* **15** (1976), 263–296.
72. D. L. KREHER. t-designs, $t \geq 3$. In *The CRC Handbook of Combinatorial Designs* (C. J. Colbourn and J. H. Dinitz, eds.), CRC Press, Boca Raton, 1996, pp. 47–66.
73. D. L. KREHER AND R. S. REES. A hole-size bound for incomplete t-wise balanced designs. *Journal of Combinatorial Designs* **9** (2001), 269–284.
74. C. W. H. LAM, L. THIEL, AND S. SWIERCZ. The nonexistence of finite projective planes of order 10. *Canadian Journal of Mathematics* **41** (1989), 1117–1123.
75. E. S. LANDER. *Symmetric Designs: An Algebraic Approach*, Cambridge University Press, Cambridge, 1983.
76. R. LIDL AND H. NIEDERREITER. *Finite Fields, Second Edition*, Cambridge University Press, Cambridge, 1997.
77. C. C. LINDNER AND C. A. RODGER. *Design Theory*, CRC Press, Boca Raton, 1996.
78. J. H. VAN LINT. *Introduction to Coding Theory, Third Edition.* Springer, New York, 1999.
79. J. H. VAN LINT AND R. M. WILSON. *A Course in Combinatorics (Second Edition)*, Cambridge University Press, Cambridge, 2001.
80. F. J. MACWILLIAMS AND N. J. A. SLOANE. *The Theory of Error-correcting Codes*, North-Holland Publishing Co., Amsterdam, 1977.

81. R. J. McEliece. *Finite Fields for Computer Scientists and Engineers*, Kluwer Academic Publishers, Amsterdam, 1987.
82. R. C. Mullin and E. Nemeth. An existence theorem for Room squares. *Canadian Mathematical Bulletin* **12** (1969), 493–497.
83. R. E. A. C. Paley. On orthogonal matrices. *Journal of Mathematical Physics* **12** (1933), 311–320.
84. D. K. Ray-Chaudhuri and R. M. Wilson. Solution of Kirkman's schoolgirl problem. In *Combinatorics* (Proceedings of Symposia in Pure Mathematics, Volume 19, T. S. Motzkin, ed.), American Mathematical Society, Providence, 1971, pp. 187–203.
85. R. Rees. Minimal clique partitions and pairwise balanced designs. *Discrete Mathematics* **61** (1986), 269–280.
86. R. Rees and D. R. Stinson. On the number of blocks in a perfect covering of v points. *Discrete Mathematics* **83** (1990), 81–93.
87. S. Roman. *Coding and Information Theory.* Springer, New York, 1992.
88. O. S. Rothaus. On "bent" functions. *Journal of Combinatorial Theory, Series A* **20** (1976), 300–305.
89. H. J. Ryser. A note on a combinatorial problem. *Proceedings of the American Mathematical Society* **1** (1950), 422–424.
90. H. J. Ryser. An extension of a theorem of de Bruijn and Erdös on combinatorial designs. *Journal of Algebra* **10** (1968), 246–261.
91. M. P. Schützenberger. A non-existence theorem for an infinite family of symmetrical block designs. *Annals of Eugenics* **14** (1949), 286–287.
92. J. Seberry and M. Yamada. Hadamard matrices, sequences, and block designs. In *Contemporary Design Theory, A Collection of Surveys* (J. H. Dinitz and D. R. Stinson, eds.), John Wiley & Sons, New York, 1992, pp. 431–560.
93. A. Shamir. How to share a secret. *Communications of the ACM* **22** (1979), 612–613.
94. S. S. Shrikhande. Affine resolvable balanced incomplete block designs: a survey. *Aequationes Mathematicae* **14** (1976), 251–269.
95. G. J. Simmons. A survey of information authentication. In *Contemporary Cryptology, The Science of Information Integrity* (G. J. Simmons, ed.), IEEE Press, New York, 1992, 379–419.
96. J. A. Singer. A theorem in finite projective geometry and some applications to number theory. *Transactions of the American Mathematical Society* **43** (1938), 377–385.
97. N. M. Singhi and S. S. Shrikhande. On the λ-design conjecture. *Utilitas Mathematica* **9** (1976), 301–318.
98. T. Skolem. Some remarks on the triple systems of Steiner. *Mathematica Scandinavica* **6** (1958), 273–280.
99. R. G. Stanton and J. G. Kalbfleisch. The λ-μ problem: $\lambda = 1$ and $\mu = 3$. In *Proceedings of the Second Chapel Hill Conference on Combinatorial Mathematics and its Applications* (R. C. Bose et al., eds.), University of North Carolina Press, Chapel Hill, 1970, pp. 451–462.
100. D. R. Stinson. Applications and generalizations of the variance method in combinatorial designs. *Utilitas Mathematica* **22** (1982), 323–333.
101. D. R. Stinson. A short proof of the nonexistence of a pair of orthogonal Latin squares of order six. *Journal of Combinatorial Theory, Series A* **36** (1984), 373–376.
102. D. R. Stinson. Frames for Kirkman triple systems. *Discrete Mathematics* **65** (1987), 289–300.

103. D. R. STINSON. Combinatorial characterizations of authentication codes. *Designs, Codes and Cryptography* **2** (1992), 175–187.
104. D. R. STINSON. Resilient functions and large sets of orthogonal arrays. *Congressus Numerantium* **92** (1993), 105–110.
105. D. R. STINSON. Combinatorial designs and cryptography. In *Surveys in Combinatorics, 1993* (K. Walker, ed.), Cambridge University Press, Cambridge, 1993, pp. 257–287.
106. D. R. STINSON AND S. A. VANSTONE. A combinatorial approach to threshold schemes, *SIAM Journal on Discrete Mathematics* **1** (1988), 230–236.
107. A. P. STREET AND D. J. STREET. *Combinatorics of Experimental Design*, Oxford Science Publications, Oxford, 1987.
108. L. TEIRLINCK. Nontrivial t-designs without repeated blocks exist for all t. *Discrete Mathematics* **65** (1987), 301–311.
109. J. A. TODD. A combinatorial problem. *Journal of Mathematical Physics* **12** (1933), 321–333.
110. V. D. TONCHEV. *Combinatorial Configurations: Designs, Codes, Graphs.* Longman Scientific & Technical, London, 1988.
111. V. D. TONCHEV. Codes. In *The CRC Handbook of Combinatorial Designs* (C. J. Colbourn and J. H. Dinitz, eds.), CRC Press, Boca Raton, 1996, pp. 517–543.
112. V. T. TRAN. Nonembeddable quasi-residual designs. In *Finite Geometries and Combinatorial Designs* (Contemporary Mathematics, Volume 111, E. S. Kramer and S. S. Magliveras, eds.), American Mathematical Society, Providence, 1990, pp. 237–278.
113. V. T. TRAN. Symmetric designs. In *The CRC Handbook of Combinatorial Designs* (C. J. Colbourn and J. H. Dinitz, eds.), CRC Press, Boca Raton, 1996, pp. 75–87.
114. W. D. WALLIS. Construction of strongly regular graphs using affine designs. *Bulletin of the Australian Mathematical Society* **4** (1971), 41–49.
115. W. D. WALLIS. *Combinatorial Designs*, Marcel Dekker, New York, 1988.
116. J. WILLIAMSON. Hadamard's determinant theorem and the sum of four squares. *Duke Mathematical Journal* **11** (1944), 65–81.
117. R. M. WILSON. Cyclotomy and difference families in elementary abelian groups. *Journal of Number Theory* **4** (1972), 17–47.
118. R. M. WILSON. An existence theory for pairwise balanced designs I. Composition theorems and morphisms. *Journal of Combinatorial Theory, Series A* **13** (1972), 220–245.
119. R. M. WILSON. An existence theory for pairwise balanced designs II. The structure of PBD-closed sets and the existence conjectures. *Journal of Combinatorial Theory, Series A* **13** (1972), 246–273.
120. R. M. WILSON. The necessary conditions for t-designs are sufficient for something. *Utilitas Mathematica* **4** (1973), 207–215.
121. R. M. WILSON. Concerning the number of mutually orthogonal Latin squares. *Discrete Mathematics* **9** (1974), 181–198.
122. R. M. WILSON. Constructions and uses of pairwise balanced designs. *Mathematisch Centrum Tracts* **55** (1974), 18–41 (Combinatorics Part 1: Theory of Designs, Finite Geometry and Coding Theory).
123. R. M. WILSON. An existence theory for pairwise balanced designs III. Proof of the existence conjectures. *Journal of Combinatorial Theory, Series A* **18** (1975), 71–79.
124. E. WITT. Über Steinersche systeme. *Abhandlungen der Mathematik Hamburg* **12** (1938), 265–275.

125. D. R. WOODALL. Square λ-linked designs. *Proceedings of the London Mathematical Society* **20** (1970), 669–687.

Index